# Combustion Science and Technology

## Recent Advances

# Combustion Science and Technology

## Recent Advances


Editors

**A.K. Agarwal**

**A. Kushari**

**S.K. Aggarwal**

**A.K. Runchal**



**Narosa Publishing House**

New Delhi Chennai Mumbai Kolkata

**Combustion Science and Technology:** ***Recent Advances***
316 pgs. | 246 figs. | 19 tbls.



*Editors*
**A. K. Agarwal**
**A. Kushari**
**S. K. Aggarwal**
**A. K. Runchal**
Department of Mechanical Engineering
Indian Institute of Technology Kanpur
Kanpur



NAROSA PUBLISHING HOUSE PVT. LTD.

22 Delhi Medical Association Road, Daryaganj, New Delhi 110 002
35-36 Greams Road, Thousand Lights, Chennai 600 006
306 Shiv Centre, Sector 17, Vashi, Navi Mumbai 400 703
2F-2G Shivam Chambers, 53 Syed Amir Ali Avenue, Kolkata 700 019

**www.narosa.com**



ISBN 978-81-8487-014-5

Published by N.K. Mehra for Narosa Publishing House,
22 Delhi Medical Association Road, Daryaganj, New Delhi 110 002.

Printed in India

# Preface

Combustion as a science has developed remarkably over the last few decades and has helped in improving the efficiencies of various processes like power generation, automobile engines, gas turbine, rocket engines and various other industrial processes. However, looking at the rapidly increasing energy demand, limited supply of fossil fuels and increased concern over global warming due to emissions, lot more needs to be done to further improve the system efficiencies, identification/invention of alternative sources of energy, development of reusable technologies and development of cleaner technologies. Fundamental research on combustion has made significant advances in modeling and diagnostics. Use of CAD & CFD models is fast emerging as an avenue for improved design of combustion systems and to validate new design concepts and ideas. Another important emerging area in combustion engineering is design of control systems for combustion processes. These techniques have helped in improving the system efficiencies, reduce emission levels and to prevent catastrophic failures. Although significant advances have been achieved in all these fields in recent times, the synergy between these different activities needs to be properly highlighted. In particular, the new emerging technologies combustion modeling, diagnostics and control as well as the advances in renewable energy and clean combustion technologies deserve a focused attention to form guideline for future research in these fields.

Extensive research on combustion has been motivated by the application of combustion in diverse areas ranging from power generation to industrial heat treatment processes and from automobile industry to aerospace propulsion and rocket technology. Fundamental research on combustion has made significant advances in experimentation, modeling and diagnostics. Availability of high-speed computers has greatly enhanced the predictive capabilities of the models while increasingly accurate diagnostics have been possible through development of sophisticated instrumentation, particularly, non-intrusive optical tools. Another important emerging area in combustion is strongly dependent on the availability of both accurate and fast-acting diagnostic tools and efficient models that can capture signal and compute data in real time. The depletion of fossil fuel reserve and increased concern of global warming has paved the way for the study of the impact of combustion on environment and development of alternative sources of energy. This book is aimed to provide a platform for the scientists and researchers to know the latest developments in combustion science and to evolve a strategy to cater to the energy needs of the future keep in mind the environmental concerns.

We, the editors along with the authors are pleased to present this book which will surely be of immense value to the students, researchers, technocrats and industrialists engaged in the field of combustion and related activities.

The editors are grateful to Mr. N.K. Mehra, publisher and managing director, Narosa Publishing House Pvt. Ltd for the keen concern taken towards faster and legible publication of the present compilation. We also thank Ms. Anupama Jauhry and the entire tam of the production team of Narosa for the technical help rendered in various ways. We welcome any further suggestion and criticism regarding the present book, which might be useful for incorporating in the next edition.

**Editors**

# Contents

# Innovative Combustion and Emissions Reduction Techniques[1]

**Ajay K. Agrawal**
Robert F. Barfield Endowed Chair Professor, Department of Mechanical Engineering,
University of Alabama, AL 35487, USA

## 1.0 INTRODUCTION

In recent years, two important factors have influenced the world's energy market. First, the increased demand for petroleum fuels has steadily raised their price to impose economic burden on importing nations. Second, the concerns for global warming have increased scrutiny on emissions of greenhouse gases such as carbon dioxide ($CO_2$) formed by the combustion of fossil fuels. The depletion of conventional fossil fuels is also a cause for concern. The above factors have prompted research on alternative energy sources. Biofuels derived from agricultural products (i.e., biomass) offer the potential for a sustainable energy future. Unlike non-renewable fossil fuels, the $CO_2$ emitted from the combustion of biofuels is recycled during cultivation of the biomass. Biofuels can increase energy security since the biomass feedstock is domestic, diffuse, and hence, secure. However, for commercial viability, the land usage and fossil energy input for biofuel production must be reduced through process integration/optimization and/or development of specialty energy crops such as microalgae.

While biofuels offer potential of a sustainable energy future, we still need to rely upon conventional resources to meet much of our near-term energy needs. For example, infrastructure to replace a sizable portion of fossil energy with alternative fuels does not exist at present, and it is unlikely that we will completely eliminate fossil energy usage in the foreseeable future. Besides the technical challenges (and opportunities) in switching to alternative energy, the energy portfolio at a given time is influenced by the prevailing economics, dependent upon the domestic infrastructure and resources. For example, coal is the most abundant fossil fuel in the U.S. and the world. This readily available domestic fuel resource can help the U.S. increase its energy security. Thus, programs to develop clean coal-based power systems with high efficiency and emissions

[1]In support of presentation at International Workshop on Advances in Combustion Science and Technology, Indian Institute of Technology, Kanpur, INDIA, December 31, 2007 – January 2, 2008.

performance and low capital cost have been embarked upon by federal and private agencies. Such initiatives for power generation will require fuel-flexible gas turbines capable of operating on syngas produced in Integrated Gasification Combined Cycle (IGCC) plants.

Regardless of the energy source (renewable or non-renewable) combustion will remain an important topic of research to address enormous challenges faced by the energy usage. In recent years, combustion research was driven by the need to reduce emissions of soot, unburned hydrocarbons (UHC), nitric oxides (NOx), and carbon monoxide (CO) emissions. Thus, conventional diffusion flame burners are being replaced with combustion systems operated in the lean premixed (LPM) mode. The LPM combustion allows greater control of the flame temperature, which can be utilized to lower the emissions. However, the LPM combustion systems introduce problems of autoignition, lean blow-off (LBO), thermoacoustic instability, and flame flashback, especially at part-load operation (Lieuwen, et al., 2006). In recent years, an extensive effort has been made to understand and mitigate problems of LPM combustion systems. Much of the research has focused on gaseous fuels, say methane or natural gas. However, low-emission combustion systems with the ability to accept different fuel streams, while meeting the increasingly stringent environmental standards without an undue economic penalty, would be essential for a sustainable energy future.

The present paper provides a summary of author's research on combustion concepts offering fuel flexibility and low-emissions. Instead of providing a comprehensive literature review, four specific topics are introduced in the paper: (1) combustion of hydrogen-enriched fuels, (2) combustion using porous inert medium (PIM), (3) combustion with annular heat recirculation, and (4) efficient liquid fuel atomization. The paper concludes with a highlight of future research involving liquid fuel combustion in miniature systems, and PIM-assisted combustion.

## 2.0 COMBUSTION OF HYDROGEN-ENRICHED FUELS

The development of advanced combustion capabilities for gaseous hydrogen and hydrogen-blended hydrocarbon fuels is an area of much current interest encompassing several needs. One need is the cost-effective utilization of alternative fuels with a wide range of heating values (Richards et al, 2001). Low and medium heating-value fuels containing hydrogen are available from gasification of coal and/or biomass, chemical waste streams, fuel cell byproducts, etc. These hydrogen-rich product gases could provide a significant source of cost-effective fuels for power generation and other applications. A second need is related to the recognition that ultra-lean premixed combustion is effective in reducing the NOx emissions. Hydrogen blended with traditional hydrocarbon fuels could improve flame stability and allow stable lean combustion at the lower temperatures needed to minimize the NOx production (Ren et al, 2001). A longer-term need is the desire to minimize and, eventually, eliminate UHC and $CO_2$ emissions. In this case, the carbon from the fuel stream (in the form of $CO_2$) is removed for sequestration, leaving hydrogen as the primary reactant. The use of hydrogen-blended fuels provides both a solution to the immediate need for NOx reduction, and a transition strategy to a carbon free energy system in the future.

Fuels containing hydrogen have been used for some time in conventional non-premixed combustion systems (Meier et al, 1986; Maden, 1998). Premixed combustion studies in spark ignition engines operating on hythane, a blend of hydrogen and natural gas, show lower exhaust emissions and leaner operating limit (Bell and Gupta, 1997; Larsen et al., 1997). An early study on LPM combustion of propane with air showed improved combustion efficiency and lower temperature at lean blow out with hydrogen addition (Anderson, 1975). Conceptual studies of large-scale gas turbine-based power plants have concluded that the hydrogen addition can be economical in meeting the stringent emissions requirements (Phillips and Roby, 1999).

Fundamental research on hydrogen-enriched fuels has been conducted using opposed jet flow configuration (Gauducheau, et al., 1998; Ren et al, 2001; Jackson et al., 2003; Som et al., 2008). Based on the results of strained flame calculations at elevated pressures, the improved lean flame stability is attributed to hydrogen's higher flame speed and increased resistance to strain (Gauducheau, et al., 1998). The presence of hydrogen substantially increases the extinction strain rate, an effect which is more pronounced in leaner mixtures (Ren et al., 2001). Similar results were obtained by Jackson et al. (2003) who found early breakdown of hydrogen increased radical production near the flame base to increase the burning rates. In the following section, selected results on combustion stability and emissions characteristics of a laboratory-scale, swirl-stabilized LPM combustor operated on hydrogen-enriched fuels are presented. Experimental details and additional results are available in a series of publications (Schefer et al., 2002; Wicksall et al., 2004; Wicksall et al., 2005; Wicksall and Agrawal, 2007).

## 2.1 Luminous Flame Shape

Figure 1 shows direct luminous photographs of the flame to gain an understanding of the flame location and size. The results are shown for fuels containing methane and hydrogen with hydrogen mole fraction, $X_{H2} = 0.0$, 0.12, 0.22, and 0.29. The burner was operated for total reactant flow rate, Q = 700 standard liters per minute (slm) at conditions near the lean stability limit or equilibrium adiabatic flame temperature, $T_{ad}$ = 1290+/-15°C. The reacting flow in this swirl-stabilized combustor consists of two flow separation zones; one located downstream of the centerbody, and another, referred to as the corner torroidal zone, is located at the intersection of the vertical combustor wall and the horizontal inlet plate. Fig 1(a) shows no visible flame in the corner torroidal zone for $X_{H2} = 0.0$. The resulting long flame extended past the downstream end of the combustor tube. Hydrogen addition ($X_{H2} = 0.12$) produces a continuous stable flame in the corner torroidal zone and a significantly shorter flame, indicating a more rapid combustion. The flame size and shape were qualitatively similar to those obtained with methane flame ($X_{H2} = 0.0$) operating at higher $T_{ad} = 1385$°C (not shown). A further increase in hydrogen mole fraction decreases the flame length although the change is modest. The results show that the flames accompanied with reactions in the corner torroidal region are shorter and more stable. Because luminous photographs provide only a qualitative description, the OH measurements using PLIF are discussed next to develop a quantitative understanding of the flame structure.

Figure 2(a) plots visible flame length as a function of $T_{ad}$ for $CH_4$ and $H_2$-enriched $CH_4$ fuels. The flame length increases with decreasing $T_{ad}$ because reactions slow down at lower temperatures and hence, a longer reaction zone is required for complete combustion. For all $T_{ad}$, the flame is shorter for the fuel with 40% $H_2$, especially near the LBO limit. Prior to the LBO, the flame length decreased slightly when the flame became unstable in the center of the combustor. In Fig. 2(a), the datum with the lowest temperature represents the LBO limit for each fuel mixture. Results show

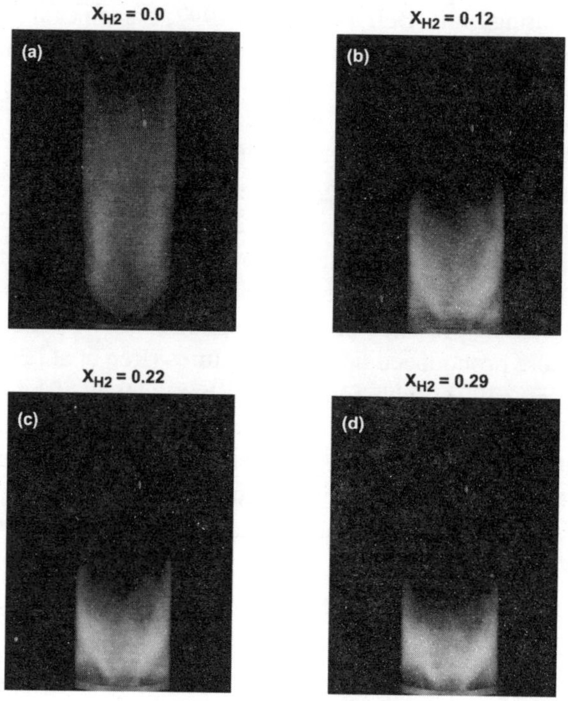


**Figure 1** Direct Flame Luminosity Photographs for Q = 700 slm and $T_{ad}$ = 1290+/-15C, (a) $X_{H2}$ = 0.0, (b) $X_{H2}$ = 0.12, (c) $X_{H2}$ = 0.22, and (d) $X_{H2}$ = 0.29

that $H_2$ in the $CH_4$ fuel decreases the temperature at LBO. Figure 2(b) shows the flame length as function of $T_{ad}$ for $C_3H_8$ and $H_2$-enriched $C_3H_8$ fuels. Results show trends similar to those observed for the previous case in Fig. 2(a), except for a greater influence of the fuel containing 20% $H_2$. The LBO temperatures in $H_2$-enriched $C_3H_8$ flames are higher compared to those in $H_2$-enriched $CH_4$ flames.

## 2.2 Hydroxyl (OH) Planar Laser Induced Fluorescence (PLIF) Measurements

Since the reaction zone contains hydroxyl (OH) radical, a set of 600 single-shot OH PLIF images was used to obtain the statistical data. Figure 3 shows the time-averaged OH PLIF images for $X_{H2}$ = 0.0, 0.12 and 0.29. The operating conditions were same as those in Fig 1, i.e., Q = 700 slm and $T_{ad}$ = 1290 C. Figure 3 shows only a modest change in the peak OH levels with increase in hydrogen addition. The location of peak OH levels is unaffected but the flame structure is significantly altered with the addition of hydrogen. At $X_{H2}$ = 0.0, the OH level in the corner torroidal region is negligible suggesting no flame in that region. This result is consistent with the luminous photograph of the flame in Fig 1(a). At $X_{H2}$ = 0.12, the OH extends into the corner torroidal zone along a layer adjacent to the inner wall of the combustor. Increase in hydrogen from $X_{H2}$ = 0.12 to 0.29 results in high OH concentrations extending across a wider portion of the corner torroidal zone.

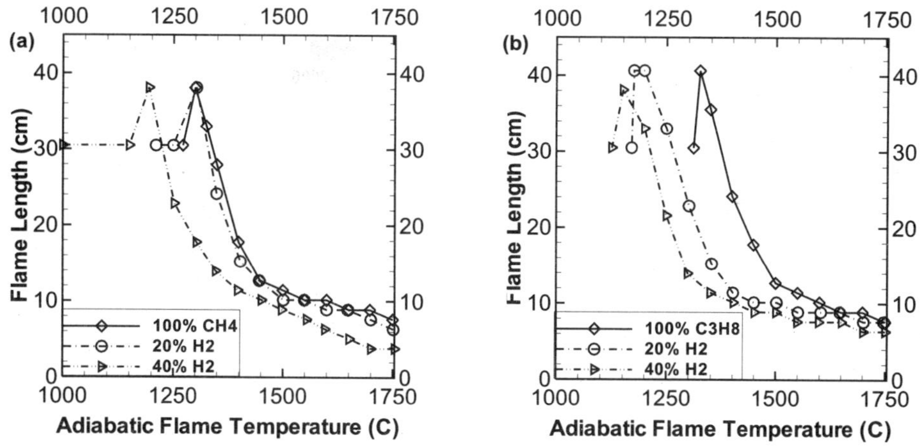


**Figure 2** Flame Length Versus Adiabatic Flame Temperature: a) $H_2$-Enriched $CH_4$ Flames, and b) $H_2$-Enriched $C_3H_8$ Flames

The corresponding RMS OH PLIF images are shown in Figure 4. Evidently, the highest fluctuation levels occur in the layer located between the incoming reactant stream and the centerbody torroidal zone. Large fluctuations in OH levels reflect the movement of the flame as it responds to the local flow turbulence. Note that the peak RMS OH levels are comparable to the peak mean OH levels. At $X_{H2} = 0.0$, the RMS OH values are near zero in the corner torroidal zone because there is no combustion. At $X_{H2} = 0.12$, the flame is present intermittently in the corner torroidal zone, resulting in large OH fluctuations between zero OH and high flame values. Finally, at $X_{H2} = 0.29$, the combustion is continuous and stable in this zone, resulting in reduced RMS OH levels.

## 2.3 Flame Stability

Flame stability characteristics were obtained by maintaining constant reactant flow rate through the combustor and incrementally decreasing the equilibrium adiabatic flame temperature until instability and blowout occurred. The flame instability refers to a weakening of the flame anchoring, which produced a tornado shaped flame in the center of the combustor. The lean stability limits of different methane-hydrogen mixtures are compared for a given equilibrium adiabatic flame temperature. Note that the actual flame temperature will be lower than the equilibrium temperature because of incomplete combustion as the lean blowout is approached. Figure 5 shows the lean stability maps for hydrogen mole fraction $X_{H2} = 0.0$, 0.12, 0.22, and 0.29. The upper data points refer to the flame instability and the lower data points represent the flame blowout. The results show that flames at higher fuel-air flow rates stabilized at higher adiabatic temperatures. This effect is attributed to the higher strain rate in the flame at larger fuel-air flow rates. At Q = 500 slm the methane flame blows out at $T_{ad} = 1175°C$. The corresponding value for $X_{H2} = 0.12$, 0.22, and 0.29 flames is, respectively, 1090°C, 1080°C, and 1070°C. Evidently, hydrogen addition extends the stability and blowout limit of the combustor.

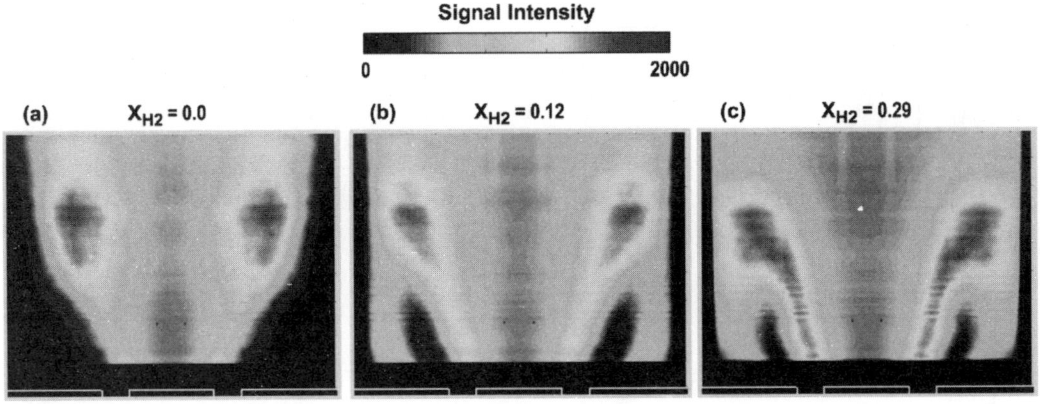


**Figure 3** Time-Averaged OH-PLIF Images for Q = 700 slm and $T_{ad}$ = 1290 +/-15C; (a) $X_{H2} = 0.0$, (b) $X_{H2} = 0.12$, (c) $X_{H2} = 0.29$

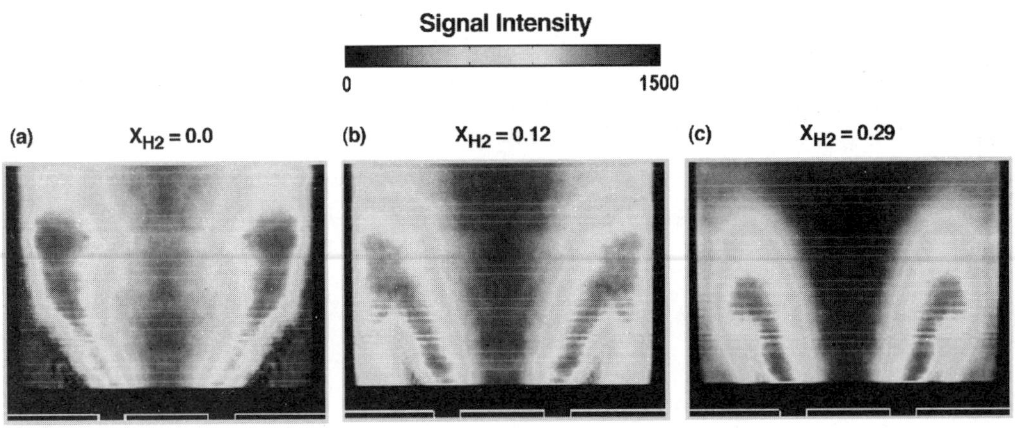


**Figure 4** RMS OH-PLIF Images for Q = 700 slm and $T_{ad}$ = 1290 +/-15C; (a) $X_{H2} = 0.0$, (b) $X_{H2} = 0.12$, (c) $X_{H2} = 0.29$

## 2.4 Pollutant Emissions

Figure 6(a) shows NOx emissions for $CH_4$ and $H_2$-enriched $CH_4$ flames. For $T_{ad} < 1550\ ^0C$, the NOx emissions for all fuels are constant to within the uncertainties of the gas analyzer. For $T_{ad} > 1550\ ^0C$, the NOx emissions increase exponentially with increase in temperature, indicating the dominance of the thermal NOx mechanism. For a given $T_{ad}$, the NOx emissions are affected by the amount of $H_2$ in $CH_4$ fuel; the highest NOx values are obtained without any $H_2$ in the fuel. Figure 5(b) shows the NOx emissions for the $H_2$-enriched $C_3H_8$ flames. Results show constant NOx emissions (within the measurement uncertainties) for $T_{ad} < 1550\ ^0C$ and an exponential increase in NOx emissions with temperature for $T_{ad} > 1550\ ^0C$. Unlike the previous case, the amount of $H_2$ in $C_3H_8$ fuel does not affect the NOx emissions at a given $T_{ad}$. The NOx emissions for $H_2$-enriched $C_3H_8$ flames are generally higher than those for $H_2$-enriched $CH_4$ flames.

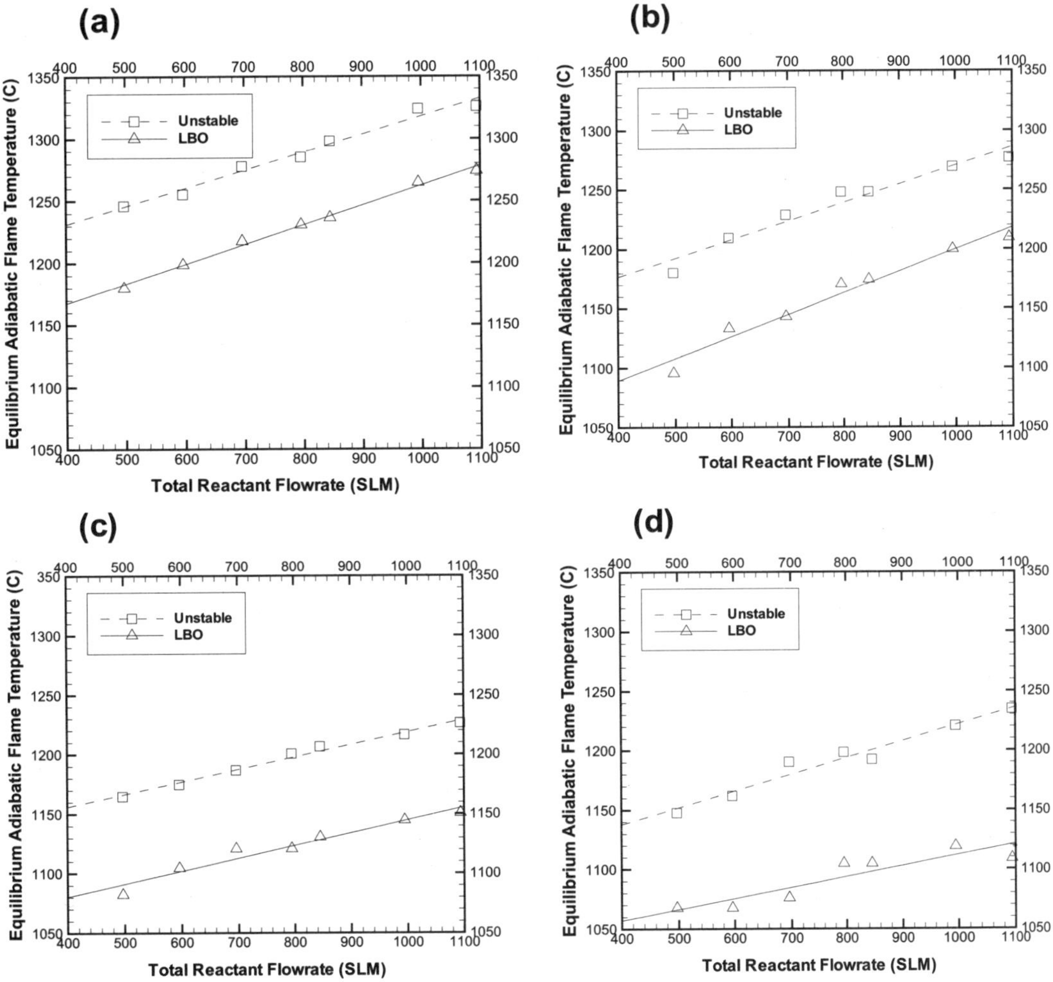


**Figure 5** Flame Stability and Lean Blow Out Limits of Hydrogen-Enriched Methane (a) $X_{H2} = 0.0$, (b) $X_{H2} = 0.12$, (c) $X_{H2} = 0.22$, and (d) $X_{H2} = 0.29$

Figure 7(a) shows CO emissions for $CH_4$ and $H_2$-enriched $CH_4$ flames. The CO emissions increase with increase in $T_{ad}$ because the oxygen concentration in the post-combustion CO burn-off zone decreases at higher $T_{ad}$. The CO emissions decrease at lower $T_{ad}$ because of the excess oxygen, until close to the LBO limit when the stable flame filling the entire cross-section of the combustor changes to a thin tornado-shaped unstable flame in the center of the combustor. Incomplete burning in the unstable flame resulted in poor combustion efficiency and very high CO emissions, exceeding the 200 ppm upper limit of the gas analyzer. Thus, the CO data near the LBO limit are not shown in Fig. 7(a). For a given $T_{ad}$, an increase of $H_2$ in $CH_4$ decreases the CO emissions. Figure 7(b) shows the CO emissions for $C_3H_8$ and $H_2$-enriched $C_3H_8$ flames. Results show a temperature dependence similar to that for $H_2$-enriched $CH_4$ flames, i.e., the CO levels decrease with a decrease in $T_{ad}$ until near the LBO limit, when the CO emissions rise suddenly (not shown). For a given $T_{ad}$,

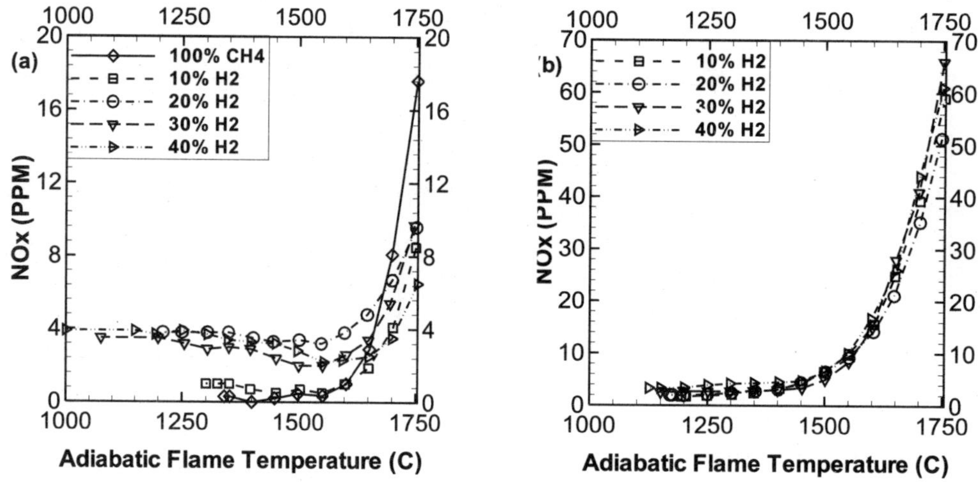


**Figure 6** NOx Emissions Versus Adiabatic Flame Temperatures: (a) $H_2$-Enriched $CH_4$ Flames, and (b) $H_2$-Enriched $C_3H_8$ Flames

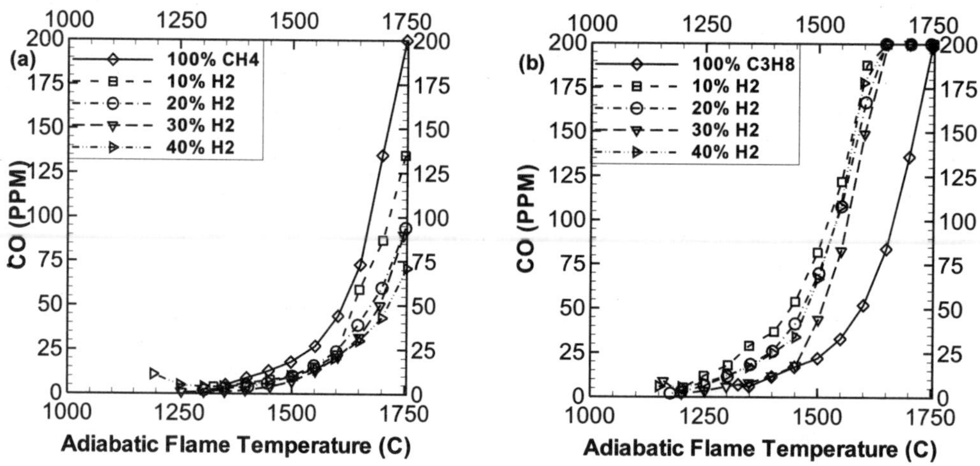


**Figure 7** CO Emissions Versus Adiabatic Flame Temperatures: (a) $H_2$-Enriched $CH_4$ Flames, and (b) $H_2$-Enriched $C_3H_8$ Flames

the CO emissions increased with the addition of $H_2$ to $C_3H_8$, a trend opposite of that for $H_2$-enriched $CH_4$ flames.

## 2.5 Combustor Acoustics

Figure 8(a) shows the total sound power in the 0-1000 Hz range versus $T_{ad}$ for $CH_4$ and $H_2$-enriched $CH_4$ flames. For all fuels, the sound levels near the LBO limit are very low in part because of the excellent fuel-air premixing in this system. For all cases, the sound level increased with an increase in the adiabatic flame temperature. The sound levels are nearly independent of the fuel

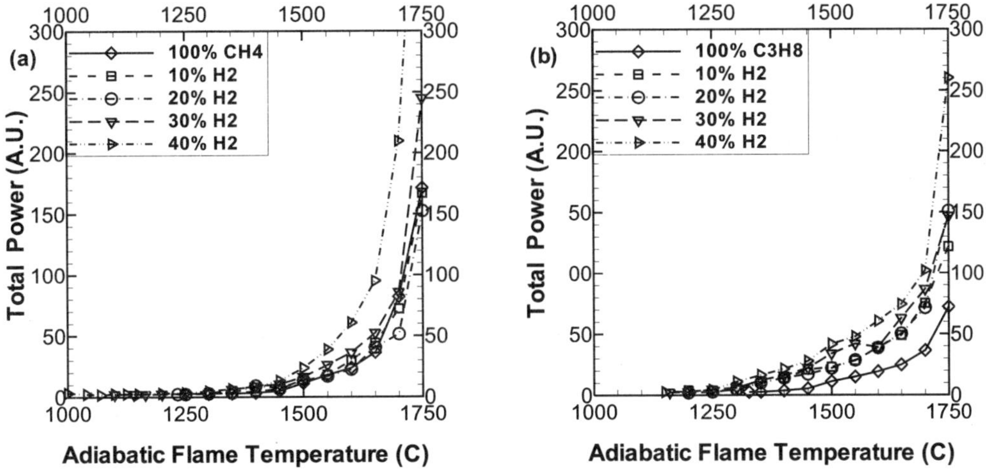


**Figure 8** Total Sound Power Versus Adiabatic Flame Temperature: (a) $H_2$-Enriched $CH_4$ Flames, and (b) $H_2$-Enriched $C_3H_8$ Flames

composition, except for the noticeably higher total sound power for the 40% $H_2$-enriched $CH_4$ flame at $T_{ad} > 1500\ ^0C$. These results suggest a switch from the stable to unstable combustion mode with the addition of $H_2$ to $CH_4$ fuel. Combustion-induced instabilities producing large pressure oscillations or sound levels can damage the system and hence, they must be avoided. Thus, the emissions benefits of $H_2$ addition must be weighed against the requirement to ensure the structural integrity of the combustor. The total sound power for $C_3H_8$ and $H_2$-enriched $C_3H_8$ flames is shown in Figure 8(b) as a function of $T_{ad.}$ For all fuel mixtures, the sound power increased exponentially with an increase in $T_{ad}$, which is similar to the previous case. For a given $T_{ad}$, the total sound power also increased with an increase in the amount of $H_2$ in $C_3H_8$ fuel. The sound levels recorded in $H_2$-enriched $C_3H_8$ fuel are lower than those in $CH_4$ and $H_2$-enriched $CH_4$ flames.

## 2.6 Summary Remarks

Lean premixed combustion of hydrogen-enriched methane or propane was investigated experimentally to determine effects of hydrogen addition on visible flame shape, reaction zones, flame stability, pollutant emissions, and combustor acoustics. For a given adiabatic flame temperature, an Increase in the reaction rates with hydrogen addition decreased the visible flame length. The flame size and shape of hydrogen-enriched fuels are qualitatively similar to those obtained with methane flame at a higher equilibrium adiabatic flame temperature. The OH PLIF measurements show that moderate amounts of hydrogen enrichment led to significant increases in the OH radical concentrations. Increasing amounts of hydrogen enrichment to the highest levels increased the radical concentrations only marginally for the conditions tested in this study. Hydrogen addition produced a broader region of high OH concentrations. Flame characteristics were most significantly affected by reactions in the corner torroidal zone. Results show that the lean stability limit was lowered with the addition of hydrogen to the fuel. The improved stability with hydrogen enrichment was postulated to be a direct result of higher OH, H, and O radical concentrations, which increase several key reaction rates. A significant reduction in CO emissions was realized with hydrogen addition by operating close of the lean stability limit of methane. The NOx and CO emissions were determined mainly by the adiabatic flame temperature. For all fuels, the total sound power increased with an increase in the adiabatic flame temperature. The sound power at a given $T_{ad}$ depended upon the fuel composition and a stable combustor could become unstable with the addition of $H_2$ to the hydrocarbon fuel. Thus, the emissions benefits of $H_2$ addition must be weighed against the requirement to ensure the structural integrity of the combustor.

## 3.0 COMBUSTION USING POROUS INERT MEDIUM (PIM)

The flame temperature (or equivalence ratio) in the combustor must be sufficiently high to prevent the LBO, and to minimize emissions. The optimum operating point can however be shifted towards lower product gas temperatures by extending the LBO limit for a given fuel. This desirable result can be achieved by preheating the incoming reactants by recirculating heat from the reaction zone, for example, in a porous inert media (PIM) burner. Figure 9 denotes the concept of "excess enthalpy flame" or heat recirculation in terms of enthalpy change along the flow direction in the combustor, introduced by Hardesty and Weinberg (1976). The dashed line indicates the normal premixed combustion. Enthalpy variation with preheating of the reactant mixture by heat recirculation is represented by the solid line. Preheating raises upstream enthalpy and thereby produces an excess enthalpy flame in the reaction zone. In terms of temperature, this is equivalent with a peak temperature in the reaction zone that exceeds the adiabatic flame temperature of the unheated reactants. Because of the finite heat loss to the surroundings, the product gas temperature decreases to sub-adiabatic values in the downstream direction.

The PIM offers an effective approach to stabilize the flame either on the surface or in the interior of the porous matrix as illustrated in Fig. 10. In either case, a portion of the heat released in the reaction zone is radiated and conducted upstream to preheat the reactants in the preheat zone.

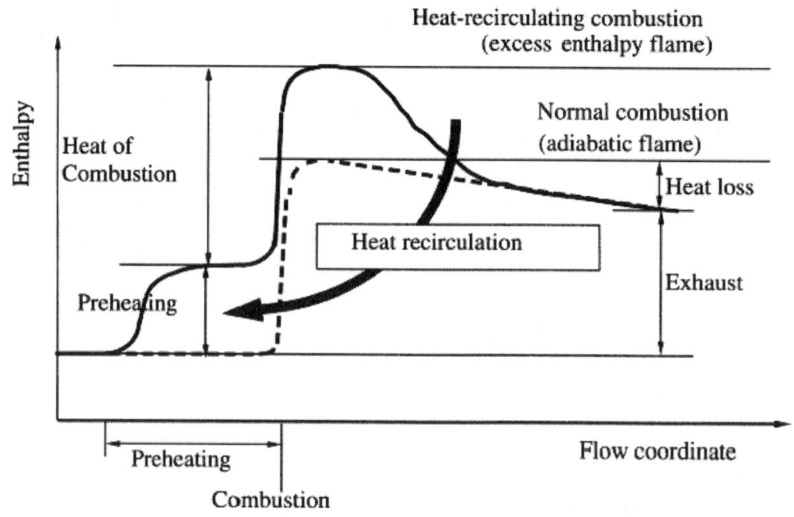


**Figure 9** Heat Recirculation Concept (Hardesty and Weinberg, 1976)

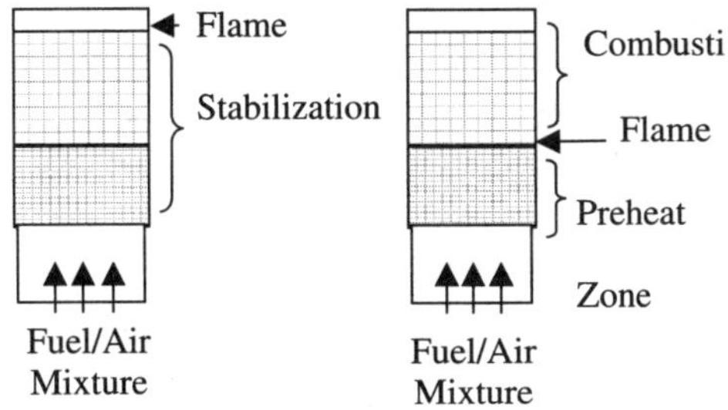


**Figure 10** Schematic Representation of Surface Combustion (left), Interior Combustion (right)

Preheated reactants undergo superadiabatic combustion, whereby the flame temperature exceeds the adiabatic flame temperature of the unheated reactants. Since the preheating increases the reaction rates, combustion at ultra-lean conditions is possible to significantly curtail the NOx and CO emissions. The PIM combustion research has been reviewed by Howell et al. (1996), Viskanta (1995), and Trimis and Durst (1996). Kotani and Takeno (1982) found that a porous burner increases the laminar flame speed by more than an order of magnitude. Subsequently, various aspects of combustion with PIM have been investigated experimentally and numerically. The typical approach for interior combustion with PIM utilizes porous blocks of two different pore sizes, such that the flame is stabilized at the interface (Hsu et al., 1993; Ellzey and Goel, 1995; Rumminger et al, 1996; Pickenaker, 1999; Smucker and Ellzey, 2004; Vogel and Ellzey, 2005, Marbach and Agrawal, 2005). This arrangements of the porous material prevents flame flashback since the flame cannot propagate through the fine-pore structure of the preheat section.

Porous burners are commonly used for space heating, wherein a thin porous combustor section is used to maximize the radiation heat transfer from the PIM to the surroundings. However, for engine applications, the energy-released must be retained within the system (or combustion products) and the heat loss to the surroundings must be kept at a minimum. Some known benefits of combustion using PIM are listed in the following.

- PIM combustion reduces NOx, CO, UHC, soot, etc, by sustaining combustion at ultra-lean conditions.
- Flow through the reticulated pores promotes lateral mixing to create uniform product gas distribution.
- PIM combustion is relatively insensitive to changes in the fuel composition and/or fuel properties.
- Pore structures can dissipate the acoustic energy to suppress the combustion noise.
- PIM combustion offers wide turndown ratio and high power density.

A critical component of PIM combustion is the porous material. Important characteristics of the porous structure are the maximum operating temperature and thermophysical properties such as absorptivity, emissivity, and thermal conductivity. Materials used in experimental porous burners are wide ranging, with ceramics being the most common. Although porous ceramics have good thermal resistance, they suffer from structural damage caused by repeated start-up and shutdown and therefore, are not suited for engine applications. In recent years, new materials have become available with advancements in the chemical vapor deposition (CVD) and other manufacturing techniques. One such example is the carbon-carbon (C-C) foam with carbon fiber reinforcement imbedded in a carbon matrix. Properly fabricated C-C foams have the highest specific strength (strength/weight) of any known material and retain their strength to over 2500°C (Sherman et al., 1990). The drawback of C-C foams is their inability to resist oxidation at temperatures above 775 K. An oxidation-resistant coating of high temperature materials such as silicon carbide (SiC), rhenium, hafnium, or iridium may be deposited on the composite structure by the chemical vapor deposition technique. These porous materials offer structural rigidity as well as thermal and oxidation resistance required for combustion at engine operating conditions.

For the first time, we have successfully used SiC coated carbon foam for PIM combustion (Marbach and Agrawal, 2005; Vijaykant and Agrawal, 2007; Alavandi and Agrawal, 2008). Figure 11 depicts a SiC coated, C-C foam material of 4 pores per cm (ppcm). Figure 12 shows photographs of surface and interior stabilized flames for the baseline configuration of 32-8-8-8, which refers to the PIM combustion section with one 32 ppcm piece (preheat zone) followed by three 8 ppcm pieces in the combustion zone. Each piece is 2.54 cm thick, with 4.04 cm by 4.04 cm flow cross sectional area. As seen in Figure 12a, surface combustion results in the characteristic blue flat flame of lean premixed combustion. However, as Fig. 12b depicts, interior combustion emits an orange glow from the radiating PIM. Audibly, surface combustion produced sound similar to that from a Bunsen burner, while interior combustion was virtually silent due to the muffling effect of the porous material between the flame front and burner exit.

## 3.1 Pressure Drop

Pressure drop resulting from cold flow through the PIM ranged from 0 to 1 percent of the operating pressure. Figure 13 presents the percent pressure drop across various individual porous pieces (each 2.54 cm thick) and the baseline configuration at a range of flow rates. As expected, the pressure drop increases with increase in the reactant flow rate. Note that the measured pressure drop across the baseline configuration is within five percent of that calculated by adding the pressure drop across the individual pieces. Figure 14 presents percent pressure drop for surface and interior combustion over a range of flow rates. With surface combustion, the pressure drop in the PIM is virtually the same as that for the cold flow. However, the pressure drop with interior combustion is

**Figure 11** SiC Coated, 4 ppcm, Carbon Foam Used as PIM

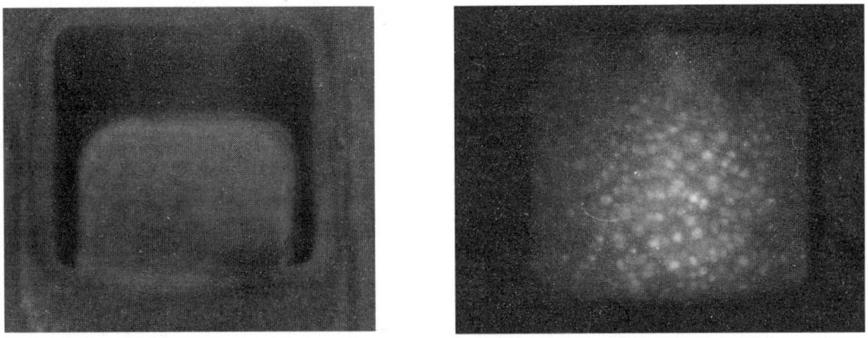

**Figure 12** a) Surface Combustion (left), b) Interior Combustion (right)

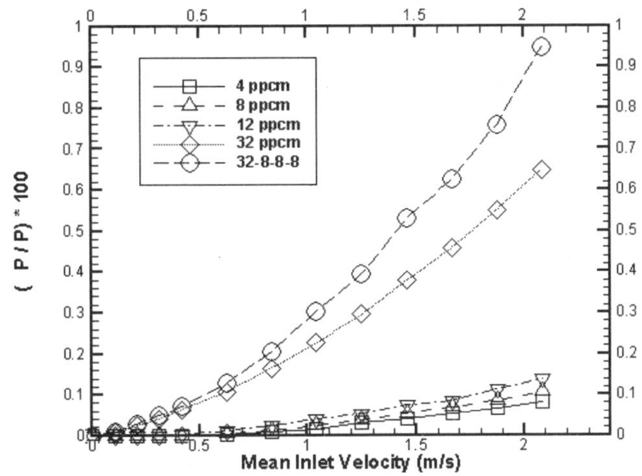


**Figure 13** Cold Flow Pressure Drop vs. Mean Inlet Velocity, Various Pore Sizes

higher, because of the higher velocity of the product gasses moving through the PIM. Results suggest the overall pressure drop for practical applications can be minimized through proper selection of the thickness and pore size of the PIM.

## 3.2 Surface versus Interior Combustion

We have made comparisons of premixed methane/air flames stabilized on the surface and interior of the porous matrix at identical operating conditions. NOx and CO concentrations vs. equivalence ratio for surface and interior combustion at various reactant flow rates are compared in Figure 15 for mean reactant flow velocity, v = 1.0 m/s. Results show that the combustion mode has little impact on NOx concentration at a given equivalence ratio. Interior combustion extends the LBO limit, offering significant NOx reduction. For the baseline PIM configuration, the equivalence ratio at the LBO is about 0.58 for surface combustion while that for interior combustion is 0.55.

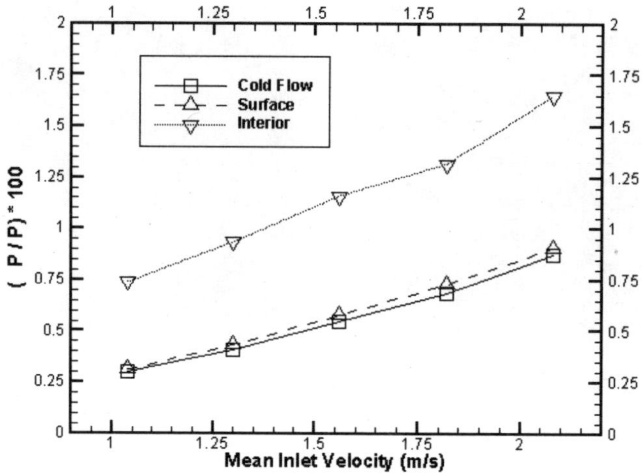


**Figure 14** Pressure Drop vs. Inlet Velocity, Various Flame Locations

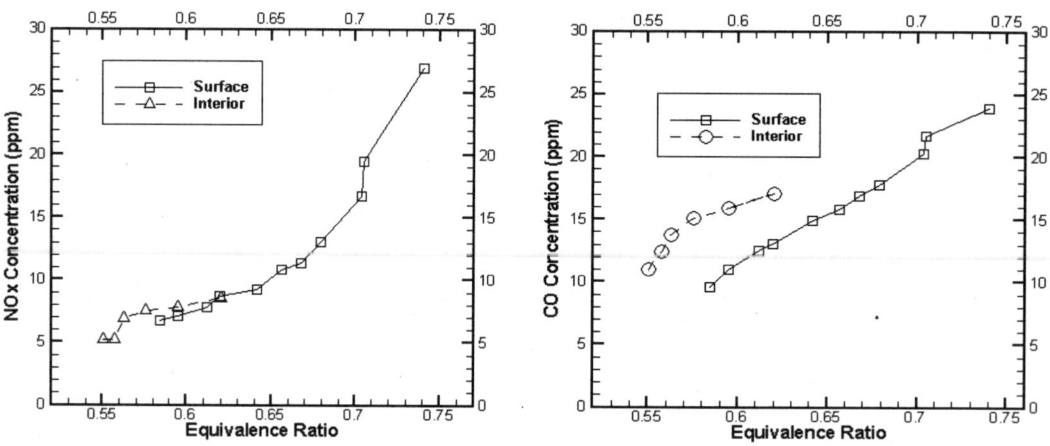


**Figure 15** a) NOx and b) CO Concentration vs. Equivalence Ratio, v = 1.0 m/s

Theoretically, this extension of the lean blow-off limit corresponds to a decrease in flame temperature of about 50 K, which can result in significant reductions in the NOx production. The CO concentrations of the two combustion modes follow a similar trend with equivalence ratio, although interior combustion results in slightly higher CO concentrations, attributed to a higher fraction heat loss from the combustor. Similar results are obtained for combustion of pre-vaorized kerosene, as shown by the plots in Figure 16, i.e., the NOx concentration is much lower for interior combustion because of the leaner operating conditions achieved. The reduction in NOx is accompanied with reduction in the CO concentration.

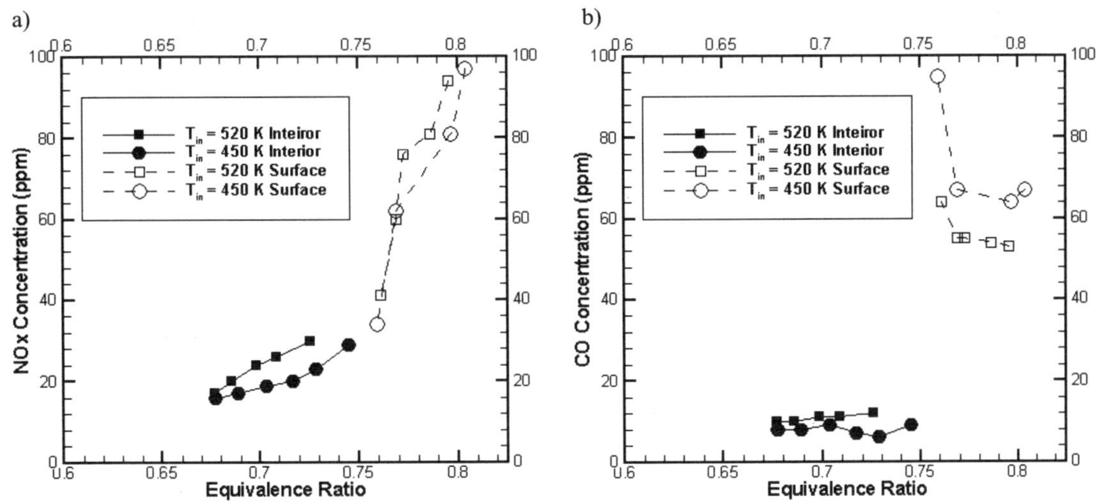


**Figure 16** Comparison of a) NOx, b) CO Emissions for Surface and Interior Combustion

## 3.3 PIM Combustion of Hydrogen-Enriched Fuels

Motivated by the need to utilize alternative fuels, we have investigated PIM combustion of gaseous fuel mixtures simulating synthetic fuels (Alavandi and Agrawal, 2008). Experiments were conducted in a SiC coated carbon foam burner using fuels containing equal amounts of hydrogen ($H_2$) and CO (by volume), while the methane ($CH_4$) percentage was varied from 100 to 0. Results showed significantly lower NOx emissions in fuels containing CO/$H_2$ as compared to $CH_4$, because of the extended lean blow-off (LBO) limit of the former fuel. Here, the LBO pertains to the equivalence ratio (expressed by the adiabatic flame temperature at the measured air and fuel flow rates) at which the flame blows off from the porous matrix. Single-digit CO and NOx emissions were obtained simultaneously in flames burning fuels containing CO/$H_2$. Figure 17 depicts the adiabatic flame temperature ($T_{ad}$) at LBO for varying percentage of $CH_4$ in the fuel, while the remaining fuel contained equal amount of CO and $H_2$ by volume. Results show only a minor reduction of $T_{ad}$ at LBO in fuels with up to 35% CO and 35% $H_2$. Further increase in CO/$H_2$ (or decrease in $CH_4$) dramatically reduced the $T_{ad}$ at LBO. These results not only demonstrate the fuel flexibility of PIM combustion, but also show that the combustor performance can be tailored for fuels containing different reactive constituents.

## 3.4 Liquid Fuel Combustion using SiC Coated Carbon Foam

Previous research on PIM combustion has focused mainly on gaseous fuels, and very few studies involving liquid fuel combustion inside porous media have been conducted. Kaplan and Hall (1995) were the first to demonstrate the feasibility of liquid fuel combustion within PIM. They used heptane supplied from an oil spray nozzle to achieve stable combustion within porous ceramics made of SiC and magnesia-stabilized zirconia. In yttrium-stabilized zirconia, the combustion was incomplete presumably because of the fuel roll-off from the surface. A PIM configuration

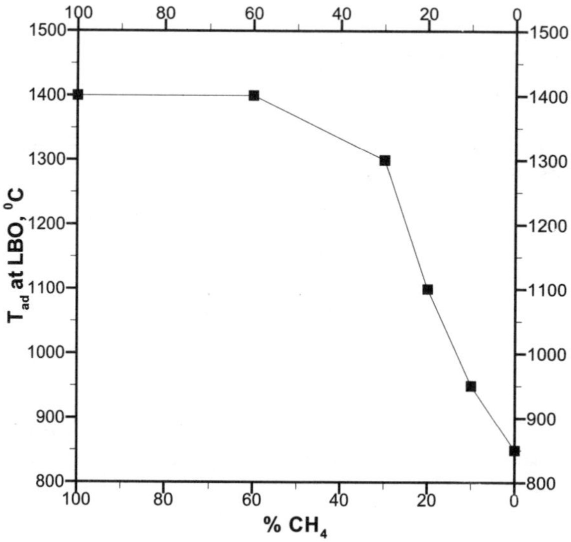


**Figure 17** Adiabatic Flame Temperature at LBO for Varying Percentage of $CH_4$ in the Fuel Containing Equal Amounts of CO and $H_2$

supporting stable combustion over the equivalence ratio range of 0.57 to 0.67 was identified after several trials. In this optimized configuration, combustion occurred inside porous media of 4 pores per cm (ppcm). A porous section of 10 ppcm was necessary in the downstream region to partly block the thermal radiation, and thus, to increase the temperature and reaction rates within the porous matrix. Kaplan and Hall (1995) reported that the distance between the injector and porous ceramic section was a critical factor to maintain stable combustion.

We have investigated kerosene combustion inside PIM using two different twin-fluid fuel injectors and several PIM configurations, with the goal of reducing the emissions, the length of the fuel-air premixing section, and the pressure drop (Vijaykant and Agrawal, 2007). Figure 18 shows the schematic diagrams of the injectors used. In the air-assist injector (Delavan Precision Nozzle WDA0.5), about 5% of the total air passed through the injector to atomize the fuel. In the swirling-air injector (Mansour, 2005a), all of the combustion air passed through the injector to create a strong swirling flow around the fuel jet to promote atomization and fuel-air premixing. Figure 19 shows the stacks of PIM configurations used for the experiments.

The effect of the injector location on CO and NOx concentrations is shown in Figure 20 for the baseline configuration of 26/4-4-4. For L > 45 cm, emissions are independent of the injector type and location because the fuel is fully pre-vaporized and premixed with air to form a homogeneous reactant mixture. The CO and NOx emissions for both injectors increase gradually for L = 35 cm to about 20 cm, but rapidly for L < 20 cm. For L < 17.5 cm, soot emitting orange flamelets, signifying droplet combustion in diffusion mode, were visually observed at the PIM exit. This result clearly indicates that the fuel did not pre-vaporize in the PIM for L < 17.5 cm. Thus, detailed measurements were not taken for L < 17.5 cm, since droplet combustion produces temperatures exceeding the maximum operating temperature of the porous material. For L > 20 cm, the fuel was pre-vaporized

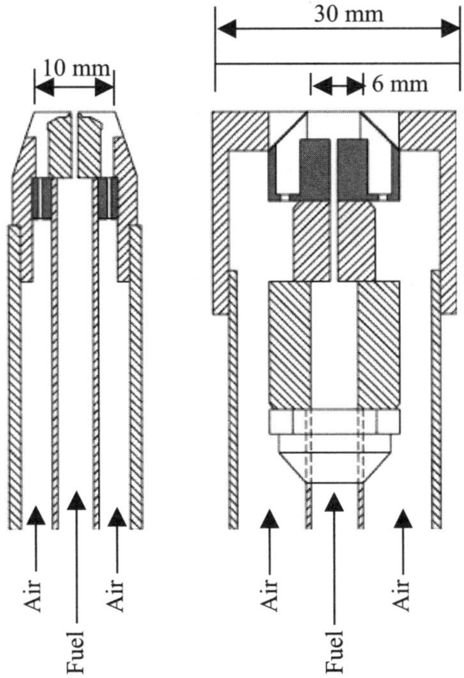


**Figure 18** Air-Blast Injector (left), and Swirling-Air Injector (right)

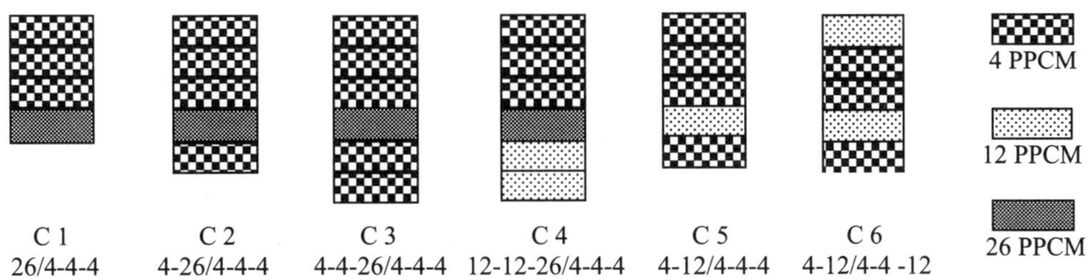


**Figure 19** Porous Inert Media Configurations

since no sooty, orange flames were observed. The rise in emissions for 20 cm < L < 45 cm is attributed to imperfect premixing between fuel and air, i.e., the spatial non-uniformities in equivalence ratio of the reactant mixture. In a non-homogeneous reactant mixture, high NOx levels are produced in zones of high equivalence ratio, and vice versa. Similarly, localized regions with equivalence ratio close to the lean blow-off (LBO) limit will generate high CO emissions. Thus, the net effect of equivalence ratio non-uniformities is to increase both CO and NOx emissions, as observed in Figure 20. Since the swirling-air injector was designed to promote premixing between fuel and air, it produced a more homogeneous reactant mixture upstream of the PIM, resulting in superior emissions performance. Figure 21 shows that the configuration 4-12/4-4-12 results in the minimum CO and NOx emissions.

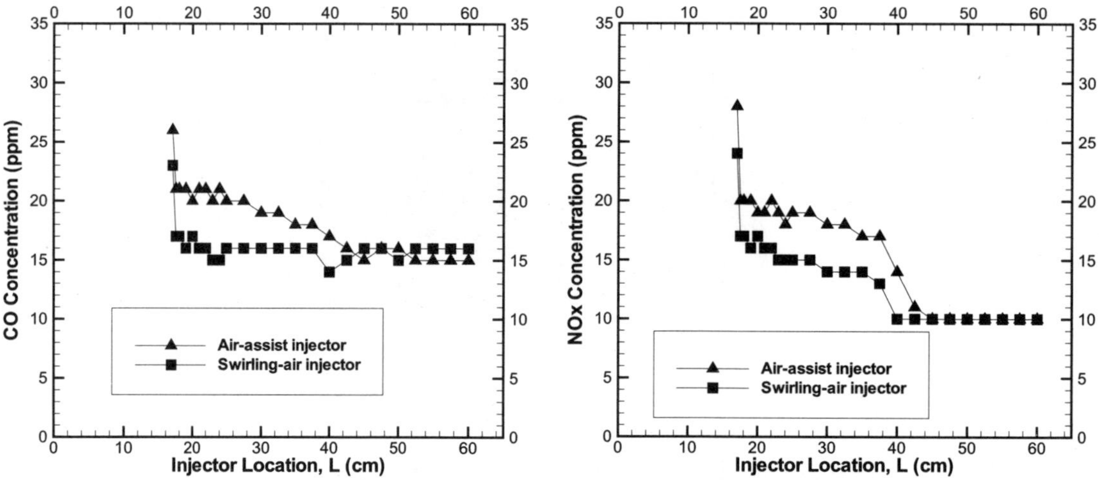


**Figure 20** Effect of Injector Location on Emissions for Swirling-Air Injector, Baseline Configuration (26/4-4-4), F = 0.67, Q = 3.3 kW, (a) CO Concentration, and (b) NOx Concentration

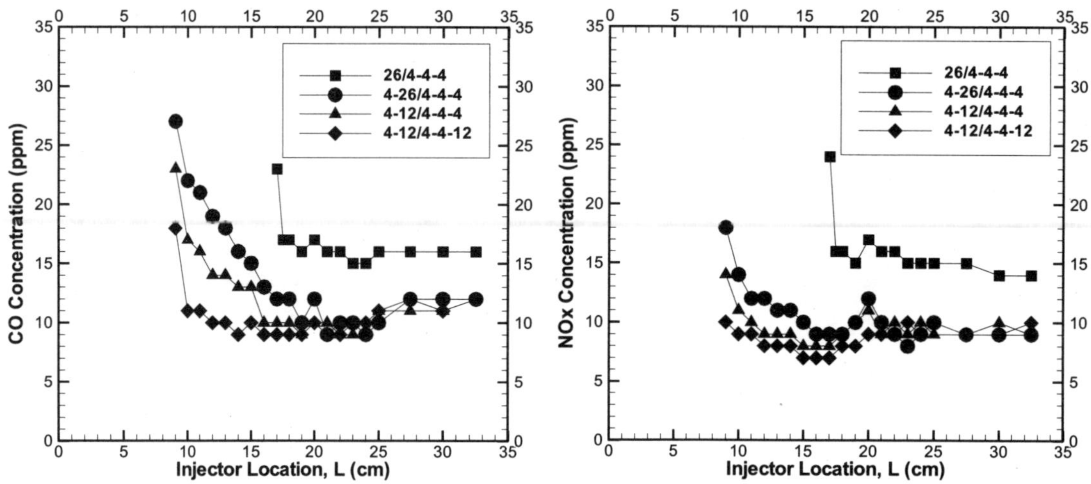


**Figure 21** Effect of PIM Configuration on Emissions for Swirling-Air Injector, $\Phi = 0.67$, Q = 3.3 kW, (a) CO Concentration, and (b) NOx Concentration

## 3.5 Summary Remarks

SiC coated, C-C foam structures offers superior structural and thermal integrity and fuel flexibility. Simultaneous NOx and CO emissions of less than 10 ppm are obtained for a range of equivalence ratios and reactant flow rates. While surface combustion produces noise typical of a flat flame burner, interior combustion is virtually silent. Surface combustion results in only slight increase in the pressure drop in the PIM compared to that for the cold flow. However, interior combustion nearly doubles the pressure drop. The NOx concentration increases exponentially with equivalence ratio, suggesting dominance of thermal mechanism of NOx formation. Interior combustion extended the LBO limit over surface combustion at all conditions tested. For liquid fuel combustion, three combustion operating regimes are identified: (a) When the injector is located far upstream of the PIM, the fuel pre-vaporizes to produce a homogeneous reactant mixture and the emissions are the lowest, (b) When the injector is at an intermediate location, the fuel pre-vaporizes to form a non-homogeneous reactant mixture and the emissions increase slightly, (c) When the injector is close to the PIM, the flow residence time is insufficient to fully pre-vaporize the fuel droplets, and hence, the combustion occurs in the diffusion mode. Thus, emissions of CO, NOx and soot increase significantly.

## 4.0 COMBUSTION WITH ANNULAR HEAT RECIRCULATION

One drawback of PIM combustion is the significant heat loss from the combustor wall(s). Thus, a design concept involving PIM filled annulus around the combustor to utilize the heat transfer through the combustor wall to preheat the reactants and/or to pre-vaporize the liquid fuel was developed by Newburn and Agrawal (2007). The PIM in the annulus is intended to increase the convective heat transfer for the reactants and also to promote the mixing between fuel and air prior to combustion. For liquid fuels, the fuel pre-vaporization and fuel-air premixing are essential to reduce emissions by approaching the fully lean-premixed, pre-vaporized (LPP) combustion. Figure 22 shows a schematic diagram of the concept with a swirl-stabilized combustor. Liquid kerosene is dripped by gravity onto an annular porous bed of stainless steel beads through twelve equidistant holes on a tubular ring positioned at the top of the annulus. The combustor is started on natural gas and subsequently, its operation is switched to the kerosene. The system produces low emissions by operating in the LPP combustion mode using room temperature air. However, fuel pre-vaporization and fuel-air premixing were incomplete at higher fuel flow rates, in part, because of an inadequate fuel injection system. Filling the PIM in the annulus increased the heat loss to the surroundings, a major disadvantage of the design.

### 4.1 Fuel Flexible PIM Combustor

Combining the advantages offered by annular heat recirculation with those of PIM combustion, a double concentric annular configuration with PIM combustion was developed (Dumetrescu, 2006). The design and operational functions are similar to those of Newburn and Agrawal (2007), but several modifications were made to improve the liquid fuel injection system and to minimize the heat loss to the surroundings. The burner system can operate on gaseous or liquid fuels or even a combination of the two fuel types. The system is essentially a counterflow heat exchanger, wherein the product gases transfer heat to the incoming reactants in the annulus by heat conduction through the combustor wall. In the following the salient features of the combustor and selected results are presented for operation with gaseous and liquid fuels.

#### *4.1.1 Operation with Gaseous Fuel*

Figure 23 shows a schematic diagram and photograph of the experimental setup. The air and fuel enter through the inlet section and then are mixed together in the mixing section. Both sections are

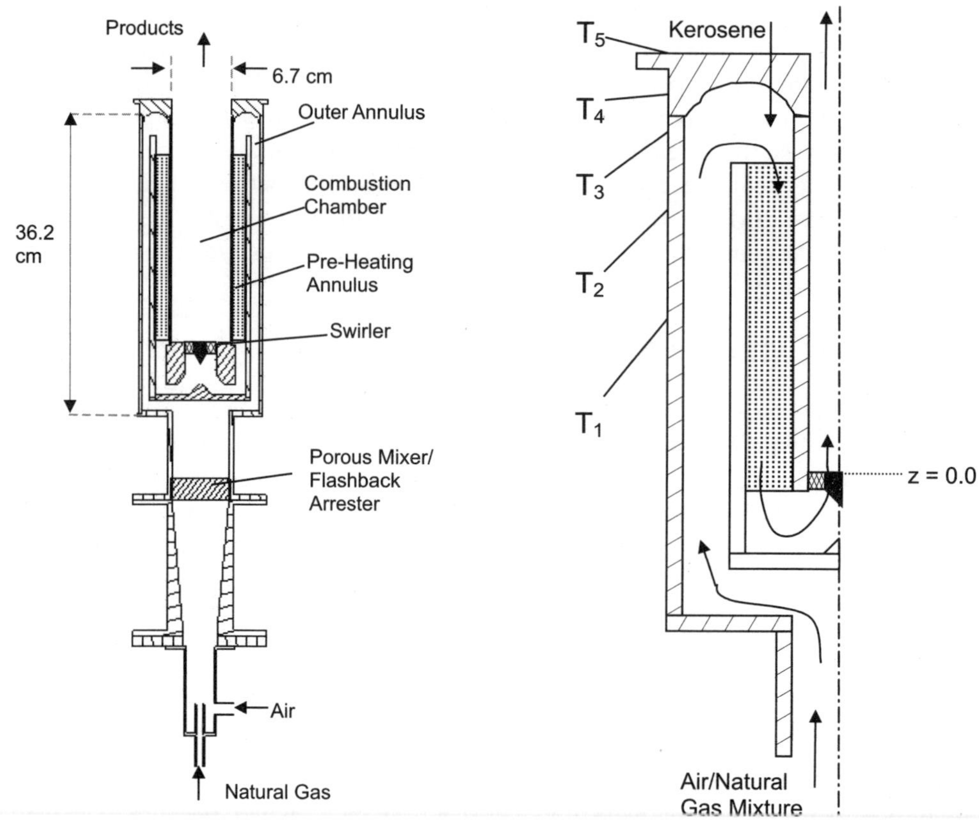


**Figure 22** Heat Recirculation Through Annular PIM; Overall System Configuration (left), Combustor Details (right)

made of stainless steel material. The burner system consists of three stainless steel concentric tubes. The main system is approximately 38 cm long, while the combustor tube is 27.5 cm long with an inside diameter (ID) of 6.7 cm. A specially designed lid covers the three tubes to form two concentric annular flow channels. The reactant mixture enters the outer annulus and then, turns 180-degress to flow through the inner annulus. The inner annulus houses a metal screen holder attached to the combustor outer wall and it is filled with 0.32 cm diameter stainless steel (SS) beads. The preheated reactants leave the inner annulus and turn again 180-degrees to enter the combustor tube containing two sections of SiC coated carbon foam. The first section consists of 1.6 cm thick piece of 40 pores per cm (ppcm) and it provides additional preheating and prevents flashback into the annulus. The second section is the combustion zone and it consists of four 1.3 cm thick porous pieces of 8 ppcm. The movement of the PIM within the combustor was constrained by a high temperature alloy holder (not shown) located downstream. Inner and outer surfaces of the two outer tubes were coated with high-temperature insulation paint to minimize the heat loss to the ambient.

Figure 24 plots radial profiles of product gas temperature at the combustor exit plane (z = 30 cm) for operation with methane fuel and different airflow rates at Ø = 0.65. Note that the measured

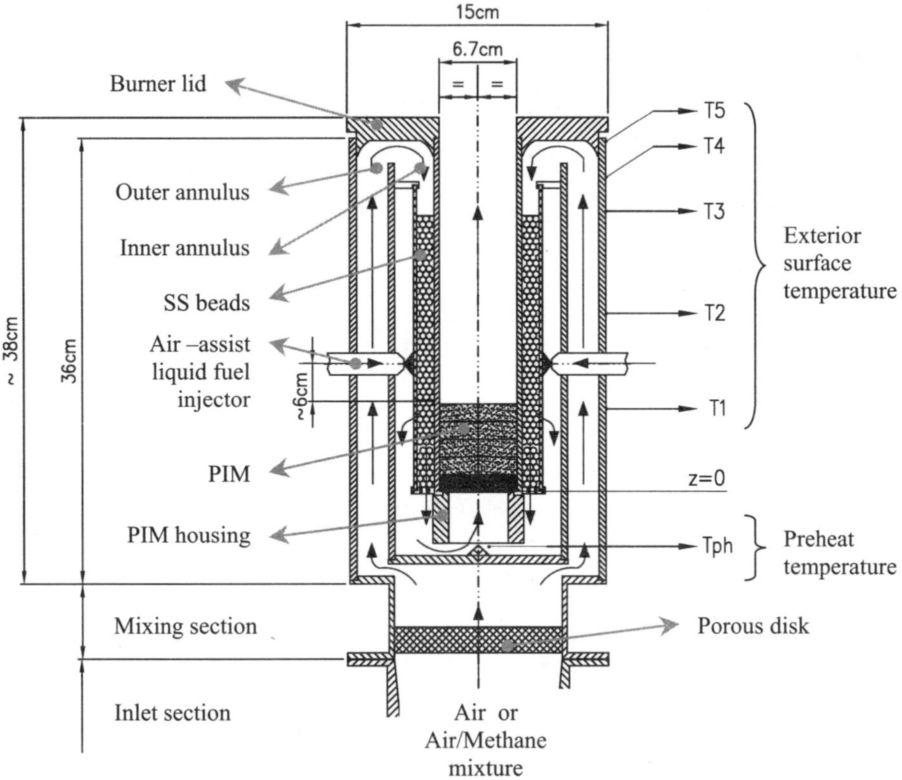


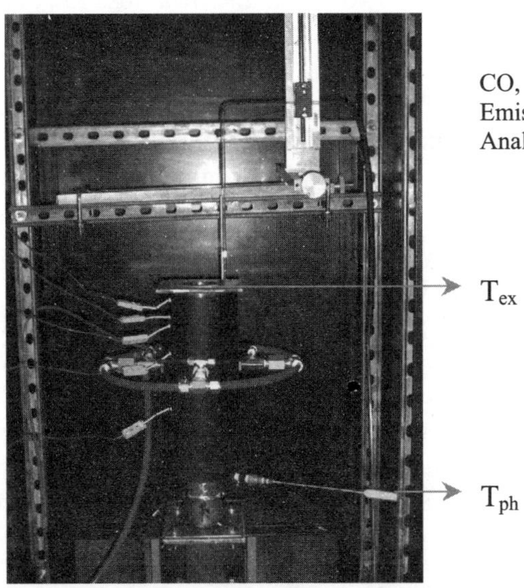


**Figure 23** Schematic Diagram of the Combustor (top); Photograph of the Combustor Setup (bottom)

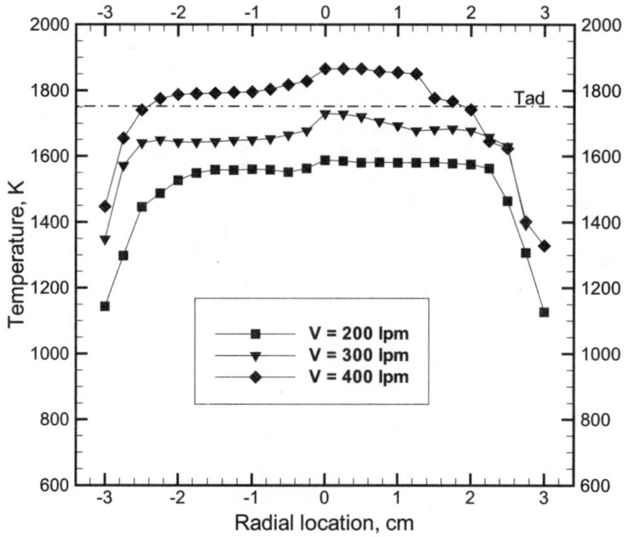


**Figure 24** Effect of Air Flow Rate on Radial Profiles of Product Gas Temperature at the Combustor Exit Plane for Surface Combustion, Ø = 0.65

temperature data were corrected for heat loss by radiation. Results show temperatures peaking in the mid-region and decreasing towards the wall. The mid-region temperature is higher than the adiabatic flame temperature for Q = 400 lpm, indicating an excess enthalpy flame. Moreover, the nearly constant temperature profile in the mid-region suggests good uniformity at the combustor exit plane. For lower airflow rates, the product gas temperature in the mid-region is lower than the adiabatic flame temperature, because of the higher heat loss to the ambient. Figure 25 summarizes thermal performance of the burner expressed heat recirculation (Fig. 25a) and heat loss (Fig. 25b), given, respectively, as percentage of heat released in the combustor. Percent heat recirculation decreases only slightly at higher equivalence ratio and higher airflow rate. Figure 25(b) shows that the percent heat loss depends upon the airflow rate, and it is relatively insensitive to the equivalence ratio. Higher fractional heat loss at lower airflow rates decreases product gas temperature as shown in Figure 24. Results show that the burner must be operated at high airflow rates to reduce heat loss to the ambient, and hence, to achieve good thermal performance.

Effect of equivalence ratio on emissions is shown in Figure 26, wherein measurements at the midpoint of the combustor exit plane are presented for different airflow rates. Results show single digit NOx emissions at low equivalence ratios and low airflow rates. The increase in NOx emissions at higher equivalence ratios signifies the importance of the thermal mechanism. For a given equivalence ratio, the CO concentration increases as the airflow rate increases, because of the reduced residence time for oxidation reactions in the combustor. The CO emissions increase with increasing equivalence ratio, a common result also observed in swirl-stabilized combustion systems.

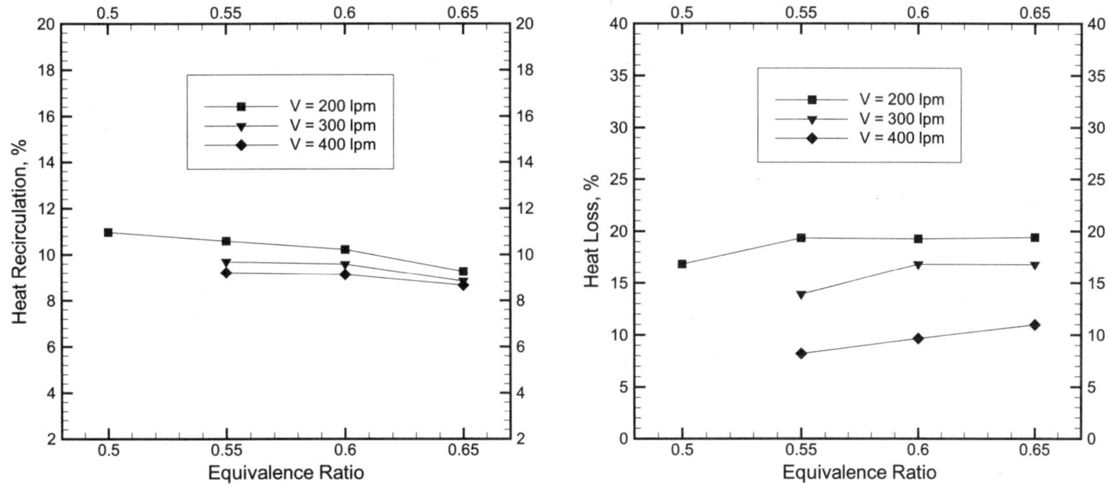


**Figure 25** Thermal Performance vs. Equivalence Ratio for Surface Combustion, (a) Percentage Heat Recirculation, (b) Percentage Heat Loss

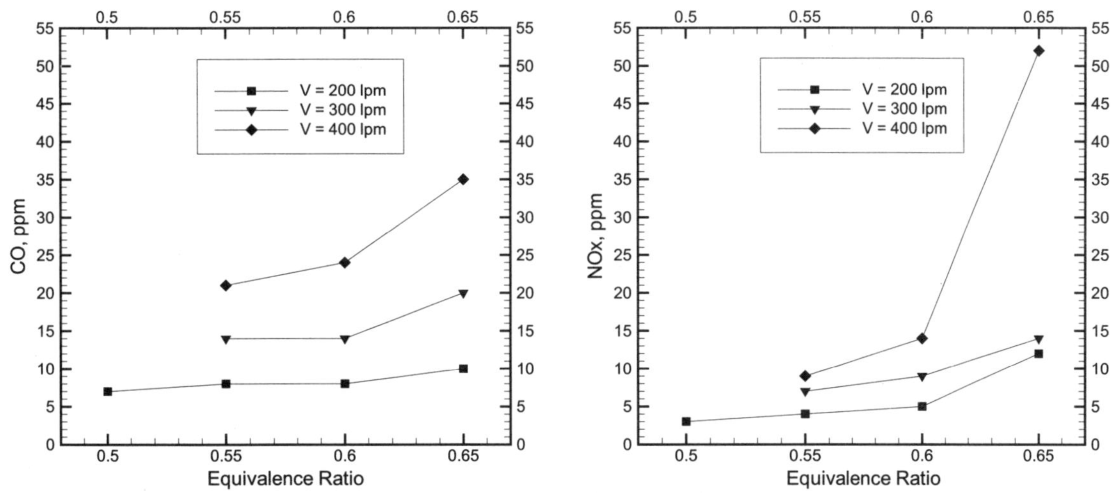


**Figure 26** Emissions vs. Equivalence Ratio at Combustor Exit Plane for Surface Combustion, r = 0, (a) CO Concentration, (b) NOx Concentration

### *4.1.2 Operation with Liquid Fuel*

The burner system, shown in Figure 23 was designed to operate on gaseous as well as liquid fuels. The liquid fuel injection system, shown schematically in Figure 27, consists of four air-assist injectors inserted through the outer and intermediate tubes. Commercially available injectors (Delavan Inc., Siphon type SNA injector) are spaced circumferentially at 90-degrees and can be adjusted so that each spray cone is projected through the holding screen around the porous bed. The

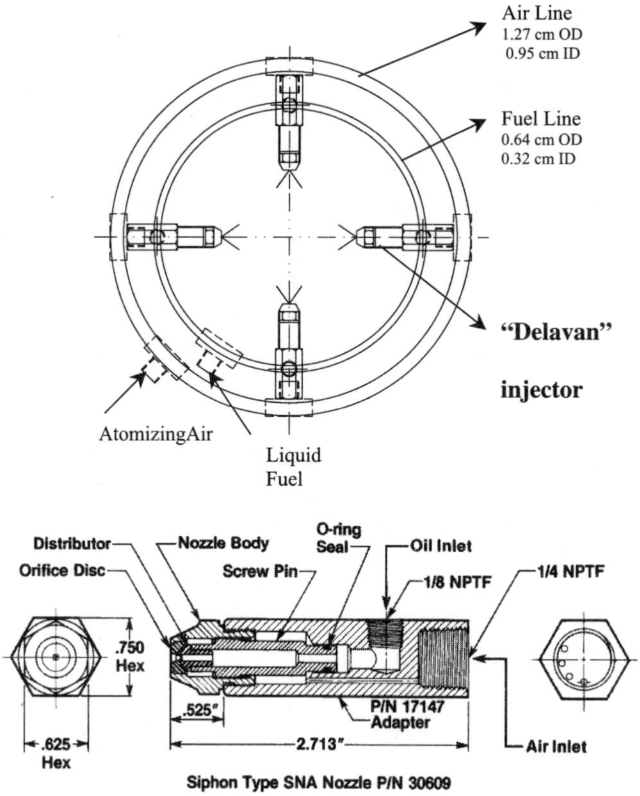


**Figure 27** Schematic Diagram of the Liquid Fuel Injection System (top) Sectional view of the Air Blast Injector, (bottom)

liquid fuel is supplied by a peristaltic pump to a 0.64 cm OD tubing connected to the fuel injectors. A pulse dampener is placed in the fuel line to maintain a relatively constant pressure supply. The secondary or atomizing air flow is supplied through a 1.27 cm OD line surrounding the fuel line.

Figure 28 shows the effect of equivalence ratio (Ø) on product gas temperature for kerosene (Figure 28a) and diesel (Figure 28b) combustion at airflow rate, Q = 335 lpm. These radial profiles show nearly constant temperature in the mid-region of the combustor, which exceeds the adiabatic flame temperature of unheated reactants. The product gas temperature increases with increasing equivalence ratio for both cases, as expected. Product gas temperatures higher than the adiabatic flame temperature pertain to significant reactant preheating by heat recirculation from the reaction zone. Figure 29 shows the thermal performance for kerosene combustion in terms of percentage heat released in the burner. Percent heat recirculation decreases slightly at high equivalence ratios and it decreases by about two percentages point with airflow rate increasing from 235 to 335 lpm (Fig. 29a). The 10-12 percent heat recirculation is comparable to that achieved with methane combustion. Figure 29b shows that the percentage heat loss is higher at lower airflow rate and higher equivalence ratios. The extra heat needed for fuel evaporation tends to reduce the heat loss

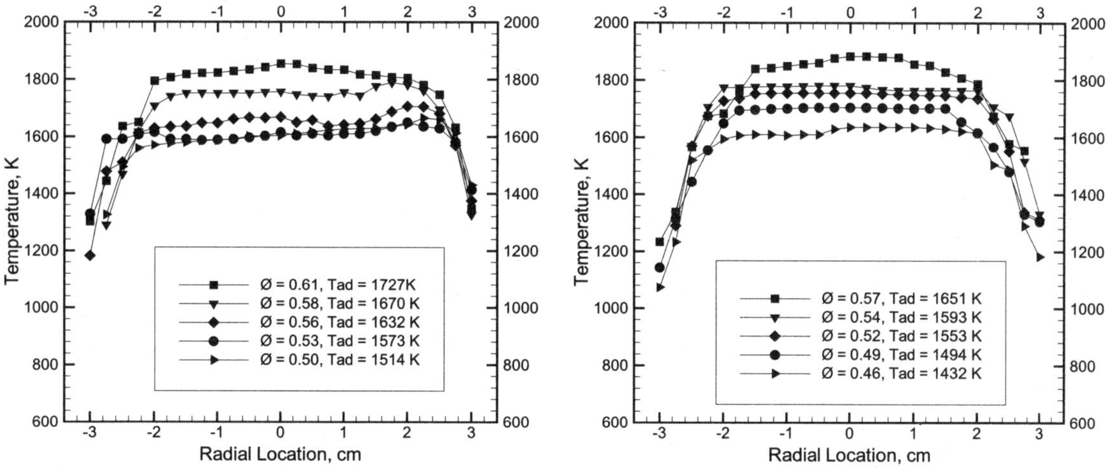


**Figure 28** Effect of Equivalence Ratio on Radial Profiles of Product Gas Temperature at the Combustor Exit Plane, Q = 335 lpm, for Surface Combustion of (a) Kerosene, (b) Diesel

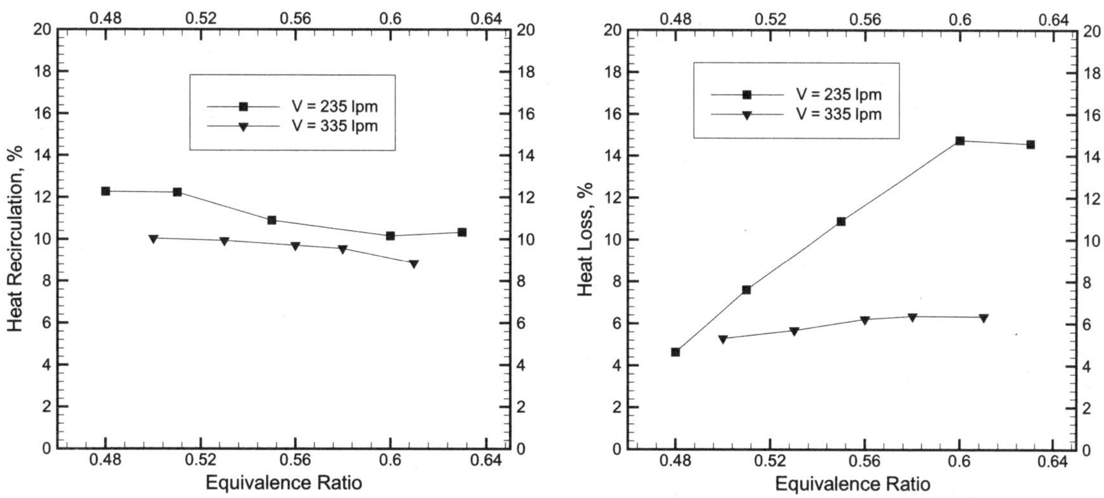


**Figure 29** Thermal Performance vs. Equivalence Ratio for Kerosene Surface Combustion, (a) Percentage Heat Recirculation, (b) Percentage Heat Loss

for liquid fuel combustion as compared to that for methane combustion. Next, emissions measurements were taken at the midpoint of the combustor exit plane for various equivalence ratios, obtained by fixing the airflow rate and varying the fuel flow rate. Figure 30 shows an increase of CO and a sharp rise in NOx concentrations at higher equivalence ratios (above 0.56) for kerosene combustion. The results also show that the emissions are nearly independent of the airflow rate for a given equivalence ratio.

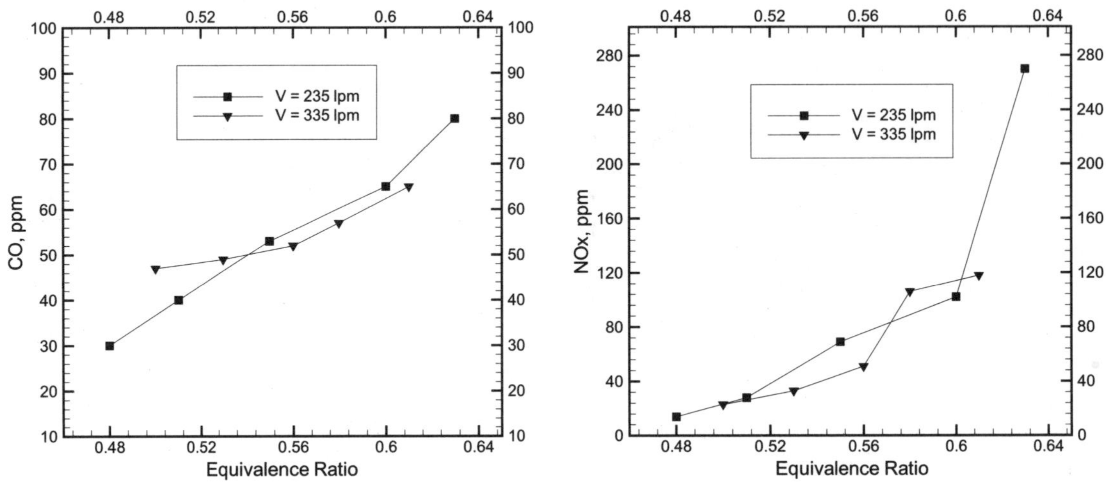


**Figure 30** Emissions vs. Equivalence Ratio at the Combustor Exit Plane, r = 0, for Kerosene Surface Combustion, (a) CO Concentration, (b) NOx Concentration

## 4.2 Miniature Combustor

The concept of recirculating combustor wall heat transfer to preheat the reactants is especially useful for miniature systems, wherein the wall heat transfer is a significant fraction of the heat released by combustion. We have experimentally demonstrated a miniature system (see Figure 31) with combustion chamber volume of 0.364 $cm^3$ to provide (a) high heat release rate of up to 90 W, (b) combustion efficiency exceeding 99.5%, (c) NOx emissions below 30 ppm, and no visible soot, and (d) heat loss of less than 15% of the heat released in the combustor (Marbach et al., 2007). Further, a detailed computational fluid dynamics (CFD) model was developed and validated to optimize the system performance (Marbach et al., 2006). The CFD model addressed the complex nature of flow through the PIM and radiation heat transfer, without the complexities of chemical kinetics processes. The CFD analysis identified two most important modes and pathways of heat loss; axial heat conduction through the combustor wall and radiation across the preheating annulus. When attempting to reduce the heat loss, a tradeoff between axial conduction and radiation is necessary because of the coupled nature of these processes.

## 4.3 Summary remarks

PIM combustion with annular heat recirculation offers the potential to achieve low-emission combustion of gaseous and liquid fuels. The combined effects of heat recirculation by conduction through the wall and upstream radiation/conduction in the PIM produce highly preheated reactants to exponentially increase the reaction rates in the PIM combustor. The high reaction rates extend the LBO limit, reduce NOx and CO emissions, and/or shorten the combustor length. Further, the thermal feedback in the annulus can be used to pre-vaporize the fuel for liquid fuel applications. The fuel and air can be premixed in the annulus to produce a homogeneous reactant mixture upstream of the reaction zone; a key requirement of low-emission combustion systems. The thermal

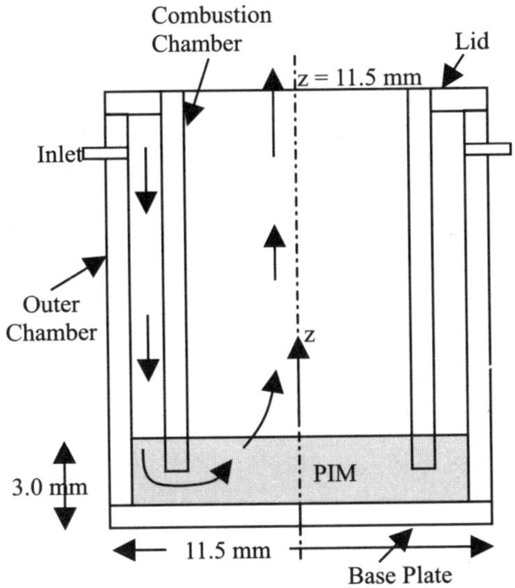


**Figure 31** Schematic Diagram of the Miniature Combustor

feedback and fuel-air premixing can be increased by strategically locating an additional porous bed around the combustor wall. The above experiments were conducted at atmospheric pressure using room temperature air. However, materials issues pertaining to high-pressure, high-temperature engine operating conditions must still be addressed. We have used SiC coated carbon foam to achieve high structural strength, which limited the maximum temperature for continuous operation to about 1200°C. Fortunately, the maximum operating temperature can be increased using CVD coating of carbon foam with materials such as hafnium oxide ($HfO_2$) and zirconium carbide (ZrC). Such efforts require simultaneous development and characterization of PIM combustion with the development of CVD coated carbon foam materials with melting point exceeding 2000°C.

## 5.0 LIQUID FUEL ATOMIZATION

A spray serves as the heart of almost every type of liquid-fueled combustion system. Large drops in a spray promote soot formation and local hot spots which cause NOx to form. Ideally, to promote combustion with maximum efficiency and minimum harmful emissions, an injector should deliver a fuel spray consisting of a fine mist with a uniform droplet size which can evaporate rapidly to yield a lean uniform mixture of gaseous fuel and air. From a practical standpoint, injectors should incur low pressure drop and be able to deliver good atomization over a large spectrum of fuel mass flow rates and with rapid mass flow changes (Lefebvre, 1989, 1999). The injector typically produces a fuel jet, which breaks up into spray in various regimes that depend upon the influence and interaction of the liquid's kinetic energy, surface tension, viscosity, and interaction with the surrounding air. Mechanism of liquid jet breakup into a spray was first broached by Rayleigh (1878), who analyzed stability of a low-speed, inviscid jet. The basic injection processes in widespread use include (1) pressure atomizer relying upon large pressure drop across the nozzle to create a high velocity fuel jet, (2) air-assist atomizer employing a low volume flow of high velocity air to break up the fuel jet, and (3) air-blast atomizer using a larger amount of air at moderate flow velocity to break up the fuel jet. The majority of fuel injection systems use combinations of these three processes in a wide variety of geometries, which can sometime become quite complex (Lefebvre, 1999). Excellent reviews concerning the preparation of fuel sprays optimized for lean combustion include those written by McDonell and Samuelsen (1991), Radzan (1998), Mansour (2005b), and Nakamura et al (2006).

A fourth process known as effervescent atomization injects high pressure air directly into the liquid fuel in a mixing chamber upstream of the discharge orifice. The injected air forms bubbles to produce a two-phase bubbly flow at the discharge orifice where the air bubbles expand rapidly to shatter the surrounding fuel into droplets. Effervescent atomization, as described by Sovani et al (2001), typically produces a spray with fine droplet sizes; however, the spray angle is small and air must be pressurized to the fuel's pumping level. The low-blurring (FB) injection concept explained in the next section also employs a form of effervescent atomization but differs considerably from current effervescent atomizers in that the two-phase bubbly flow is produced at the tip of the fuel supply line and, consequently, it requires a relatively small amount of low pressure atomization air.

## 5.1 Flow Blurring Injector Concept

The discovery of the FB injection concept was reported recently by Ganan-Calvo (2005). As illustrated in Figure 32, it consists of a liquid supply tube and an orifice plate situated downstream. The tube outer wall is tapered at the outlet, and the sharpened edge of the orifice in the plate is the same diameter as the inner diameter, d, of the tube. The gas flows through the lateral cylindrical passageway (LCP) between the tube end and the exit orifice. When the axial distance between the tube end and the orifice (H) is small (H < 0.25 d), some of the gas flowing radially through the LCP

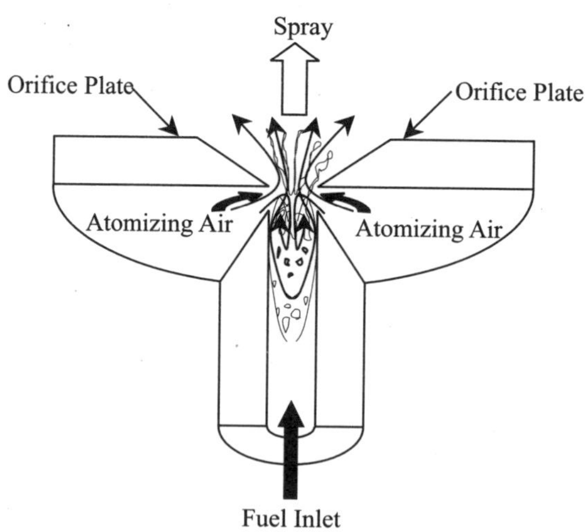


**Figure 32** Working Principle of the Flow Blurring Injector Concept

is forced a short distance upstream into the tube. Flow separation and vigorous mixing of the liquid and the gas occur in this short tube length. This process results in a fine spray forming downstream of the orifice plate. Ganan-Calvo reports that the FB configuration creates about five to fifty times more liquid surface than any other pneumatic atomizer of the plain jet air-blast type. For $H/d > 0.25$, the liquid flow forms a typical microjet, which breaks down by processes similar to those in air-assist or air-blast atomizers. Ganan-Calvo (2005) performed experiments with water and ethanol as liquids, and air, nitrogen, or argon as the atomizing gas. He developed a correlation to predict the droplet size in terms of H/d, Weber number, Ohnesorge number, and gas to liquid mass ratio. Ganan-Calvo (2005) states that the extremely simple geometry of the FB injector can be implemented down to micrometric scales, thus opening the door to improvement of current atomizers and to unprecedented designs. In the next section, we present our experimental data that indeed demonstrate that the FB injector results in significantly lower CO and NOx emissions in diesel and kerosene flames, as compared to those produced in flames formed by a commercial air-blast atomizer operated at identical conditions.

## 5.2 Combustion Experiments with the FB Injector

We performed atmospheric pressure experiments in a 46 cm long quartz combustor of 8.0 cm ID. The primary-air entered the vertical combustor through a swirler with six vanes positioned at 28° to the horizontal. The theoretical swirl number is 1.5, assuming that the flow exited tangentially from the swirler vanes. The bulk axial inlet velocity of the primary air is 1.9 to 2.1 m/s, which resulted in Reynolds number varying from 5960 to 6750. The liquid fuel was supplied to the injector with separate concentric tube inlets for fuel and atomization air. The comparison injector is a commercial air-blast atomizer (Delavan Siphon type SNA nozzle). The commercial version creates a swirling flow of atomizing air to breakdown the fuel jet after it exits the orifice plate. The FB injection concept was implemented in the same injector hardware, simply by replacing the air swirler with a spacer tube. In this latter configuration, air and fuel mix together before exiting through the orifice plate to form the spray downstream. One notable difference between the two injectors is the exit diameter of the fuel jet; approximately 0.4 mm for the air-assist injector versus 1.5 mm for the flow-blurring injector.

Figure 33 offers a quantitative evaluation of the injector performance in terms of the emissions profiles in the axial direction for kerosene flames. Results show that: (i) the CO emissions initially increase and then decrease in the axial direction, (ii) the NOx is formed within a short reaction zone near the injector exit, (iii) the CO profiles at the combustor exit display asymmetry for low atomizing airflow rates but they are uniform at higher atomizing airflow rates when combustion occurs in the premixed mode, (iv) the NOx emission profiles are uniform at the combustor exit plane, (v) both CO and NOx emissions decrease with increase in the atomizing airflow rate, and (vi) for a given atomizing airflow rate, the CO and NOx emissions for the FB injector are always lower than those for the AB injector. Experiments for diesel fuel also resulted in the same conclusion, i.e., for a given $Q_a$, the CO and NOx emissions for the FB injector were always lower than those for the AB injector. Measurements indicted that the pressure drop in the atomizing air supply line of the FB injector was less than 10% of that in the AB injector, even though no design optimization was carried out. Detailed results are reported in a recent paper by Panchasara et al. (2007).

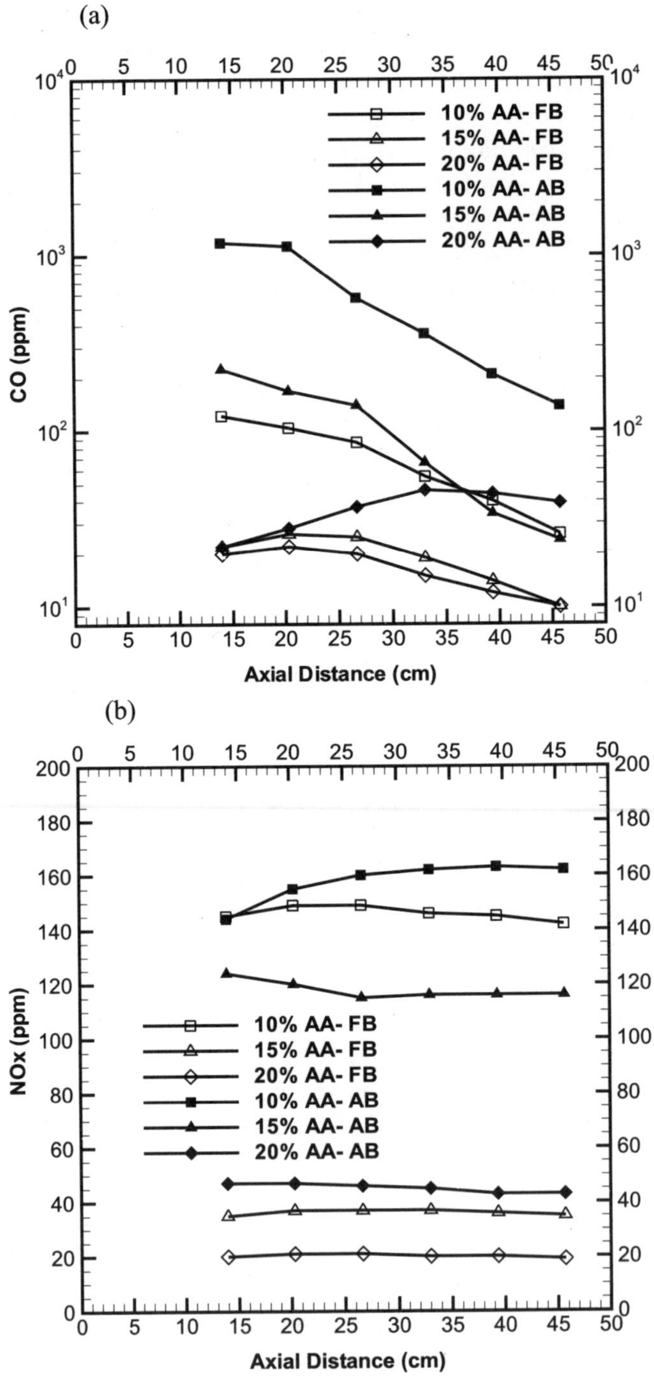


**Figure 33** Axial Profiles of Emissions in Kerosene Flames; (a) CO Concentration, (b) NOx Concentration. [Open symbols represent FB Injector, Closed Symbols represent AB Injector]

### 5.3 Summary Remarks

Our experiments show that the FB injector always results in significantly lower emissions of nitric oxides (NOx), carbon monoxide (CO) and soot in comparison to those from a commercial air-blast atomizer operated at identical fuel (kerosene and diesel), atomizing air, and total air flow rates. Results show that the FB injector concept is less susceptible to changes in the fuel physical properties, fuel flow rate, or impurities in fuel that could otherwise clog the injector. Moreover, the FB injector could be scaled up or down for applications in combustion systems with a broad power range. The simplicity of the FB injector makes it a prime candidate for a multitude of combustion applications.

## 6.0 FUTURE DIRECTIONS

Various concepts presented above offer exciting opportunities for combustion research in the future. Two examples are highlighted in this section; miniature liquid fuel combustion, and PIM-assisted combustion.

The potential for creating chemically-powered small-scale devices has intrigued combustion researchers. However, fulfilling the promise of miniature power devices requires combustion systems operated with high energy density liquid fuels. Liquid fuel combustion at small scales introduces the requirements of fuel dispersion, fuel vaporization, and fuel-air mixing, all within the short flow residence time, and preferably before combustion to minimize the emissions of soot, unburned hydrocarbons, CO, and NOx. Sirignano et al (2002) and Pham et al (2007) present the concept of injecting fuel as a liquid film on the combustor surface to redirect the wall heat transfer for fuel vaporization. However, experiments have shown incomplete fuel vaporization and inadequate fuel-air mixing, resulting in poor combustion efficiency. Kyritsis et al (2004a, 2004b) were the first to utilize an electrospray to disperse the liquid fuel and mix it with air before combustion occurred in a meso-scale catalytic reactor. Gomez et al (2007) present the next step of integrating a multiplexed electrospray-catalytic combustor-heat recuperator system with a Stirling engine to produce 42.4 W of electric power at fuel to electric efficiency of 22%. For heat release rate of about 250 W, the combustor/recuperator volume was 400 $cm^3$, and evaporation/mixing chamber volume was 6.6 $cm^3$.

We have developed a novel meso-scale combustion concept utilizing flow-blurring injector to produce fine fuel droplets, a counter-flow heat exchanger to preheat reactants using product gas enthalpy, and porous inert medium to homogenize reactants and stabilize the flame (Vijaykant, 2008). The overall system is 30 mm long and 17 mm in diameter, with combustor diameter of 10 mm. For atmospheric pressure operation on kerosene fuel, the combustion system achieved heat release rate of up to 460 W, pertaining to energy density of 90 $MW/m^3$ based on the total volume and 230 $MW/m^3$ based on the combustor volume. No soot or coking problems were experienced during or after combustor operation on kerosene fuel. Nearly 94 % of the heat released was retained by the products, thus, less than 6 % of the heat released was lost to the ambient. The combustor maintains good performance over a range of air and fuel flow rates. The next step is to integrate this system with a power generator to produce electricity.

Next, we introduce the term "PIM-assisted combustion" to refer to applications wherein the combustion with free flames is assisted by PIM combustion. The concept is akin to plasma-assisted combustion, which has received considerable attention in recent years. Unlike the plasma-assisted combustion requiring energy input from a high-voltage electrical source, the PIM-assisted combustion depends upon heat-released by chemical reactions. The PIM combustion can be strategically employed to assist the free flames to improve the overall performance. The underlining premises of this concept are: (i) the porous media creates an intense region of combustion which can be used to improve the stability of free flames, and (ii) the porous media homogenizes the flow field by disintegrating the large vortical structures. Thus, it can be employed in regions of large non-uniformities to locally homogenize the flow field. The PIM combustion implemented in localized regions can improve the stability of free flames, and most importantly, it could suppress the thermoacoustic coupling and combustion noise, both of which are serious problems in advanced combustion systems. Since PIM combustion is used together with free flames, the concept can be implemented without major hardware changes. However, much research remains to be done to achieve this objective.

## Acknowledgements

The author would like to thank his former/current graduate students whose work is reported in this paper: Sandeep Alavandi, Cristina Dumitrescu, Ryan Heatly, Timothy Marbach, Ryan Newburn, Heena Panchasara, Daniel Sequera, Benjamin Simmons, S. Vijaykant, and Donald Wicksall. The author would also like to thank sponsors who provided financial support for the research.

## References

Alavandi, S., and Agrawal, A.K., 2008, "Experimental Study of Combustion of Hydrogen-Syngas/Methane Fuel Mixtures in a Porous Burner," *International Journal of Hydrogen Energy*, vol. 33, pp. 1407-1415.

Anderson, D. N. (1975). "Effect of Hydrogen Injection on Stability and Emissions of an Experimental Premixed Prevaporized Propane Burner," *NASA TM X-3301*.

Bell, S. R., and Gupta, M. (1997). "Extension of the Lean Operating Limit for Natural Gas Fueling of a Spark Ignited Engine Using Hydrogen Blending," *Combustion Science and Technology* vol. 123, pp. 23-47.

Dumitrescu, C., (2006), "Experimental Study of Combustion of Gaseous and Liquid Fuels Using Porous Inert Media with Heat Recirculation," M.S. Thesis, Department of Mechanical Engineering, University of Alabama.

Ellzey, J.T., and Goel, R., (1995), "Emissions of CO and NO from a Two Stage Porous Media Burner," Combustion Science and Technology, Vol. 107, pp. 81-91.

Gañán-Calvo, A. M. (2005), Enhanced liquid atomization: From flow-focusing to flow-blurring, *Applied Physics Letters*, 86, 2141.

Gauducheau, J. L., Denet, B. and Searby, G. (1998). "A Numerical Study of Lean CH4/H2/Air Premixed Flames at High Pressure," *Combustion Science and Technology,* vol. 137, pp. 81-99.

Gomez, A., Berry, J.J., Roychoudhury, S., Coriton, B., and Huth, J., (2007), "From Jet Fuel to Electric Power using a Meso-scale, efficient Stirling Engine," Proceedings of the Combustion Institute, Vol. 31 (2007) 3251-3259.

Hardesty, D.R. and Weinberg, F.J. (1976). "Converter Efficiency in Burner Systems Producing Large Excess Enthalpies," Combustion Science and Technology, Vol. 12, pp. 153-157.

Howell, J.R., Hall, M.J., and Ellzey, J.L. (1996). "Radiation Enhanced/Controlled Phenomena of Heat and Mass Transfer in Porous Media," Progress in Energy and Combustion Science, Vol. 22, pp. 121-145.

Hsu, P.F., Evans, W.D., and Howell, J.R. (1993). "Experimental and Numerical Study of Premixed Combustion within Nonhomogeneous Porous Ceramics," Combustion Science and Technology, Vol. 90, pp. 149-172.

Jackson, G. S., Sai, R., Plaia, J. M., Boggs, C. M., and Kiger, K. T. (2003). "Influence of $H_2$ on the Response of Lean Premixed $CH_4$ Flames to High Strained Flows," *Combustion and Flame*, Vol. 132, pp. 503-511.

Kaplan, M., and Hall, M.J., (1995), "The Combustion of Liquid Fuels within a Porous Media Radiant Burner," Experimental Thermal and Fluid Science, Vol. 11, 1995, pp. 13-20.

Kotani, Y., and Takeno, T. (1982). "An Experimental Study on Stability and Combustion Characteristics of an Excess Enthalpy Flame, Proceedings of the Combustion Institute, Vol. 19, pp.1503-1509.

Kyritsis, D. C., Roychoudhury, S., McEnally, C. S., Pfefferle, L. D., and Gomez, A. (2004a). "Mesoscale Combustion: A First Step Towards Liquid Fueled Batteries," Experimental Thermal and Fluid Science, Vol. 28, pp 97-104.

Kyritsis, D.C., Coriton, B., Faure, F., Roychoudhury, S., and Gomez, A., (2004b), "Optimization of a Catalytic Combustor using Electrosprayed Liquid Hydrocarbons for Mesoscale Power Geenration," Combustion and Flame, Vol. 139, (2004), pp. 77-89.

Larsen, J. F., and Wallace, J. S. (1997). "Comparison of Emissions and Efficiency of a Turbocharged Lean-Burn Natural Gas and Hythane-Fueled Engine," *Journal of Engineering for Gas Turbines and Power*, vol. 119 pp. 218-226.

Lefebvre, A. H., 1989, Atomization and Spray, Hemisphere Publishing, New York.

Lefebvre, A.H. (1999). Gas Turbine Combustion, 2nd Ed. Taylor and Francis, Ann Arbor, MI.

Lieuwen, T., McDonnell, V., Petersen, E., and Santavicca, D., 2006, "Fuel Flexibility Influences on Premixed Combustor Blowout, Flashback, Autoignition, and Stability, *ASME Paper GT2006-90770.*

Maden, K. H. (1998). "Fuel Flexibility in Industrial Gas Turbines," Turbomachinery Technology Seminar, Solar Turbines, San Diego, CA.

McDonell, V. G. and Samuelsen, S. (1991). "Gas and drop behavior in reacting and non-reacting air-blast atomizer sprays", *J. Propulsion*, 7, 684.

Mansour, A., Parker Hannifin, Personal Communication, (2005a)

Mansour, A. (2005b). Gas turbine fuel injection technology, ASME Paper No. GT2005-68173.

Marbach, T. L and Agrawal, A. K. (2005). "Experimental Study of Surface and Interior Combustion Using Composite Porous Inert Media," Journal of Engineering for Gas Turbines and Power, Vol. 127, pp. 307-313.

Marbach, T.L., and Agrawal, A.K., 2006, "A Computational Study of Heat Recirculating Combustor using Porous Inert Media," *AIAA Paper-2006-1443*, 43rd AIAA Aerospace Science Meeting, Reno, Nevada.

Marbach, T.L., S. Vijaykant, and Agrawal, A.K., 2007, "Investigation of a Miniature Combustor Using Porous Media Surface Stabilized Flame," *Combustion Science and Technology*, vol. 179, pp. 1901-1922.

Meier, J. G., Hung, W. S. Y., and Sood, V. M. (1986). "Development and Application of Industrial Gas Turbines for Medium-BTU Gaseous Fuels," *Journal of Engineering for Gas Turbines and Power*, vol. 108, pp. 182-190.

Nakamura, S., McDonell, V., and Samuelsen, S. (2006). The effect of liquid-fuel preparation on gas turbine emissions, ASME Paper No.2006-90730.

Newburn, E.R., and Agrawal, A.K., 2007, "Liquid Fuel Combustion using Heat Recirculation through Annular Porous Media," *Journal of Engineering for Gas Turbines and Power*, Vol. 129, pp. 914-919.

Panchasara, H., Sequera, D., Schreiber, W., and Agrawal, A.K. (2007), Combustion performance of a novel injector using flow-blurring for efficient fuel atomization, 5$^{th}$ US Combustion Meeting paper # G12.

Pham, T.K., Dunn-Rankin, D., and Sirignano, W.A., (2007), "Flame Structure in Small-Scale, Liquid Film Combustors," Proceedings of the Combustion Institute, Vol. 31, (2007), pp. 925-931.

Phillips, J. N., and Roby, R. J. (1999). "Enhanced Gas Turbine Combustor Performance Using $H_2$-Enriched Natural Gas," *ASME Paper 99-GT-115.*

Pickenaker, O., Pickenacker, K., Wawrzinek, K., Trimis, D., Pritzkow, W.E.C., Muller, C., Goedtke, P., Papenburg, U., Adler, J., Standke, G., Heymer, H., Tauscher, W., and Jansen, F., (1999). "Innovative Ceramic Materials for Porous-Medium Burners," Interceram, Vol. 48, No. 5 and 6, pp. 1-12.

Rayleigh, W. S. (1878). On the instability of jets, *Proc. Lond. Math. Soc.*, 4, 10.

Razdan, M. K., (1998). Fuel/air preparation in the design of low emissions gas turbine combustion systems, NATO RTO Meeting Proceedings 14 on Gas Turbine Combustion Emissions and Alternative Fuels, Paper No. 34.

Ren, J. Y., Qin, W., Egolfopoulos, F. N., and Tsotsis, T. T. (2001). "Strain Rate Effects on Hydrogen-Enhanced Lean Premixed Combustion," *Combustion and Flame*, Vol. 124, pp. 717-720.

Richards, G. A., McMillian, M. M., Gemmen, R. S., Rogers, W. A., and Cully, S. R. (2001). "Issues for Low-Emission, Fuel-Flexible Power Systems," *Progress in Energy and Combustion Science*, Vol. 27, pp. 141-169.

Rumminger, M.D., and Dibble, R.W., (1996) "Gas Temperature above a Porous Radiant Burner: Comparison of Measurements and Model Predictions," Proceedings of the Combustion Institute, Vol. 26, pp. 1755-1762.

Schefer, R. W., Wicksall, D. M., and Agrawal, A. K. (2002). "Combustion of Hydrogen-Enriched Methane in a Lean Premixed Swirl-Stabilized Burner," *Proceedings of The Combustion Institute*, Vol. 29, pp 843-851.

Sherman, A.J., Williams, B.E., DelaRosa, M.J. and LaFerla, R., (1990). "Characterization of Porous Cellular Materials Fabricated by CVD", Proceedings of the 1990 MRS Fall Meeting, Boston.

Sirignano, W.A., Pham, T.K., and Dunn-Rankin, D., (2002). "Miniature Scale Liquid-Fuel Film Combustor," Proceedings of the Combustion Institute, Vol. 29, pp. 925-931.

Smucker M.T. Ellzey JL., (2004), "Computational and Experimental Study of a Two-Section Porous Burner," *Combustion Science and Technology*, 2004; Vol. 176, pp. 1171-1189.

Som, S., Ramirez, A.I., Hagerdorn, J., Saveliev, A., Aggarwal, S.K., (2008), "A numerical and experimental study of counterflow syngas flames at different pressures," *Fuel*, vol. 87, n 3, March, 2008, p 319-334.

Sovani, S. D., Sojka, P. E., and Lefebvre, A. H. (2001) Effervescent atomization, *Progress in Energy and Combustion Science*, 27, 483.

Trimis, D., and Durst, F., (1996). "Combustion in a Porous Medium- Advances and Applications," Combustion Science and Technology, Vol. 121, p. 153.

Vijaykant, S., and Agrawal, A.K., 2007, "Liquid Fuel Combustion within Silicon-Carbide Coated Carbon Foam," *Experimental Thermal and Fluid Science*, vol. 32, pp 117-125.

Vijaykant, 2008, "Meso-scale Combustion of Liquid Fuels using Porous Inert Media," Doctoral Dissertation, Department of Mechanical Engineering, The University of Alabama, 2008.

Viskanta, R., 1995, "Interaction of Combustion and Heat Transfer in Porous Inert Media," in Transport Phenomena in Combustion, Ed. S.H. Chan, pp. 64-87, Taylor and Francis.

Vogel, B.J. and Ellzey, J.L., (2005), "Subadiabatic and Superadibatic Performance of a Two-Section Porous Burner," *Combustion Science and Technology*, 2005; Vol. 177, pp. 1323-1338.

Wicksall, D.M., Agrawal, A.K., Schefer, R.W., and Keller, J.O., 2005, "Influence of Hydrogen Addition on Flow Structure in Enclosed Swirling Methane Flame," *AIAA Journal of Propulsion and Power*, vol. 21, pp. 16-24.

Wicksall, D.W., Agrawal, A.K., Schefer, R.W., and Keller, J.O., 2005, "The Interaction of Flame and Flow Field in a Lean Premixed Swirl-Stabilized Combustor Operated on $H_2/CH_4$/Air, *Proceedings of the Combustion Institute*, vol. 30, pp. 2875-2883.

Wicksall, D.M., and Agrawal, A.K., 2007, "Acoustics Measurements in a Lean Premixed Combustor Operated on Hydrogen-Hydrocarbon Fuel Mixtures," *International Journal of Hydrogen Energy*, Vol. 32, pp. 1103-1112.

# Laser Ignition of Engines

J. Tauer[1], H. Kofler[1], K. Iskra[2], G. Tartar[1] and E. Wintner[1]
[1]Institut für Photonik, Technische Universität Wien, Gusshausstrasse 25-29, A-1040 Wien, Austria
[2]Institut für Experimentalphysik, Technische Universität Graz, Petersgasse 16, A-8010 Graz, Austria

## 1.0 INTRODUCTION

A steadily growing fraction of electricity production is based on combined heat and power plants where the mechanical energy of internal combustion engines is employed to drive electric generators, and the exhaust gas is used for the exploitation of thermal energy. Consequently, the efficiency of such power plants and hence of the engines must be as high as possible while the pollutant emissions should be kept low. Such requirements cannot be realized satisfactorily any more when using conventional engine ignition techniques. Spark plugs reach their limits at the necessary high ignition pressures demanding for excessively high voltages. However, there are several alternative concepts like plasma ignition, high-frequency ignition, Diesel micro-pilot ignition, corona ignition and laser ignition which might contribute to an improvement of the overall efficiency. To our knowledge, laser ignition represents the most promising future ignition concept out of a number of reasons [1-4]. The main advantages of laser ignition, among many others, are performance enhancing high effective mean pressures in the combustion chamber as well as the feasibility of very lean mixtures lowering the flame temperature and consequently the $NO_X$ emissions.

Another promising field of potential application is direct fuel injection in modern gasoline engines. In this case, very high precision and constant shape of the injection plume would be required producing curls at its edges to guarantee spark ignition under all operation conditions. Also the injection pressure generally has to be higher than in other concepts. Large droplets of fuel have to be avoided because they might cause electrical shorts and/or carbonaceous layers at the electrodes. As these conditions hardly can be met, conventional spark plugs face severe drawbacks which might be overcome by laser ignition. Even varying injection cones can be ignited because plasma formation happens at the right position nearly under all circumstances where the right fuel/air concentration is met. This eventually leads to reduced fuel consumption and most reliable ignition [5].

Furthermore, first tests of application of laser ignition to HCCI (homogeneous charge compressed ignition) have shown a useful potential to make this technique applicable to realistic general purpose engine operation. This engine operation scheme is based on reaching self-ignition of lean mixtures by applying the necessary pressures and temperatures to the load dependent fuel/air mixtures. This represents a very delicate requirement associated with enormous efforts to guarantee the precise adjustment of these parameters. If an external source of inflammation is added, and laser ignition would optimally represent such a trigger, precise timing and a much larger range of tolerable parameters can be achieved. Besides, even for lean mixtures rather low laser energies are required (compare ignition of lean mixtures below) [6].

## 2. BACKGROUND AND COMBUSTION CHAMBER RESULTS

In general, the mechanism of laser ignition is based on non-resonant gas breakdown [7] of the tightly focused pulsed (ns) laser beam. Initial electrons absorb photons to gain energy via the inverse bremsstrahlung process. These energetic electrons can ionize gas molecules leading to the breakdown in the focal region via the electron cascade growth. It is important to note that this process requires initial seed electrons [8]. These electrons might be produced from thermally heated or linearly ionized impurities like soot or dust in the gas mixture. The plasma formed by the mentioned mechanism can ignite the combustible mixture. Plasma diagnostics revealing shock waves, ignition kernels and flame propagation was observed by high-speed Schlieren photography [9]. The different phases of laser ignition can be defined in chronological order [10]:

i. Electric breakdown and energy transfer from laser to plasma
ii. Shock-wave generation and propagation
iii. Gasdynamic effects
iv. Chemical induction of branching chain reactions of radicals leading to ignition
v. Turbulent flame initiation

In order to illustrate plasma formation and the onset of combustion, the combined Fig. 1 is depicted. The left image (a) therein shows a multi-exposure image enabling the observation of a shockwave in 500-ns steps. The temporal expansion of the flame kernel is also illustrated using Schlieren photography (b) and PLIF imaging (c). The laser pulse entering from the right had an energy $E_p \approx 140$ mJ, the initial pressure of the mixture amounted $p_{in} = 4.3$ bar and the relative air/fuel ratio $\lambda = 1.3$ [11].

Non-resonant breakdown in gases requires intensities of several $10^{11}$ W/cm² in the focal region [1,2,12-15] to be provided by a focused laser pulse with minimum energy $E_p = 0.1$ mJ and pulse durations $\tau_p \leq 10$ ns. For pulses down to several ps the breakdown threshold increases according $\tau_p^{-1/2}$ [8]. Experiments revealing the minimum pulse energies MPE for ignition of different fuel-air mixtures (methane-air, hydrogen-air, methane-hydrogen-air) under engine-like conditions have been reported in former work, partly by our group [1,2,12,16-19]. Furthermore, laser ignition for biogas applications has also been studied successfully [20]. While a plasma can be formed by pulse energies of only a few 100 μJ @ 8 ns, the development of a flame kernel requires the more energy in order to ignite the mixture the higher $\lambda$ and the lower the temperature T is [21]. This circumstance is clearly illustrated in Fig. 2. Moreover, for stoichiometric mixtures MPE is independent of T, as it can be seen in the figure.

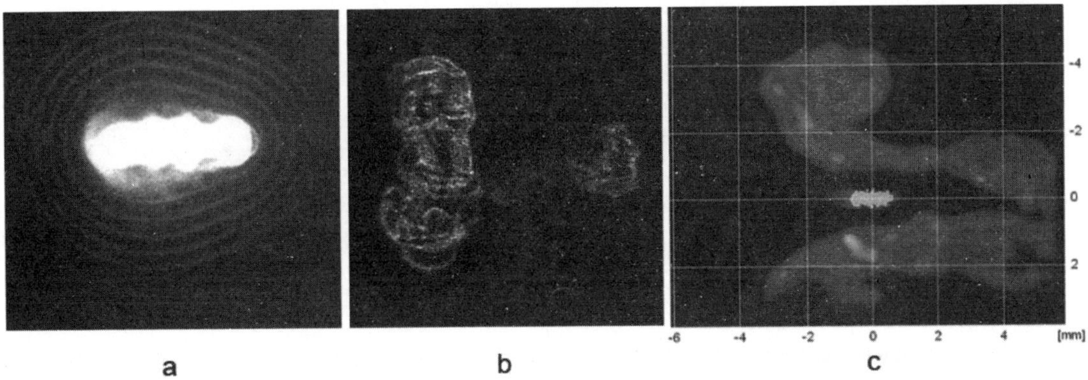


**Figure 1** (a) Multi-exposure Schlieren Image of Shockwave Emission Around a Plasma Spark in air at 10 Bar in 500-ns Steps after Ignition. (b) Schlieren Image of a Stoichiometric Methane-air Mixture at 10 Bar, 500 μs after Ignition. (c) Planar Laser-Induced-Fluorescence (PLIF) Image of OH-radicals; in all Cases, the Laser Beam Enters from the Right Hand side [9,11]

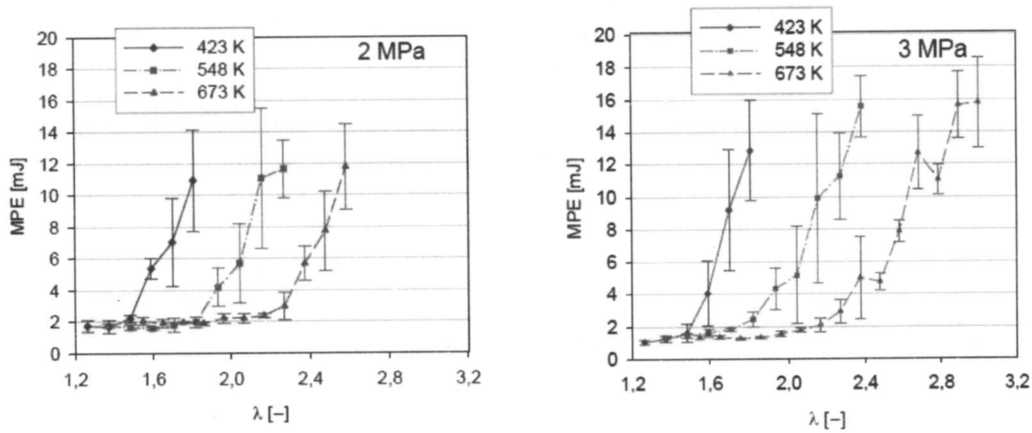


**Figure 2** The Minimum Pulse Energy (MPE) Versus the Relative Air/Fuel Ratio (λ) for Different Temperatures (T) and Initial Pressures ($p_{in}$) of 20 Bar (left) and 30 Bar (right). The Measurements were taken in a Constant Volume High-Pressure Vessel [21]

The higher $p_{in}$ and λ are, the higher is the gas engine efficiency and the lower are the $NO_X$ emissions, $\lambda \geq 2$ is desirable. In order to ensure reliable ignition for various operation conditions of the gas engine, a ns-laser pulse with $E_p$ = 5–10 mJ is necessary. Furthermore, we could prove that shorter ns-pulses (≈ 1 ns) are more effective than long ns-pulses (> 10 ns) as the transmission of a plasma decreases with $\tau_p$ down to several 100 ps.

It is interesting to compare these results of laser ignition depending on λ and p with well-known flammability limits out of the literature [22] shown in Fig. 3.

In Fig. 4, a constant volume test of ignition probability was carried out over values of λ from 1.40 to 2.30 and ignition pressures ranging from 10 to 40 bar. A clear advantage of laser ignition can be observed expressed by the red shaded area representing 100% ignition probability.

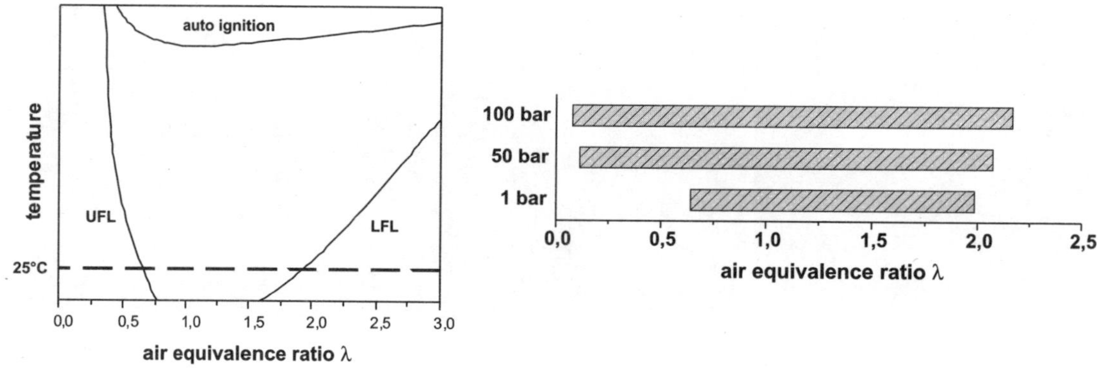


**Figure 3** a) Schematic Drawing of the Temperature Dependence of Flammability Limits. UFL – Upper Limit, LFL – Lower Limit. Autoignition starts at a Certain Temperature. b) Methane Flammability Limits for Different Ignition Pressures at 20°C [22]

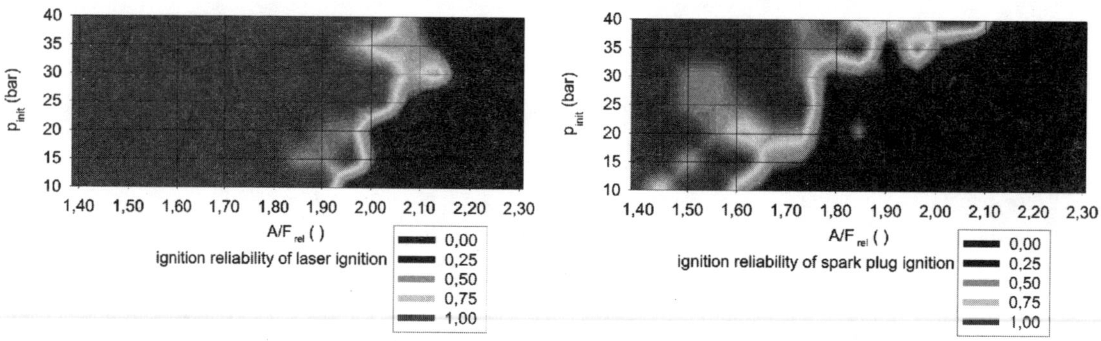


**Figure 4** Comparison of Ignition Probability between Laser and Spark Plug in a Combustion Chamber of Constant Volume. Methane/air, $T_{gas}$ = 200°C; Laser Quality Factor $M^2$ < 1.2; Laser Pulse Energy $E_p$ = 25 mJ Constant, well above the Plasma Breakdown Threshold for all Conditions; overall Ignition Attempts for Statistics: 1201 for Spark Plug, 642 for Laser. Currently: Maximum λ = 1.77 and BMEP = 19 Bar on the Engine for Reliable Run, Typical Spark Duration = 400 – 500 μs [23]

## 3. LASER IGNITION CONCEPTS

The concept of an optical ignition system emerged in the late 1960ies and has been successively studied by several groups [8,10,14,24,25]. The coupling of laser radiation into the combustion chamber can be realized by various approaches [25], two of them are:

- A separate ignition laser mounted on every cylinder head being supplied by an external pump source (laser diode), termed laser spark plug.
- Laser unit and engine separated and the ignition pulses transported via optical fibers to the cylinders.

The conceptual ideas (compare Fig. 5) and experiments dealing with the development of a compact ignition laser can be found in a recent publication by us [26]. Such a laser system, if mounted on the cylinder as an optical spark plug, must be small in size and has to be manufactured

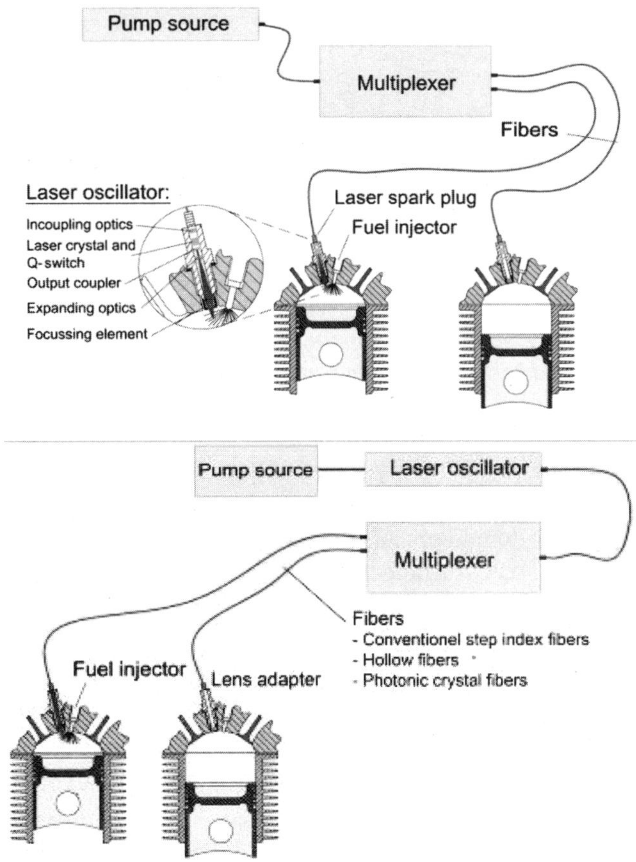


**Figure 5** Two Possible Laser Ignition System Concepts: Common Pump Source (above) or Common Oscillator (below)

in a robust form in order to withstand the adverse influences on the cylinder head. The regime of short ns-pulses can be well covered by passivelyQ-switched lasers which are, due to their nature, robust, simple in design and cheap. The pump pulse generating the inversion in the laser active medium can be propagated through a conventional optical step-index fiber from an external laser diode to the solid-state laser. In Fig. 6 the concept of an optical spark plug is depicted.

The transportation of ignition pulses via optical fibers implicates some plain advantages, but as a matter of fact, the restrictive factor is the damage threshold of the fiber material. Such pulses may induce intensities in a small core far beyond $1.5 \times 10^9$ W/cm² which is the break-down threshold of fused silica for ns-laser pulses [27,28]. An enlargement of the solid fiber core area, where the beam is guided, in order to reduce the intensity is not suitable since for fixed wavelengths higher core diameters imply multi-mode beam profiles. The focusability of the ignition pulse depends strongly on the beam profile, hence disadvantageous multi-mode radiation would increase MPE.

However, there exist some concepts for the transportation of ignition laser pulses via hollow-core optical fibers [29-31]. The employment of Photonic Band Gap (PBG) fibers would bring some

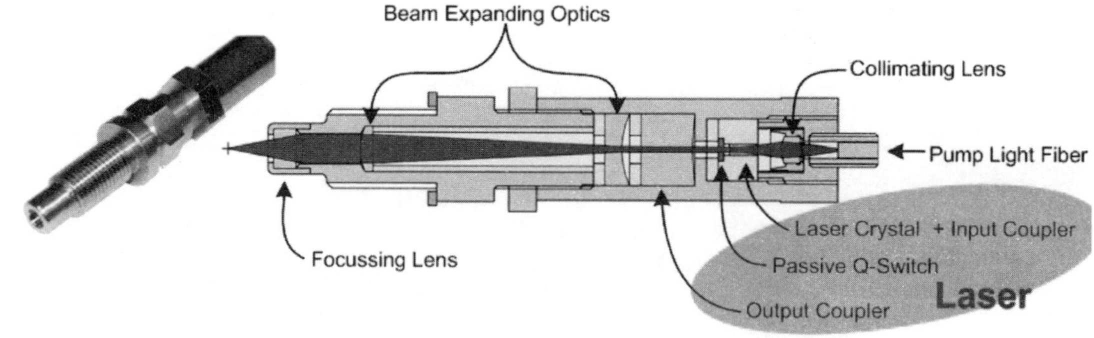


**Figure 6** Preliminary Concept of an Optical Spark Plug Comprising a Diode-pumped, Q-switched, Solid-state Laser and Suitable Focusing Optics

advantages like single-mode propagation. As reported in a recent paper by our group [32], the highest transmitted single pulse energy of approximately 0.8 mJ is not adequate for laser ignition. The further improvement of hollow fibers has to be successfully achieved before an increase of the delivered pulse energy up to a level sufficient for laser ignition can be realized.

In contrast to the ignition pulse, the pump beam can be guided via conventional optical fibers since the intensity is lower by about six orders of magnitude. In addition, the pump beam generated from diode bars generally is not of single mode nature and therefore only multi-mode fibers are appropriate.

A further cornerstone of laser ignition is the window representing the gateway between the hot combustion chamber and the laser oscillator. There is a number of good candidates for window materials whose properties are enlisted in Table 1. So far, nearly all experiments involved either sapphire or silica. The former has a higher bandgap and hence is more robust against electro-optic damage as well as it is also mechanically the hardest one. However, it is most expensive making it very hard to produce lenses of every desired shape. The latter, silica, is much cheaper, however less robust. Its lower heat conductivity, however, may even be advantageous allowing to maintain gradients of temperature between the inner and outer window surfaces.

Previous investigations on the window include combustion-sided deposits and their formation as well as laser-induced deposit formation. Even a few deposits on the window lead to substantial absorption of laser radiation. Deposits on the combustion window are not well defined substances ranging from nutty, carbonaceous matters to small inorganic particles [33].

The ignition laser system must fulfil the aforementioned requirements in order to ensure reliable operation under engine-like conditions. Unfortunately, such small ns-pulsed laser systems are commercially not available. Moreover, customary laser systems fulfilling the needs of laser ignition are large in size, heavy, expensive and consume several kilowatts of electrical power. Thus the need to develop an ignition laser of our own emerged.

An end-pumped fiber-coupled passively Q-switched solid-state laser seems to be the most promising candidate. Such a laser system satisfies the requirements like compactness, robustness, low cost and the pumpability via optical fibers. Fig. 5 (upper scheme) shows the schematic concept

**Table 1** Properties of Some Suitable Window Substrates [34,35]

| | *Sapphire* $Al_2O3$ | *Fused Silica* $SiO_2$ | *ALON* $Al_{23}O_{27}N_5$ | *Spinel* $MgAl_2O_4$ | *Magnesia* $MgO$ | *YAG* $Y_3Al_5O_{12}$ |
|---|---|---|---|---|---|---|
| Refractive index n | 1.77 | 1.55 | 1.79 | 1.72 | 1.74 | 1.83 |
| Window reflection loses | 0.14 | 0.09 | 0.15 | 0.13 | 0.14 | 0.16 |
| Band gap (eV) | 9.9 | 8.4 | - | - | 7.8 | - |
| Flexure strength (MPa) | 1200 (25°C) | 50 (25°C) | 310 (25°C) | 170 (25°C) | 130 (25°C) | 280 (25°C) |
| | 200-600 (800°C) | 75 (900°C) | 270 (1000°C) | | | |
| Fracture toughness (MPa/m$^2$) | 3 | 0.8 | 1.4 | 1.5 | 1 | 1.0/1.4 |
| Hardness (kg/mm$^2$) | 1500 - 2200 | 600 | 1850 | 1140 | 690 | 135 |
| Melting point (K) | 2300 | 1900-2000 | 2300 | 2400 | 3100 | 2200 |
| Thermal conductivity (W/mK) | 58 | 8 | 13 | 30 | 30 | 15 |
| | | 13 | | 25 | 21 | 13 |
| Thermal expansion (10$^{-6}$K) | 6,7 | 12.46.9 | 5,7 | 7 | 10.6 | 7.7 |

of laser ignition applying a laser spark plug with an external pump source. Besides this advantageous concept, other research groups favour the idea of a transversally-pumped solid-state laser for ignition. To our knowledge, transversally-pumped laser systems are worse in beam quality and pulse duration. Moreover, such side-pumped systems are more complex and expensive than end-pumped ones. The following section introduces the schematic setup of the laser system as well as the experimental realization.

## 4. LASER EXPERIMENTS

### 4.1 Basics of the Setup

A suitable laser ignition system covering the requirements like robustness, simple design and compactness is a fiber coupled, end-pumped passively Q-switched solid-state laser. Fig. 7 depicts the schematic setup of such a laser unit. Beside the financial aspect the availability of the components is a major point within the selection criteria. Therefore $Nd^{3+}$:YAG as laser active medium and $Cr^{4+}$:YAG as passive absorber have been chosen.

The Nd-doping of the laser crystal was around 1.0 at. %, and the fluorescence lifetime of the upper laser level $^4F_{3/2}$ amounts to $\tau_{fl} \approx 255$ µs. The absorption line of $Nd^{3+}$:YAG is centered around $\lambda_{diode} = 808$ nm being suitable for a GaAs high power laser diode as pump source. Pump pulses up to a power $P_{pump} = 300$ W @ 500 µs can be generated by this laser diode. Since the laser diode shows a temperature gradient of 0.3 nm/K, temperature stabilization was necessary. Furthermore, as shown in Fig. 8, temperature stabilization enables the exact adjustment of the diode emission line and the crystal absorption wavelength with the aim of highest conversion efficiency. A step-index fiber for transportation of the pump pulse with a core diameter of 600 µm and a numerical aperture NA = 0.22 were employed.

Different aspheric collimating lenses with effective focal lengths $l_{f,eff} = 1.5$-18 mm were tested, however, the lens with $l_{f,eff} = 8$ mm and NA = 0.5 turned out to be the most suitable for in-

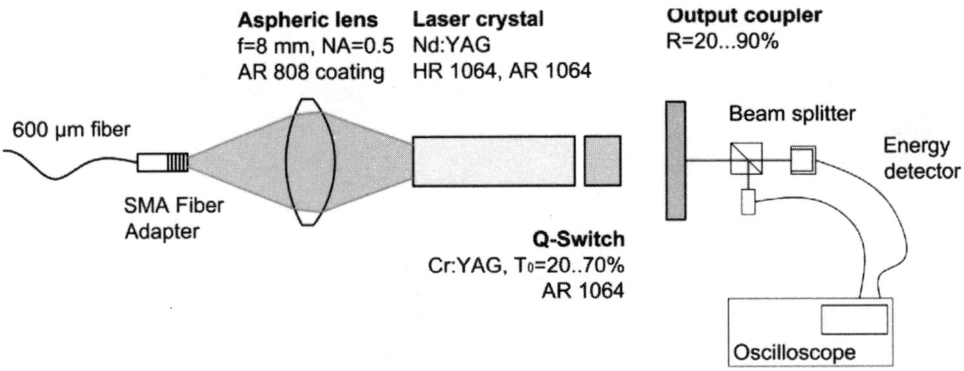


**Figure 7** Schematic view of the Developed Laser System. The Pulse Length was Measured by a Photodiode while for the Energy Measurement a Pyroelectric Sensor was used

coupling the pump beam into the laser crystal. The in-coupling mirror of the resonator was directly coated onto the crystal in order to reduce the length of the system and, as a result, the pulse duration. Moreover, the end face of the crystal was coated with an anti-reflection (AR) layer for the 808 nm pump beam reducing the reflection losses. The output-coupler was separated from the crystals since the adjustment of the mirror allows flexibility in pulse energy and beam profile.

Passive absorbers with initial transmissions $T_0$ = 25%-70% and output-couplers with reflectivities R = 25%-90% cover the field of laser operation. All passive absorber crystals were coated with an AR layer for the laser wavelength $\lambda_{em}$ = 1064 nm. In order to achieve highest output power the in-coupling optics were always adapted with respect to the resonator conditions (length, reflectivity, initial transmission). The results are summarized in the next section.

## 4.2 Resulting Laser Performance

The main parameters strongly influencing the maximum output power are the pump duration $\tau_{pump}$, $T_0$ and R. At a pump power $P_{pump}$ = 70 W, $\tau_{pump}$ = 300 μs, R = 50% and $T_0$ = 40% the maximum pulse energies of approximately 2.1 mJ were yielded. This result corresponds to an optic-to-optic efficiency $\eta_{optic}$ = 7.1%. The length of the resonator was about 15 mm while the length of the Nd-doped laser rod having ø = 2 mm thickness was 5 mm. Pulse durations $\tau_p \approx$ 1.0 ns were achieved leading to a peak power $P_{p,peak}$ > 2 MW.

An additional parameter influencing $E_p$ is the crystal doping being typically between 0.8-1.4 at.%. However, some manufacturers offer 2.5 at.%-doped Nd:YAG crystals which result in a higher cross-section and, therefore in higher absorption efficiency, but the fluorescence life-time $\tau_{fl}$ decreases dramatically. For our purposes, a compromise between the absorption efficiency and the storage ability of the upper laser level had to be found. It turned out that Nd-doping around 1.4 at.% led to the maximum output power.

An increase of $P_{pump}$ = 300 W and $\tau_{pump}$ = 300 μs lead to $E_p \approx$ 6 mJ at $\tau_p \approx$ 1.0 ns using the in-coupling lens shown in Fig. 7, but with an effective focal length $l_{f,eff}$ = 4.6 mm. We found out, that $E_p$ is very sensitive to $\tau_{pump}$ due to the thermal load of the laser crystal. Fig. 9 depicts the experimental data of $E_p$ versus $\tau_{pump}$.

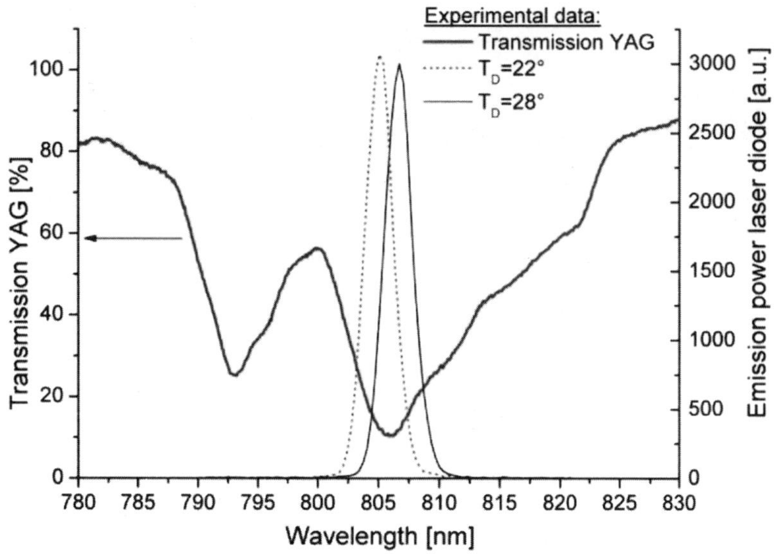


**Figure 8** Emission and Absorption Spectrum of the Laser Diode and the Laser Crystal, Respectively. At a Emission Wavelength $\lambda_{diode}$ = 805.7 nm, Corresponding to a Diode Temperature ($T_{diode}$) of Approximately 26°C, the Maximum Absorption (Minimum Transmission) was Noted

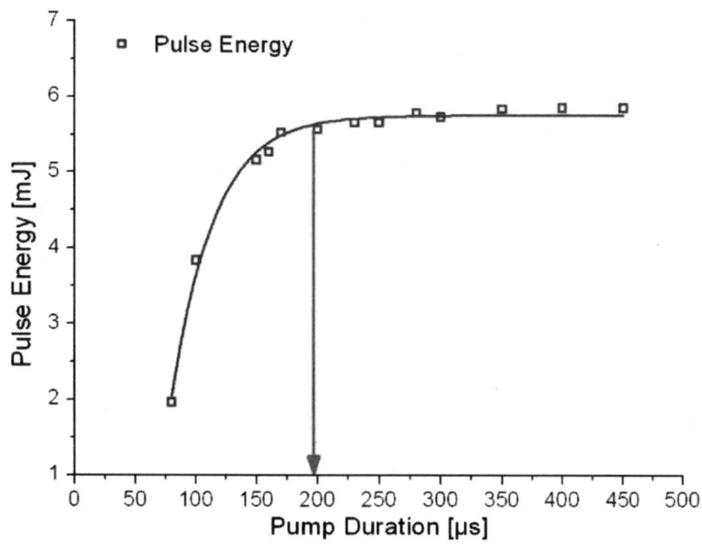


**Figure 9** The Ignition Laser Pulse Energy $E_p$ vs. Pump Duration $\tau_{pump}$. At a Certain $\tau_{pump}$ » 200 µs $E_p$ does not Rise any Further

An improvement in performance can be yielded by a thicker laser crystal, on the one hand, and by a more efficient in-coupling process, on the other hand, resulting in up to 10 mJ pulse energies. A further interesting aspect is the intensity profile of the outgoing beam. While the intensity

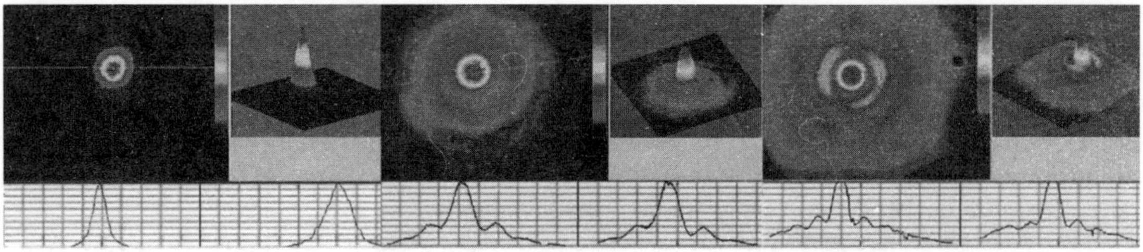

a b c

**Figure 10** Observed Transversal Modes of the Outgoing Laser Beam. For Pulse Energies $E_p$ in the Order of 2 mJ (a) the Characteristic Gaussian Mode can be seen. For Higher Energies (in (b), $E_p$ = 6 mJ, in (c), $E_p$ = 10 mJ)) also Higher Modes are Present

distribution for a pulse with $E_p$ = 2 mJ is quasi-Gaussian (Fig. 10a), the profile for pulses with energies $E_p$ = 6 and 10 mJ is somewhat different. In addition to the fundamental Gaussian mode higher $TEM_{lm}$ modes can oscillate [30]. In our case, we observed $TEM_{10}$ for $E_p$ = 6 mJ (Fig. 10b) and $TEM_{20}$ (Fig. 10c) for $E_p$ = 10 mJ.

## 4.3 Discussion of the Laser Experiments

Although passively Q-switched solid-state lasers using Nd:YAG as active medium are well-known and field tested with respect to many applications [37-41], we spent various investigations towards an optimization of $E_p$ at pulse length in order the of 1 ns. The special requirements of laser ignition with respect to pulse energy could be satisfactorily fulfilled by our system. Taking the maximum pulse energy of 10 mJ at a pulse duration of approximately 1.5 ns into account, the creation of a plasma in air under atmospheric condition is possible. Therefore, also fuel/air mixtures can be ignited with this laser unit since the plasma breakdown thresholds of combustible gases are in the same order of magnitude as air. Furthermore and most importantly, the output power of 10 mJ @ 1.5 ns is sufficient to also feed the flame kernel of the combustion in case of very lean mixtures.

This improvement of pulse energy is mostly a consequence of optimized in-coupling optics and thicker laser crystals. To our knowledge, these output parameters $E_p$ = 10 ns @ $\tau_p$ = 1.5 ns corresponding to an optic-to-optic efficiency $\eta_{optic}$ > 11% are unique due to the fact that no low-repetitive, self-Q-switched, high-peak power and compact lasers have been developed and published up till now. In [38] a compact monolithic passively Q-switched Cr,Nd:GSGG microlaser with a single-pulse energy of several 100 μJ at $\tau_p$ = 1.5 ns was reported. Further experimental studies on compact, longitudinal diode-pumped Nd:YAG laser employing $V^{3+}$:YAG as saturable absorber have been discussed in [40]. The maximum observed pulse energy therein was $E_p$ = 131 μJ at $\tau_p$ = 6 ns and an $\eta_{optic}$ = 2.7%. Another small Nd-doped laser has been published in [41] using a 600 W quasi-cw diode. The achieved output parameters of the 11 mm long resonator were pulse durations of about $\tau_p$ = 4 ns and pulse energies $E_p$ = 4.5 mJ. An interesting pump arrangement, where the laser rod was pumped at both end-faces leading to better absorption efficiency can be found in [39]. Nevertheless, such a laser setup is not useful for our purposes since the ignition laser should be assembled into a small, rod-like housing.

Recently, a first monolithic and hence mechanically very robust version of this Q-switched Nd:YAG laser could be realized, with Ep ≈ 6 mJ being already sufficient for gas engine application. Details of this approach will be published elsewhere.

## 5. SOME ENGINE RESULTS

Already several years ago, the first laser ignition engine tests were carried out on one independently controllable cylinder of a several hundred kW gas engine fabricated by Jenbacher AG[1]. These test bed experiments revealed the potential for substantial $NO_x$ reduction very clearly as illustrated in Fig. 11. Additionally, and not of least importance, a clear increase of effectivity can be expected which is currently ≤ 46%. Details still underlie privacy of our partner company. After reaching maturity of the laser ignition system, longer service intervals due to the avoidance of the spark plug electrode erosion problem can be expected.

We believe that large gas engines as depicted in Fig. 12 are the first and best candidates for successful economic performance of the laser ignition principle which may be realized in a few years.

Another technically realistic field of application for laser ignition is direct injection gasoline engines, i.e. spray-guided combustion, as already mentioned in the introduction (Fig. 13).

A major advantage is the reliable ignition at the edge of the combustion spray cone which can vary according to the changing load. Separate optics allow to use a cheaper planar incoupling window which has only to fulfil the requirements with respect to pressure, chemical and photochemical erosion. Additionally, the spot size faces a rather high fluence for a given laser pulse

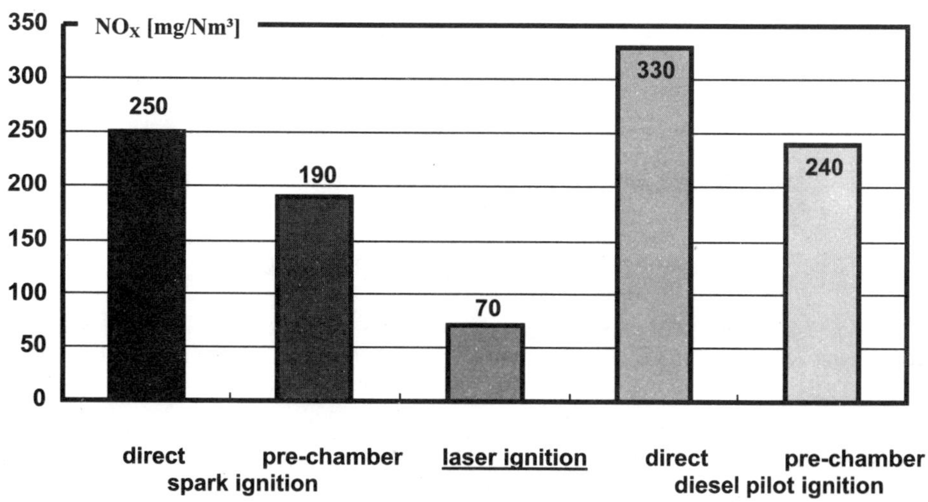


**Figure 11** Comparison of $NO_x$ Emissions with Different Ignition Sources for Gas Engines Operated with Methane/air Mixtures

[1]Due to a take over 4 years ago, currently GE Jenbacher is a branch factory of General Electrics, division GE Energy.

Figure 12 20-cylinder, 3-MW Gas Engine of GE Jenbacher

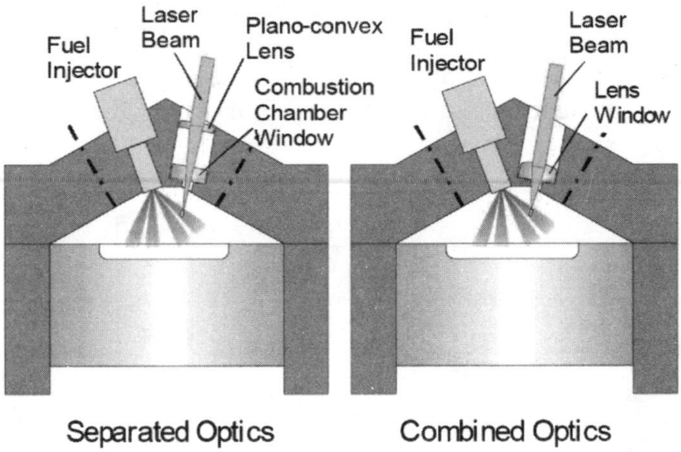


**Figure 13** Spray-guided Combustion of Gasoline Engines. Laser Ignition can be Realized with 2 Different Set-ups for the Optics: Separated and Combined. For the same Beam Parameters the Fluence at the Entrance Spot is Different having Substantial Impact on the Window Deposition Behaviour

energy due to the converging beam which is able to ablate some deposits. In case of the combined optics, a more complicated lens-window has to be applied. If it is kept above 350°C and the laser fluence is low enough no deposition will take place. Table 2 [42] provides an overview on the details:

For a Otto engine, operated under homogenous burn conditions, typical data [5] as shown in Fig. 14 have been collected: (left upper) COV (coefficient of variation) under laser ignition is lower

**Table 2** Experimentally Measured Fluences for Different Effects of the Ignition Laser on Organic Depositions [42]

| *Material* | | *Laser fluence [mJ/mm²]* | |
|---|---|---|---|
| Suprasil | d" 0.25 | 0.30 – 2.60 | $^{3}$ 3.50 |
| Sapphire | d" 0.15 | 0.15 – 1.60 | $^{3}$ 2.00 |
| ALON | (no evaluation) | 0.20 – 3.00 | $^{3}$ 3.00 |
| | no effect | laser-induced deposition | laser ablation |

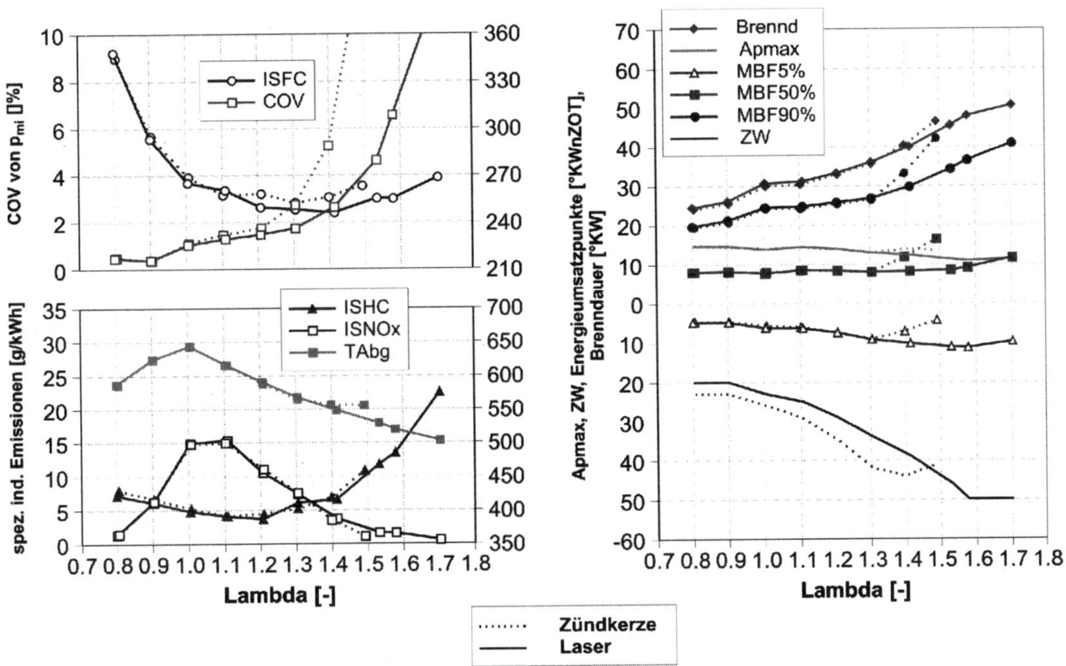


**Figure 14** Comparison of Spark Plug and Laser Ignition of a Gasoline Engine: n = 2000 min$^{-1}$, $p_{mean}$ = 4 bar; Homogenous; w/o EGR (Exhaust Gas Recycling). (Courtesy J. Graf, Vienna University of Technology 2006)

indicating a more smooth operation, ISFC (indicated specified fuel consumption) is clearly lower, both for lean mixtures. (left lower) The emissions of HC and $NO_x$ are reduced slightly under laser ignition at higher leanness. (right picture) All combustion processes are somewhat faster with laser ignition, as it can be recognized.

Finally, a third field of laser application might be HCCI in order to improve ignition timing and to achieve much more tolerance for the change of the ignition parameters being pressure, temperature and composition. Fig. 15 shows measurements of the first laser-triggered HCCI engine tests [13]. With laser ignition the maximum cylinder pressure is higher and occurs earlier (left), CA50 (the crank angle where 50% of the fuel/air mixture is burned) occurs earlier under laser ignition than without and shows smaller fluctuations in the first case. Although it was shown that to

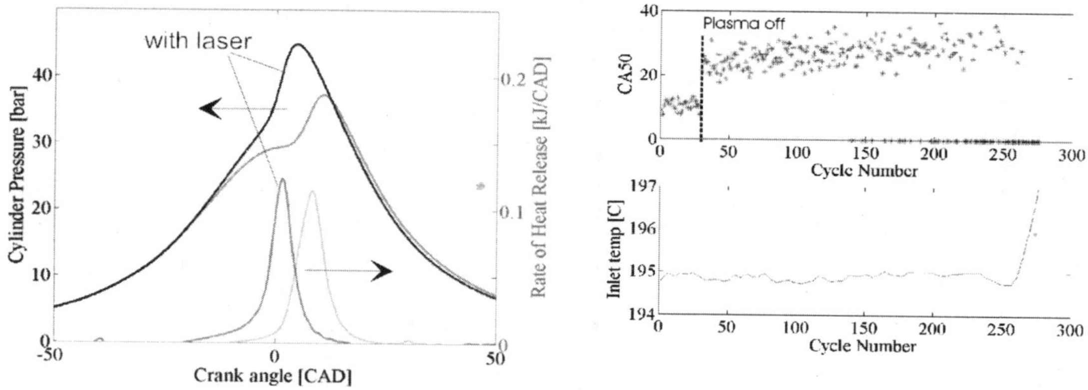


**Figure 15** (left) Influence of Laser Plasma on HCCI Combustion; Weak Color: without Plasma, Intense Color: Laser Plasma at 40° BTDC; λ= 2.3, 90% Methane, 10% Isooctane. (right) Laser-stabilized HCCI Engine Operation at Reduced Inlet Temperature Fades away after Plasma is Turned off (Zero Value for CA50: no Combustion at all; CA50: crank Angle at 50% Burn Point; λ = 2.04, 90% Methane, 10% Isooctane, Plasma at 25° BTDC)

some extent the outstanding properties of HCCI, i.e. no flame front and hence lowest burn temperatures, are somewhat deteriorated, the advantage of laser application is proven. If this ever will come to standard technical application cannot be assessed today.

## 6. CONCLUSION AND OUTLOOK

This paper reviews the milestones of laser ignition in short, being based on laser-induced plasma formation, and reveals the relevant ignition parameters. These facts give information about the required ignition laser specifications. The important parameters which are pulse energies around 10 mJ at pulse durations of about 1 ns could be realized by a laser development of our own. For a realistic application to combustion engines, the laser system must be assembled in small housing with dimensions of a conventional spark plug. A prototype has been realized, even with monolithic setup which is highly robust against mechanical interactions.

Most likely fields of application of laser ignition have been pointed out: i) large scale MW gas engines, ii) spray-guided direct fuel injection gasoline engines, and iii) HCCI operation with laser triggering. The first one also has the chance for economic implementation within the next few years.

### Acknowledgements

G. Herdin deserves special thanks for his continuing helpful discussions over many years. Financing by the company GE Jenbacher GmbH & Co OHG, Austria, and by the Austrian Industrial Research Promotion Fund under the project "*Adaptierung von Gasmotoren zur Nutzung der Potenziale der Laserzündung*" is gratefully acknowledged. The fruitful cooperation with the Institute of Internal Combustion Engines and Automotive Technology (Head: Prof. B. Geringer) is remarked with gratitude.

## References


1. H. Kopecek, et al., *"Laser ignition of methane-air mixtures at high pressures and diagnostics"*, Journal of Engineering for Gas Turbines and Power, Vol. 127(1), pp. 213-219, (2004).
2. H. Kopecek, et al., *"Laser-induced ignition of methane-air mixtures at pressures up to* 4MPa*"*, Laser Physics Letters, Vol. 13(11), pp.1-5, (2003).
3. H. Kopecek, et al., *"Laser-stimulated Ignition in a Homogeneous Charge Compression Ignition Engine"*, SAE International, (2003).
4. P. D. Ronney, *"Laser versus conventional ignition of flames"*, Optical Engineering, Vol. 33(2), pp. 510-521, (1994).
5. J. Graf, "*Laserzündung – Potenzial bei ottomotorischen Brennverfahren und Anforder-ungen für die praktische Umsetzung*" ("Laser ignition - potential under Otto engine-like combustion schemes and requirements for practical realization"), Ph.D. Thesis, Vienna University of Technology, November 2006.
6. M. Weinrotter, E. Wintner, K. Iskra, T. Neger, J. Olofsson, H. Seyfried, M. Aldén, M. Lackner, F. Winter, E. Vressner, A. Hultqvist and B. Johansson, "*Optical Diagnostics of Laser-Induced and Spark Plug Assisted HCCI Combustion*", SAE Technical Paper Series, Vol. 1963, 129-140, (2005).
7. L. J. Radziemski, et al., "*Laser-Induced Plasmas and Applications*", Thompson Brian J., 1st ed., New York, Marcel Dekker Inc., pp. 445, ISBN: 0-8247-8078-7, (1989).
8. T. X. Phuoc, *"Laser-induced spark ignition fundamental and applications"*, Optics and Lasers in Engineering, Vol. 44(5), pp. 351-397, (2006).
9. M. Lackner, et al., *"Investigation of the early stages in laser-induced ignition by Schlieren photography and laser-induced fluorescence spectroscopy"*, Optics Express, Vol. 12, pp. 4546-4557, (2004).
10. D. Bradley, et al., *"Fundamentals of high-energy spark ignition with lasers"*, Combustion and Flame, Vol. 138(1-2), pp. 55-77, (2004).
11. M. Lackner, et al. "*Optical Diagnostics of Laser Ignition for Future Advanced Engines*", Proc. of ASME Fall Technical Conference, (2004).
12. H. Kopecek, et al., *"Laser ignition of methane-air mixtures at high pressures"*, Experimental Thermal and Fluid Science, Vol. 27, pp. 499-503, (2003).
13. H. Kopecek, et al. "*Laser-induced ignition characteristics of methane- and hydrogen-air mixtures at high pressures*", Proc. of SPIE, Vol. 5506, (2004).
14. T. X. Phuoc, *"Laser spark ignition: Experimental determination of laser-induced break-down of combustion gases"*, Optics Communications, Vol. 175(4), pp. 419-423, (2000).
15. M. Weinrotter, et al., *"Application of laser ignition to hydrogen-air mixtures at high pressures"*, International Journal of Hydrogen Energy, Vol. 30(3), pp. 319-326, (2004).
16. M. Weinrotter, et al., *"An extensive comparison of laser-induced plasma ignition and conventional spark plug ignition of lean methane-air mixtures under engine-like conditions"*, SAE Technical Paper Series, Vol. 1972, pp. 248-256, (2005).
17. M. Weinrotter, et al., *"Laser ignition of ultra-lean methane-hydrogen-air and hydrogen-air mixtures under engine-like conditions"*, Experimental Thermal and Fluid Science, Vol. 29(5), pp. 569-577, (2005).
18. J. L. Beduneau, et al., *"Measurements of minimum ignition energy in premixed laminar methane/air flow using laser induced spark"*, Combustion and Flame, Vol. 132, pp. 653-665, (2003).

19. D. H. McNeil. "*Minimum Ignition energy for laser spark ignition*", Proc. of Combustion Institute, Vol. 30, (2005).

20. C. Forsich, et al., *"Characterization of Laser-Induced Ignition of Biogas-Air Mixtures"*, Biomass & Bioenergy, Vol. 27(3), pp. 299-312, (2004).

21. M. Weinrotter, et al. "*Laser ignition of engines – a realistic option!*" Proc. of SPIE, Vol. 6053.

22. R. Günther, *"Verbrennung und Feuerungen"*, Springer, 1st ed., 432, ISBN: 3-540-13256-2, (1984).

23. G. Ast, "*Vergleich Laserzündung mit Zündkerzenzündung*"; Photonics Institute, Supervisor: E. Wintner; 2004.

24. J. D. Dale, et al., *"Laser Ignited Internal Combustion Engines - An Experimental Study"*, SAE Technical Paper Series, Vol. 780329, (1978).

25. M. Weinrotter, et al., *"Laser ignition of engines"*, Laser Physics, Vol. 15(7), pp. 947-953, (2005).

26. H. Kofler, et al., *"An Innovative Solid-State Laser for Engine Ignition"*, Laser Physics Letters, Vol. 4(6), pp. 322-327, (2007).

27. B. C. Stuart, et al., *"Laser-Induced Plasma in Dielectrics with Nanosecond to Subpicosecond Pulses"*, Physical Review Letters, Vol. 74(12), pp.2248-2251, (1995).

28. R. Wood, *"Laser damage in optical materials at 1.06μm"*, GEC Journal of Science & Technology, Vol. 45(3), pp.109-115, (1979).

29. Y. Matsura, et al. "*Hollow fibers for delivery of high-power lasers*", Proc. of SPIE, Vol. 4065, (2000).

30. A. Yalin, et al., *"Use of hollow-core fibers to deliver nanosecond Nd:YAG laser pulses to form sparks in gases"*, Optics Letters, Vol. 30(16), pp. 2083-2085, (2005).

31. A. Yalin, et al. "*Fiber Delivered Systems for Laser Ignition of Natural Gas Engines*", Proc. of ASME Internal Combustion Engines Division, (2006).

32. J. Tauer, et al., *"High-throughput of single high-power laser pulses by hollow photonic band gap fibers"*, Laser Physics Letters, Vol. 4(6), pp. 444-448, (2007).

33. H. Kofler, et al. "*Fenster zur Verbrennungsdiagnostik: Herabsetzung der Verschmutz-ungsneigung für einen störungsfreien Langzeitbetrieb*", Proc. of VDI, Vol. 1888, (2005).

34. M.J. Weber, "Handbook of Optical Materials", Laser & Optical Science & Technology Vol. 19, New York etc., CRC Press, ISBN: 0-8493-3512-4, (2003).

35. D.C. Harris, "Durable 3-5 μm transmitting infrared window materials", Infrared Physics & Technology, Vol. 39(4), pp. 185-201, (1998).

36. W. Koechner, et al., "*Solid-State Laser, A Graduate Text*", Springer, pp. 409, ISBN: 0-387-95590-9, (2003).

37. G. Hennig, et al. "*Laser engraving in gravure industry*", Proc. of SPIE, Vol. 6157, (2005).

38. R. L. Schmitt, *"Monolithic passively Q-switched Cr:Nd:GSGG microlaser"*, Proc. of SPIE - The International Society for Optical Engineering, Vol. 5871, 2005, Article No. 587105, pp. 1-8.

39. S. K. Sudheer, et al., *"Set temperature dependence of fractional thermal loading and effective stimulated emission cross section of a diode double end pumped $TEM_{00}$ Nd:YAG laser for micromachining application"*, Laser Physics Letters, Vol. 3(5), pp. 244-248, (2006).

40. J. Sulc, et al., *"Nd:YAG/V:YAG microchip laser operating at 1338 nm"*, Laser Physics Letters, Vol. 2(11), pp. 519-524, (2005).
41. W.-Q. Wen, et al., *"Short cavity passive Q-switched Nd:YAG laser end-pumped by QCW-LDA "*, Journal of Optoelectronics Laser, Vol. 13(9), pp. 900-903, (2002).
42. H. Ranner, *"Beiträge zur Entwicklung der Laserfunken-Zündung von Ottomotoren: Plasmacharakterisierung, Analyse von Brennraumfenster-Ablagerungen und Kombinierte Laser-Funkenzündung"* ("Contributions to the development of laser-spark ignition of Otto engines: Plasma characterization, analysis of combustion chamber window depositions and combined laser-spark plug ignition"), PhD thesis, Vienna University of Technology, October 2007.

# Entropy Generation and Exergy Destruction in Combustion

**Amitava Datta**
Department of Power Engineering, Jadavpur University, Salt Lake Campus, Kolkata – 700 098
E-mail: amdatta_ju@yahoo.com

## INTRODUCTION

The impact of energy crisis and the growing concern for the environment have resulted in a specific thrust in the efficient and clean utilization of natural resources. In all the conventional work producing systems, energy in the form of thermal energy or heat is converted to work using a thermodynamic engine. The heat is obtained from the primary resource or fuel. Even with the advent of alternative energy resources, fossil fuels continue to serve as the principal source of energy in the bulk of power industries and transportation and aviation applications. The chemical energy stored in the fossil fuels is released upon combustion. Combustion is a rapid chemical reaction, where the fuel is oxidized, often using the atmospheric air, to produce thermal energy. However, combustion is a very complex phenomenon involving several physical and chemical processes simultaneously. There occurs momentum, heat and mass transfers, chemical reaction and thermal radiation all at a time in the combustor, where the fuel is burnt.

During the energy conversion that takes place during combustion, energy changes from one form to the other. According to the first law of thermodynamics energy can never be destructed. Therefore, it is justified to state that energy conversion processes do not have permanent energy destructions, except for the losses from the system to the environment. However, the second law of thermodynamics has also to be considered, which states that the irreversible nature of the energy conversion processes lead to the generation of entropy and destruction of exergy. Exergy (or available energy, or availability) is the maximum useful work that can be extracted from a quantity of energy and refers to the quality of energy. Thus, though the energy is conserved in the process of conversion, its quality may degrade reducing the work potential with each conversion. The various irreversible processes encountered within the combustor leads to certain degree of exergy destruction. Several studies have indicated that the conventional combustion involves inherent thermodynamic irreversibility, which significantly limits the conversion of fuel energy into useful

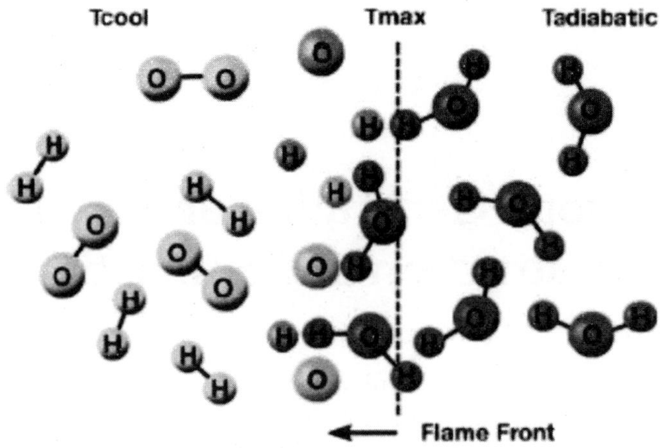


**Figure 1** Schematic of how entropy is generated in flame front due to internal heat transfer. Energetic product molecules dissipate its energy in collision with surrounding cool reactant molecules [22]

work [1,2]. For typical atmospheric combustion systems, about 1/3$^{rd}$ of the fuel energy becomes unavailable and is discharged into the environment as heat. Most of this irreversibility is associated with the internal heat transfer within the combustor between the products and reactants. Such heat transfer becomes inevitable in both premixed and diffusion flames, where highly energetic product molecules are free to exchange energy with unreacted fuel and air molecules. As depicted in Figure 1, the product and reactant molecules have large energy difference (i.e. temperature difference) and considerable entropy is generated when they interact. Internal heat transfer within the combustor is often difficult to be recognized as an efficiency problem, because it does not result in a direct energy loss from the combustion zone to the surrounding. Instead, internal heat transfer only degrades the exergy of the product flue gas and reduces its ability to produce useful work. Therefore, the actual energy penalty does not become apparent until the work generation step, which is separated from the combustion step, is taken into account.

It is increasingly recognized that exergy analysis must be added to the conventional energy accounting analysis during the conception, analysis, development and design of the energy systems, of which combustor is a part. However, most of these analyses are nowadays conducted on the system level development, by evaluating the exergy values of the component inlet and outlet streams and energy interactions. While this can identify the points of exergy destruction in a system, it does not deliver the detailed information regarding the specific process phenomena, often space and time dependent, which causes the exergy changes. The actual process may include several phenomena like heat transfer, mass transfer, fluid friction, chemical or nuclear reaction, presence of gravitational, electrical or magnetic field etc. To find the root cause of entropy generation and exergy destruction attention has to be paid on each of the individual processes and its rate. A recent review conducted by Som and Datta [23] has focused on such studies for different flames with gaseous, liquid or solid fuels.

## EXERGY ANALYSIS FOR THERMAL POWER PLANTS

In a recent work (Sengupta et al., [4]), a thermodynamic analysis based on first and second laws have been done on the performance of a working 210 MW thermal power unit (Figure 2) to obtain the energy (first law) and exergy (second law) efficiencies using the plant operating parameters at different loads. The major streams entering and leaving the components of the plant are shown in the figure. To identify the sources of the availability destruction, the entire plant has been split into three different control volumes, viz. (1) only the turbo-generator with its inputs and outputs, (2) turbo-generator, condenser and regenerative system and (3) the entire cycle with boiler, turbine, generator, condenser and regenerative system. This would help to find out the individual contribution towards the gross irreversibility of the plant, as well as, the response of different parts towards exergy destruction with varying operating conditions. In the analysis, the second law efficiency in a control volume is evaluated as,

$$\eta_2 = \frac{P_{net}}{\dot{E}_{in} - \dot{E}_{out}} \times 100\% \qquad ...(1)$$

where, $\dot{E}$ represents the exergy flow rate across the control surface and $P_{net}$ is the net power output. Two types of exergy are taken into account, thermo-mechanical exergy for all the streams and chemical exergy associated with the fuel stream. The total rate of exergy in a stream is obtained from its specific value as,

$$\dot{E}_j = \dot{m}_j e_j \qquad ...(2)$$

where, $\dot{m}_j$ is the mass flow rate of the stream. The specific thermo-mechanical exergy (neglecting kinetic and potential energy) is evaluated from the following equation:

$$e_j = (h_j - h_0) - T_0(s_j - s_0) \qquad ...(3)$$

Chemical exergy evaluation, required for the fuel only, has been evaluated separately.

Restricting to two control volumes out of the three for present, i.e. without (C.V.#2) and with (C.V.#3) the boiler, the following formulation can be envisaged.

Considering the thermo-mechanical exergy for streams in the case of control volume without the boiler, i.e. CV 2 in figure 2, the exergy flow rate entering and leaving the control volume are (refer figure 2):

$$\dot{E}_{in} = \dot{E}_1 + \dot{E}_3 + \dot{E}_{13} \qquad ...(4)$$

and

$$\dot{E}_{out} = \dot{E}_{11} + \dot{E}_{12} + \dot{E}_{14} \qquad ...(5)$$

Because of the unavailability of required data for the calculation of exergy destruction in the condenser under all the operating conditions, the the exergy destruction inside the condenser is neglected considering that the heat transfer in the condenser takes place with a very low temperature difference. Therefore, an exergy balance of the condenser gives,

$$\dot{E}_{14} - \dot{E}_{13} = \dot{E}_6 - \dot{E}_{15} \qquad ...(6)$$

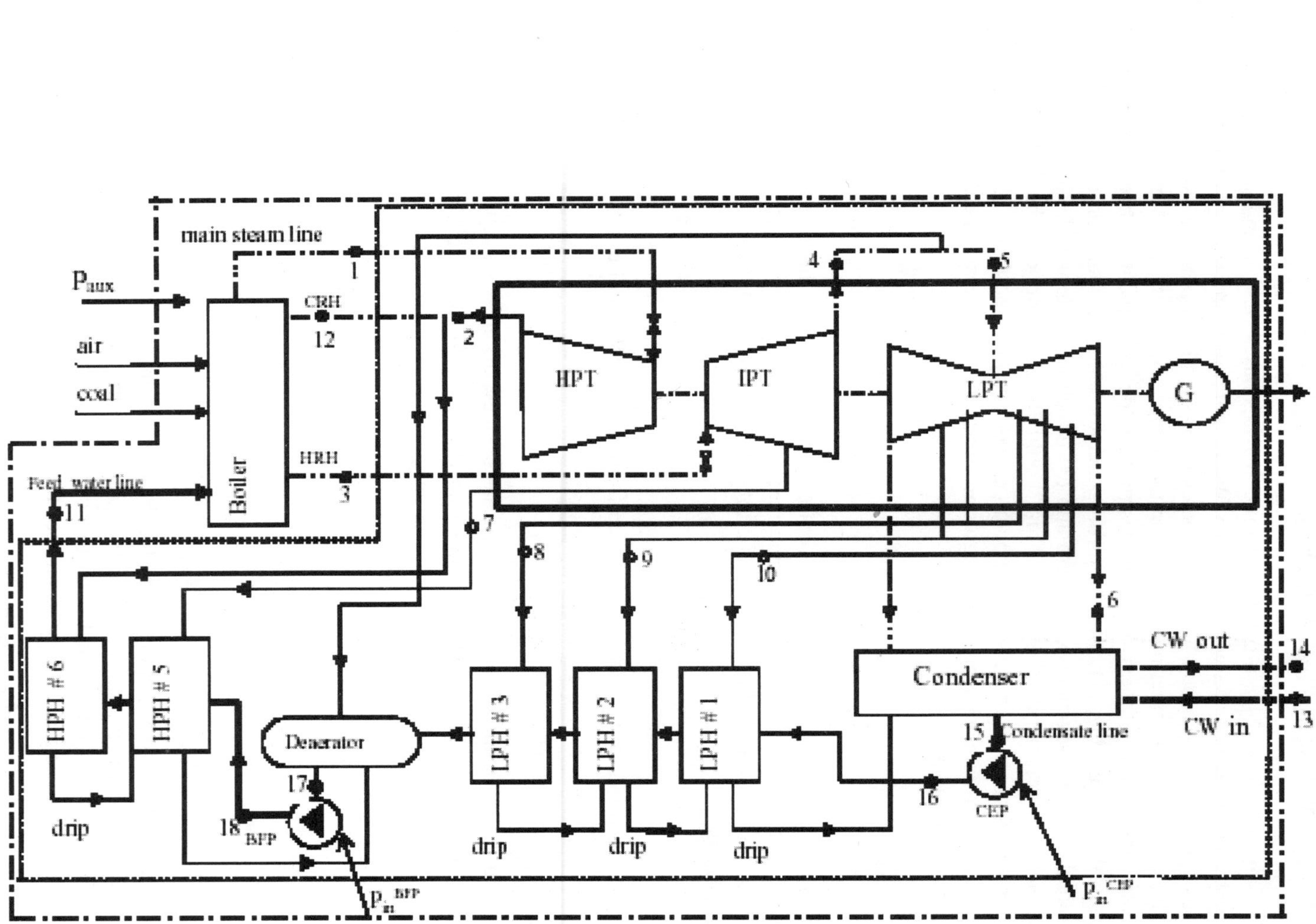


**Figure 2** Schematic layout of a thermal power plant showing the different control volumes under study.——— Control Volume 1,............ Control Volume 2,— · — · — Control Volume 3

The above equation is used to take care of the net exergy transfer rate with the cooling water.

The net power output from the control volume is:

$$P_{net} = G - P_{in}^{CEP} - P_{in}^{BFP} \quad \text{...(7)}$$

The second law efficiency of the control volume 2 ($\eta_2^{cycle}$) is calculated using equations (1) to (7).

When the entire plant, including the boiler, is considered as the control volume (CV 3), considering the exergy input with fuel and circulating water, the total exergy flow rate entering the control volume is

$$\dot{E}_{in} = \dot{E}_{in}^{f} + \dot{E}_{13} \quad \text{...(8)}$$

Therefore, the rate of exergy entering the control volume with fuel:

$$\dot{E}_{in}^{f} = \dot{m}_{in}^{f} n_C \frac{\bar{e}_{in}^{f}}{M_C} \quad \text{...(9)}$$

The total exergy flow rate leaving the control volume including that with circulating water is:

$$\dot{E}_{out} = \dot{E}_{out}^{g} + \dot{E}_{14} \quad \text{...(10)}$$

Out of it the exergy flow rate leaving the control volume with flue gas is:

$$\dot{E}_{out}^{g} = \sum \dot{m}_j^g e_j^g \quad \text{...(11)}$$

where, the summation applies over all the components of the flue gas.

The net power output from the control volume is:

$$P_{net} = G - P_{aux} \quad \text{...(12)}$$

The auxiliary power is considered as a consolidated manner at different load for the plant using the standard plant data.

Figures 3a and 3b show the variation of the second law efficiencies for the two cases with CV-2 and CV-3 respectively. In both the cases, it is observed that the efficiency increases with the load and is the maximum at full load. The same is the case for the energy efficiency evaluated and shown for the entire plant in Figure 3b. The reason can be attributed to the increased throttling that is required at part load operation for the plant. It is noteworthy to mark the difference in the second law efficiency values for the cases without and with the boiler. While the second law efficiency remains within 80-90% for a variation of load between 60–100% in control volume 2, the same resides only between 32-35% in the given load range for control volume 3. This shows the large irreversibility in the boiler, which is the added component for the case of control volume 3.

Although the analysis is indicative of the fact that boiler is the most important source of irreversibility in a thermal power plant, the drawback of the gross analysis is that it could not identify the causes of the irreversibilities. In the boiler the major sources of irreversibility may be the combustion process in the furnace as well as the heat transfer between the high temperature flue gas and the steam/water flowing through the tubes. However, the present results fail to evaluate the contribution of each of these towards the loss in exergy.

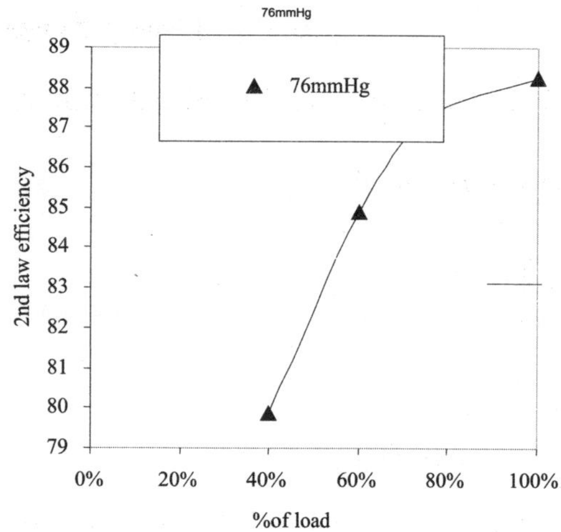


**Figure 3a** Variation of Second Law Efficiency with % of Full Load for Control Volume#2

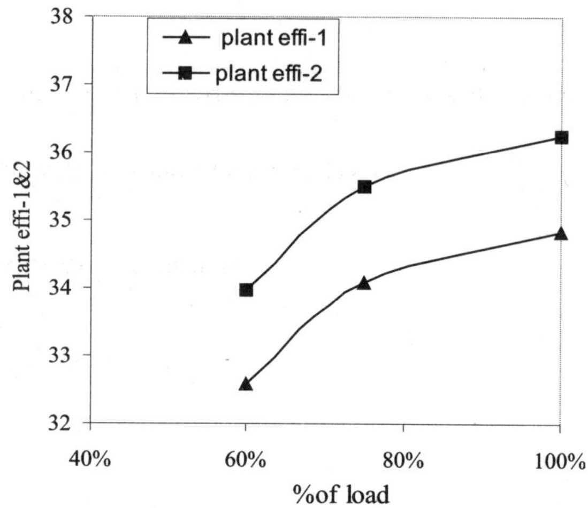


**Figure 3b** Variation of First and Second Law Efficiencies with % of Full Load for Control Volume#3

**Table 1** Exergy and energy analysis in thermal plant

| *Item* | *Exergy balance (%)* | *Energy balance (%)* |
|---|---|---|
| Fuel input (coal) | 100 | 100 |
| Combustion and heat transfer loss | –46 | –1 |
| Stack gas loss | –3 | –8 |
| Condensation loss | –2 | –46 |
| Miscellaneous loss | –9 | –5 |
| Power output | –40 | –40 |

The case of a typical thermoelectric plant for fossil fuel and at an ambient temperature of 20°C.

Taniguchi et al. [5] presented the exergy and energy analysis data of a thermal power plant, though in their work the details of the plant is not given. The results of their analysis are summarized in Table 1. It shows the combustion and heat transfer loss occurring in the boiler to be the major source of irreversibility (46%). They have mentioned that out of the 46% loss only 1% is towards combustion and 45% is due to heat transfer. The energy loss due to combustion is referred as 1% (given under the combustion and heat transfer loss in the table). This result is probably a misrepresentation as the parameters are evaluated only at a gross level and 1% exergy loss is attributed to the loss due to the unburnt fuel and incomplete combustion, which also gives rise to the energy loss to the same tune.

The most important and useful objective of exergy analysis are: (i) identification of the specific phenomena/processes that have large exergy losses or irreversibilities, (ii) understanding why these losses occur (iii) evaluation how they change with any changes in the process parameters and configuration, and (iv) as a consequence of all these, suggestions on how the process could be improved. As shown in the two examples cited above, a large fraction of the publications in the field of exergy analysis deal only with (i) above, and are therefore not totally conclusive for the betterment of the system [3].

## CHEMICAL EXERGY OF FUEL

It is observed in the analyses shown above that a chemical exergy has been associated with the fuel. In order to define the thermomechanical exergy it is required to define a reference state characterized by the reference pressure ($p_o$) and reference temperature ($T_o$). A stream at ($T_o$,$p_o$) condition has no thermomechanical exergy. However, a fuel stream at ($T_o$,$p_o$) condition has the capacity to do useful work, as it can react with oxygen at reference condition to produce energy. The energy released in the form of heat may be transformed to useful work in a heat engine. This phenomenon can be explained through the definition of another form of exergy- chemical exergy. The chemical exergy is defined at the reference pressure and temperature conditions. However, in addition to the reference pressure and temperature, a reference environment consisting of species of defined concentration is needed. The exergy reference environment is assumed to consist of 20.35% oxygen, 3.03% $CO_2$, 0.03% $H_2O$ and the balance $N_2$ by volume.

For a hydrocarbon fuel ($C_aH_b$), the chemical exergy is evaluated by considering its reaction with stoichiometric quantity of oxygen from exergy reference environment resulting in complete combustion and leaving product species again at the condition of the reference environment (Figure 4).

The chemical reaction may be expressed as,

$$C_aH_b + \left(a + \frac{b}{4}\right)O_2 \rightarrow a\,CO_2 + \frac{b}{2}H_2O \qquad \ldots(13)$$

The steady state energy balance for a control volume across the combustor per mole of fuel gives

$$w_{cv} = q_{cv} + \bar{h}_F + \left(a + \frac{b}{4}\right)\bar{h}_o - a\bar{h}_c - \frac{b}{2}\bar{h}_w \qquad \ldots(14)$$

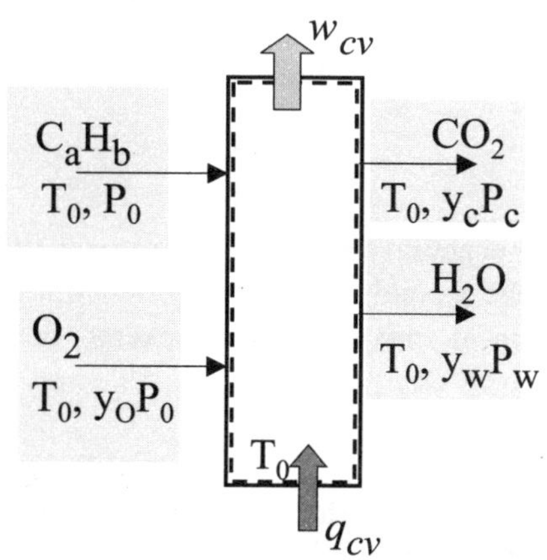


**Figure 4** Illustration used to introduce the fuel chemical exergy concept

An entropy rate balance across the combustor takes the form

$$\frac{q_{cv}}{T_0} + \bar{s}_F + \left(a + \frac{b}{4}\right)\bar{s}_o - a\bar{s}_c - \frac{b}{2}\bar{s}_w + \bar{s}_g = 0 \qquad ...(15)$$

where, $\bar{s}_g$ is the entropy generation due to irreversibility per mole of fuel being burnt. Combining equations (14) and (15), the work output from the control volume may be expressed as,

$$w_{cv} = \left[\bar{h}_F + \left(a + \frac{b}{4}\right)\bar{h}_o - a\bar{h}_c - \frac{b}{2}\bar{h}_w\right] - T_0\left[\bar{s}_F + \left(a + \frac{b}{4}\right)\bar{s}_o - a\bar{s}_c - \frac{b}{2}\bar{s}_w\right] - T_0\bar{s}_g \qquad ...(16)$$

The maximum theoretical value for the work developed is obtained when no irreversibilities are present in the process. Under that condition, the following expression for chemical exergy of the fuel per unit mole results.

$$\bar{e}_{ch} = \left[\bar{h}_F + \left(a + \frac{b}{4}\right)\bar{h}_o - a\bar{h}_c - \frac{b}{2}\bar{h}_w\right] - T_0\left[\bar{s}_F + \left(a + \frac{b}{4}\right)\bar{s}_o - a\bar{s}_c - \frac{b}{2}\bar{s}_w\right] \qquad ...(17)$$

In the above equation, the specific molar entropy of oxygen, carbon dioxide and water vapor are expressed at their respective partial pressures at the reference condition. In order to express all the enthalpy and entropy terms at ($T_o$, $p_o$) condition, the following equation is used:

$$\bar{s}_i(T_o, y_i p_o) = \bar{s}_i(T_o, p_o) - \bar{R}\ln(y_i) \qquad ...(18)$$

Applying the above equation (18), equation (17) becomes,

$$e_{ch} = \left[\bar{h}_F + \left(a + \frac{b}{4}\right)\bar{h}_o - a\bar{h}_c - \frac{b}{2}\bar{h}_w\right]_{T_0, P_0} - T_0\left[\bar{s}_F + \left(a + \frac{b}{4}\right)\bar{s}_o - a\bar{s}_c - \frac{b}{2}\bar{s}_w\right]_{T_0, P_0}$$

$$+ \bar{R}T_0 \ln\left[\frac{(y_o)^{a+b/4}}{(y_c)^a (y_w)^{b/2}}\right] \qquad ...(19)$$

Equation (19) represents the specific molar chemical exergy of a substance, e.g. the fuel, which is used to determine the exergy input through the fuel stream into the combustor.

## ENTROPY GENERATION

The irreversibility present in the actual processes occurring in the combustor results in the generation of entropy following the second law of thermodynamics. The rate of entropy generation may be used to calculate the rate of irreversibility or exergy destruction using the well known Gouy-Stodola equation,

$$\dot{I} = T_o \dot{S}_g \qquad ...(20)$$

where, $T_o$ is the reference temperature. Calculating the total rate of entropy generation is an involved process in an intrinsic analysis in the combustor. This is done by separately calculating the rate of entropy generation in each physical and chemical process that take place.

In the flow field of a combustor, the nonequilibrium conditions are due to the exchange of momentum, energy and mass of different species (multicomponent) within the fluid and at the solid boundaries. This nonequilibrium phenomenon causes a continuous generation of entropy in the flow field. The entropy generation is due to the irreversible nature of heat transfer, mass diffusion, viscosity effects within the fluid and at the solid boundaries, chemical reaction, coupling effects between heat and mass transfers and body force effects. Every irreversible process can be viewed as the relevant flux driven by the corresponding potential, e.g. the flux of heat is driven by the temperature gradient. The entropy production rates in different processes can be explicitly obtained considering the conservation of multicomponent species mass in presence of chemical reaction, conservation of momentum and energy along with the local entropy balance equation. The local entropy generation per unit volume has been derived by Hirschefelder et al. [6] and is given by,

$$\dot{s}_g''' = \frac{\tau_{ij}\frac{\partial v_i}{\partial x_j}}{T} - \frac{q_i}{T^2}\frac{\partial T}{\partial x_i} - \frac{\sum_{k=1}^{K} j_{k_i}\frac{\partial \mu_k}{\partial x_i}}{T} - \frac{\sum_{k=1}^{K} s_k j_{k_i}\frac{\partial T}{\partial x_i}}{T} + \frac{\sum_{k=1}^{K} j_{k_i} f_{k_i}}{T} - \frac{\sum_{r=1}^{R} \dot{\omega}_r \sum_{k=1}^{K}(\gamma''_{kr} - \gamma'_{kr})\mu_k}{T} \quad ...(21)$$

where, $t_{ij}$ is the stress tensor, $q_i$ is the heat flux vector per unit area and $j_{k_i}$, $\mu_k$ and $s_k$ are the species mole flux per unit area, chemical potential and partial molal entropy of the kth species, respectively. $\gamma'_{kr}$ and $\gamma''_{kr}$ are the stoichiometric coefficients on the reactant side and product side, respectively, for the kth species in the rth reaction, and$\dot{\omega}_r$ is the reaction rate of the rth reaction. In the above equation, the first term is due to fluid friction, the second due to heat transfer, the third due to mass transfer, the fourth is a result of the coupling between heat and mass transfer, the fifth term is due to the body force effects and the last term results from chemical reaction. Almost all the existing intrinsic exergy analysis in combustion systems had employed the above equation for computing the entropy generation rate [7-12].

Observation of the equation (21) shows that knowledge of the field parameters like velocity (and hence $t_{ij}$), temperature (and hence $q_i$), species concentration (and hence also $j_k$, $m_k$), of the state equations for the materials used to determine the thermophysical properties (density, viscosity and thermal conductivity), and of the chemical reactions involved (to determine the rate) allows the computation of the exergy, exergy destruction, entropy, and entropy generation fields. All of these field parameters are obtained from the solution of the full field and state equations, consisting of the Navier-Stokes, energy, species conservation, entropy generation, and thermodynamic properties equations, combined in a combustion process with the reaction kinetics equations, all tightly coupled.

## SIMPLIFIED ANALYSIS BASED ON HYPOTHETICAL SUBPROCESSES

In the literature of exergy analysis in combustion systems, the work of Dunbar and Lior [1] receives a special mention because they for the first time evaluated the primary causes for irreversibility using some heuristic finite increment exergy analysis method for a simple hydrogen and methane combustor. However, the method adopted in the work was not fully intrinsic. The authors did not solve the full conservation equations using the reaction kinetic equations in the whole combustion chamber to obtain the flux and gradient terms for the different processes involved. Instead they divided the entire combustion phenomenon into a number of hypothetical sub-processes and apply them along prescribed process paths. The sub-processes considered by the authors are:

(i) a diffusion process where the fuel and oxygen molecules are drawn together

(ii) a chemical reaction process leading to oxidation of the fuel

(iii) an internal thermal energy exchange between high temperature product and the unburned reactant

(iv) a physical mixing process where the system constituents mix uniformly.

The sub-processes are arranged differently in the process paths resulting in different types of combustion phenomena.

For example, in one case (Path 1), the authors considered an incremental quantity of fuel to be mixed with its corresponding stoichiometric quantity of oxygen to result the chemical reaction, internal energy exchange and product mixing. Such incremental process continues along the length of the combustor in small compartments till the gas constituents reach the fuel ignition temperature when the fuel oxidation is assumed to be instantaneous. The path according to the authors corresponds to the behaviour of a diffusion flame.

In an alternative path (Path 2), the entire fuel and air are first mixed thoroughly. Subsequently, incremental amount of the mixture is reacted followed by the internal energy exchange and mixing. Such a path may exhibit the behaviour of a premixed flame.

Dunbar and Lior performed their analysis separately with hydrogen and hydrocarbon (methane) fuels. Despite the simplification, the analysis yielded consistent results. It was observed that in each of the hypothetical paths, the major share of exergy destruction took place due to the internal energy exchange. In the process path 1, the overall exergetic efficiency was reported to be within 66.5% and 77.3% in hydrogen combustion for a range of excess air of 0-100%, decreasing with the increasing amount of excess air. Out of the total exergy destruction, 72-77% was reported due to the internal heat exchange, 15-18% due to chemical oxidation reaction and 8-10% due to gas mixing. In the process path 2, 66-73% of the total exergy destruction was accounted to internal heat exchange, 18-25% to chemical reaction and 8-10% to the mixing processes. Therefore, Dunbar and Lior reported no perceptible difference in exergy destruction due to the premixing of the reactants. The difference between hydrogen and methane combustion towards exergy destruction was that in case of the latter a relatively higher percentage of irreversibility resulted due to chemical reaction than that in the former. However, even with methane combustion the largest share of irreversibility resulted due to heat transfer only.

## DETAILED ANALYSIS OF COMBUSTION

Studies have shown that the combustion process typically destroys about one third of the exergy of the fuel. The significance is even clearer when one notes (cf. [3]) that of the many processes occurring in the typical electricity-producing power plant, combustion of the fuel is the single largest contributor of exergy losses in power production. It is thus of great interest to determine the magnitudes and causes of irreversibility in this process through a detailed intrinsic analysis. The full intrinsic analysis of entropy generation and exergy destruction has been reported in the literature in case of certain specific flame situations. The key existing literatures in specific combustion applications are as below:

| *Applications* | *References* |
|---|---|
| Laminar gaseous jet diffusion flame | Datta [9], Stanciu et al. [13], Datta and Leipertz [14], Nishida et al. [10], Datta [11] |
| Laminar gaseous premixed flame | Arpaci and Selamet [15], Nishida et al. [10] |
| Turbulent gaseous jet flame | Yapici et al. [16,17] |
| Liquid droplet combustion | Puri [18], Hiwase et al. [7], Lior et al. [3] |
| Spray Combustion | Datta and Som [8], Som and Sharma [12] |
| Coal Combustion | Som et al. [19], Lior et al. [3] |

All the above work solved the conservation equations of mass, momentum, energy and species concentration with an assumed chemical kinetics of reaction to predict the variables like velocity, temperature, species concentration etc in the combustion zone. Subsequently the rate of entropy generation is evaluated using equation (21) and then the irreversibility in the process is obtained from the Gouy-Stodola equation (20).

As an example of the above cases, let us take the exergy analysis in a laminar jet diffusion flame.

## ANALYSIS IN A LAMINAR JET DIFFUSION FLAME

A diffusion or non-premixed flame is the type of flame where the fuel and oxidizer remains separated as they approach the reaction front and reacts as they mix. This is one of the two types of deflagarating flames (the other being premixed) having wide applications in practical burners, furnaces and combustors. Though, most of the diffusion flames employed in practice are turbulent, laminar diffusion flames have never lost their importance as they are better tractable, both theoretically and experimentally, than their turbulent counterparts. Moreover, understanding the transport and reaction processes in a laminar diffusion flame serves as a necessary precursor for understanding the structure of the more complicated turbulent flames. In many cases, the turbulent flames are studied through the laminar flamelet concept, establishing a direct correspondence between the two flame types.

The exergy analysis in laminar jet diffusion flame had been analysed by Datta [9,11] in a confined environment with a co-flowing burner (Figure 5). The combustion process is simulated by employing a detailed numerical model, solving the governing equations for a laminar, axi-symmetric, reacting flow with appropriate boundary conditions. The flow is vertical through the reaction space and the gravity effect is included in the axial momentum equation. The conservation equations for mass and momentum are as follows:

Mass :

$$\frac{\partial \rho}{\partial t} + \frac{1}{r}\frac{\partial}{\partial r}(r\rho v_r) + \frac{\partial}{\partial z}(\rho v_z) = 0 \qquad ...(22)$$

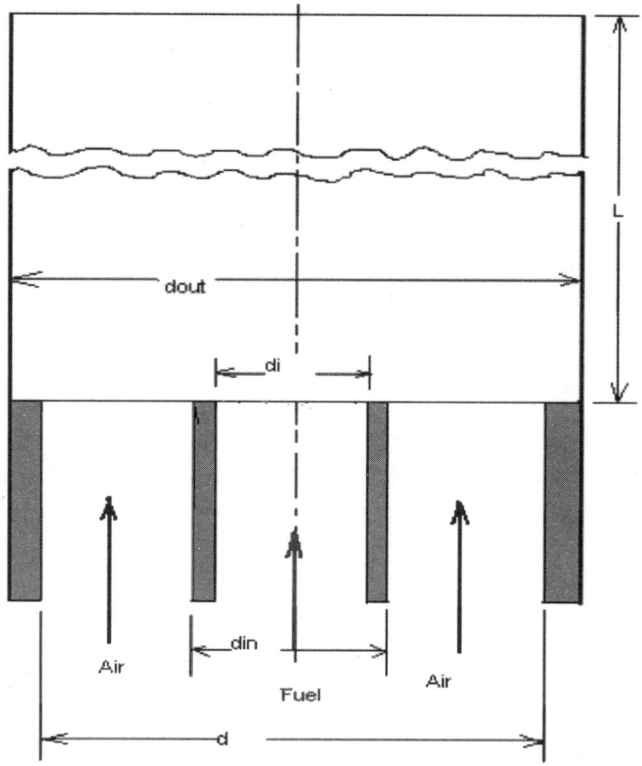


**Figure 5** Physical model : Schematic of the Geometry

Radial Momentum :

$$\frac{\partial}{\partial t}(\rho v_r) + \frac{1}{r}\frac{\partial}{\partial r}(r\rho v_r^2) + \frac{\partial}{\partial z}(\partial v_r v_z) = -\frac{\partial p}{\partial r} + \frac{2}{r}\frac{\partial}{\partial r}\left(r\mu\frac{\partial v_r}{\partial r}\right) - \frac{2}{r}\mu\frac{v_r}{r^2} +$$

$$\frac{\partial}{\partial z}\left\{\mu\left(\frac{\partial v_z}{\partial r} + \frac{\partial v_r}{\partial z}\right)\right\} - \frac{2}{3}\frac{\partial}{\partial r}\left(\mu\left(\frac{\partial v_r}{\partial r} + \frac{v_r}{r} + \frac{\partial v_z}{\partial z}\right)\right) \quad ...(23)$$

Axial Momentum :

$$\frac{\partial}{\partial t}(\rho v_z) + \frac{1}{r}\frac{\partial}{\partial r}(r\rho v_r v_z) + \frac{\partial}{\partial z}(\rho v_z^2) = -\frac{\partial p}{\partial z} + \frac{1}{r}\frac{\partial}{\partial r}\left(r\mu\left(\frac{\partial v_z}{\partial r} + \frac{\partial v_r}{\partial z}\right)\right) + 2\frac{\partial}{\partial r}\left(\mu\frac{\partial v_z}{\partial z}\right)$$

$$-\frac{2}{3}\frac{\partial}{\partial z}\left\{\mu\left(\frac{\partial v_r}{\partial r} + \frac{v_r}{r} + \frac{\partial v_z}{\partial z}\right)\right\} + \rho g \quad ...(24)$$

In the axial momentum equation, 'g' represents the acceleration due to gravity. At normal gravity, the value of g is 9.81 m/s$^2$. This is reduced at reduced gravity levels and will come to zero at zero gravity.

The combustion reaction of methane and air is assumed to proceed through two step global reaction chemistry as

$$CH_4 + 1.5\,O_2 \rightarrow CO + 2\,H_2O \qquad ...(25)$$

$$CO + 0.5O_2 \rightarrow CO_2 \qquad ...(26)$$

The reaction rates for the above reactions are obtained following an Arrhenius type rate equation, given as

$$\omega = A\,\rho^{(a+b)}\,\frac{C_1^a\,C_2^b}{M_1^a\,M_2^b}\exp\left(-\frac{E_a}{\bar{R}\,T}\right) \qquad ...(27)$$

where, A and E are the pre-exponential factor and activation energy, respectively, for the respective equations and a and b are the reaction orders in terms of fuel and oxygen respectively. The values for these parameters are taken from the work of DuPont et al. [20].

The conservation equation for chemical species is solved for five species, viz. $CH_4$, $O_2$, $CO_2$, CO and $H_2O$. The concentration for the sixth species $N_2$ is obtained by difference. The governing equation for the species conservation is as follows:

$$\frac{\partial}{\partial t}(\rho C_j) + \frac{1}{r}\frac{\partial}{\partial r}(r\rho\varpi_r C_j) + \frac{\partial}{\partial z}(\rho v_z C_j) = \frac{1}{r}\frac{\partial}{\partial r}\left(r\rho D_{jm}\frac{\partial C_j}{\partial r}\right) + \frac{\partial}{\partial z}\left(\rho D_{jm}\frac{\partial C_j}{\partial z}\right) + \dot{S}_{cj} \qquad ...(28)$$

where, $C_j$ is the mass fraction of the jth species. $D_{jm}$ is the diffusion coefficient of the species in a binary mixture of that species and nitrogen (following Katta et al., [21]). The source term $S_{cj}$ is the rate of production or destruction of the species j per unit volume due to chemical reaction. The source terms are calculated as

$$\dot{S}_{cj} = \sum_{k=1}^{n}(\gamma''_{jk}\dot{\omega}_k - \gamma'_{jk}\dot{\omega}_k)M_j \qquad ...(29)$$

where, $M_j$ is the molecular weight of the jth species. $\gamma''_{jk}$ and $\gamma'_{jk}$ are the stoichiometric coefficients on the product side and reactant side respectively for the jth species in the kth reaction. Obviously, for the present simple reaction mechanism, k is either 1 for the reaction given by Eq. (25) or 2 for the reaction given by Eq. (26).

The density of the species mixture is calculated using the equation of state considering all the species as ideal gases. The enthalpy for chemically reacting flows is given as the weighted sum of each mass fraction

$$h = \sum_{j=1}^{n} C_j h_j = \sum_{j=1}^{n}\left(h_{fj}^0 + \int_{T_o}^{T} c_{pj}\,dT\right) \qquad ...(30)$$

where, $h^0_{fj}$ is the heat of formation of the jth species at the reference temperature $T_0$ and the integral part is the contribution of the sensible heat.

The energy equation is written as

$$\frac{\partial}{\partial t}(\rho h) + \frac{1}{r}\frac{\partial}{\partial r}(r\rho v_r h) + \frac{\partial}{\partial z}(\rho v_z h) = \frac{1}{r}\frac{\partial}{\partial r}\left(r\frac{\lambda}{\partial r}\frac{\partial h}{\partial r}\right) + \frac{\partial}{\partial z}\left(\frac{\lambda}{c_p}\frac{\partial h}{\partial z}\right) +$$

$$\frac{1}{r}\frac{\partial}{\partial r}\left[r\frac{\lambda}{c_p}\sum_{j=1}^{n} h_j(LE_j^{-1} - 1)\frac{\partial C_j}{\partial r}\right] + \frac{\partial}{\partial z}\left[\frac{\lambda}{c_p}\sum_{j=1}^{n} h_j(Le_j^{-1} - 1)\frac{\partial C_j}{\partial z}\right] \quad ...(31)$$

$Le_j$ in the above equation is the local Lewis number of the jth species defined as

$$Le_j = \frac{\lambda}{c_p}\frac{1}{\rho D_{jm}} \quad ...(32)$$

The last two terms in the energy equation (31) are caused by the preferential diffusion of species in the radial and axial directions respectively. The Soret effect, caused by the thermal-diffusion effect, is neglected in the present analysis. The specific heat $c_p$ is a strong function of temperature and is locally calculated for each species at the respective temperature. The mixture specific heat is then calculated considering an ideal gas mixture. The temperature of the gas mixture is implicitly calculated by solving equation (30) using a Newton-Raphson method. The solution is refined until the accuracy within the prescribed criteria is achieved.

The conservation of mass, momentum, energy and species conservation equations are solved simultaneously with their appropriate boundary conditions for the domain using a suitable numerical method, which is described in detail in the original papers. The model adopts a finite difference discretization technique for the variables, which are defined in the domain at discrete locations given be a staggered grid arrangement. The transport and thermodynamic properties have been locally evaluated based on the temperature and concentrations. The ignition is made by a spark, which is simulated by raising the temperatures of a few cells to a high temperature for a certain time denoting the spark duration. The final solution obtained from the numerical model is the steady state diffusion flame for the particular inlet conditions denoted by the fuel and air flow rates and temperatures. The distribution of velocity, pressure, temperature and species concentrations in the entire solution domain is obtained at discrete locations into which the domain is divided.

The entropy generation rate due to the transport of heat, mass and momentum along with the chemical reaction is determined from the general entropy transport equation [equatoion (21)]. The equation used the discrete values of the variables found from the numerical solution to evaluate the entropy generation rate. Each term of the equation (21) enables to find out the contribution of the individual process like fluid friction, heat transfer, mass transfer, chemical reaction etc. towards the entropy generation and irreversibility of the process. The second law efficiency is defined as the fraction of the incoming exergy lost during the process and is given by

$$\eta_{II} = 1 - \frac{\dot{I}}{\dot{E}_{in}} \quad ...(33)$$

The objective of the exergy analyses in diffusion flame was to identify the contribution of the different processes towards entropy generation and exergy destruction in a laminar diffusion flame as well as under various conditions like different gravity level, fuel jet velocity, air inlet temperature.

***Results: Entropy generation and exergy loss in diffusion flame at different gravity levels*** [11]

Figure 6a and 6b show the structure of a laminar diffusion flame predicted by the numerical model. Figure plots the temperature distribution in a part of the confined domain near the core, where the flame exists. The plot near the wall is not shown as there is almost no variation there. The flame contour is also plotted in Figure by drawing the heat release zone. Figure , on the other hand, shows the velocity distribution inside the domain. The figures illustrate the structure of the flame. The highest temperature zone is located within the flame in an annular region somewhat above the burner port. The high temperature accelerates the flow in the flame zone and thereby entrains air from the surrounding as is depicted in Figure . The resulting flow field generates the over-ventilated structure of the flame with the tip rounded on the axis. Downstream to the flame zone, the temperature gradually reduced due to the transport of heat to the fluid through internal heat exchange.

Figure 7a-e show the concentration distributions of the major species like fuel ($CH_4$) and $O_2$, which are the reactants and CO, $CO_2$ and $H_2O$, which are the intermediate and final products. The distributions are the result of the transport process of the species mass in a multicomponent environment as well as chemical reaction. The fuel coming from the central jet is consumed at the flame and its concentration gets rapidly reduced. Therefore, the fuel species exist only inside the

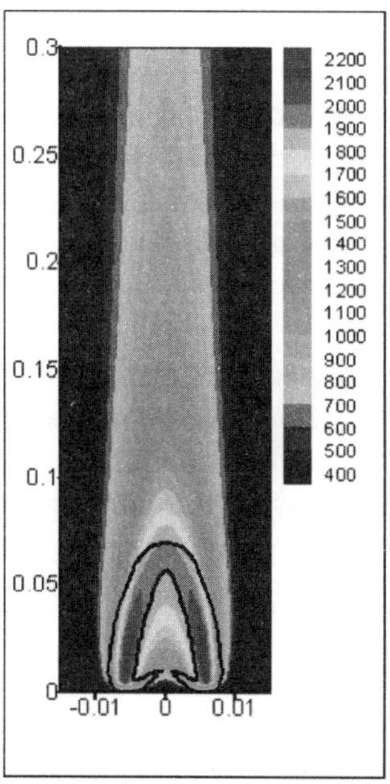


**Figure 6a** Flame (black line) and temperature distribution in a part of the computational domain showing diffusion flame at normal gravity

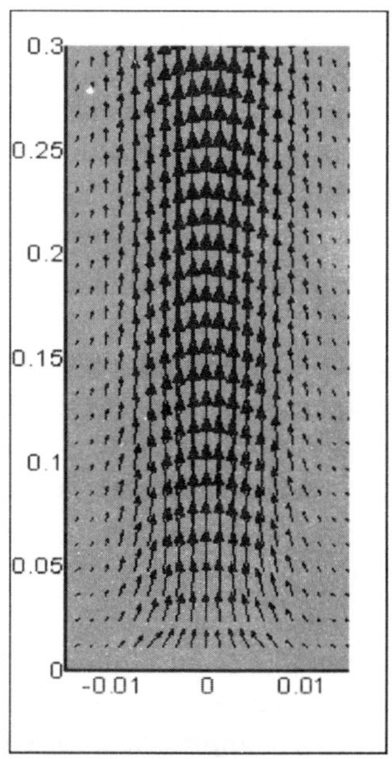


**Figure 6b** Velocity distribution in a part of the computational domain showing diffusion flame at normal gravity

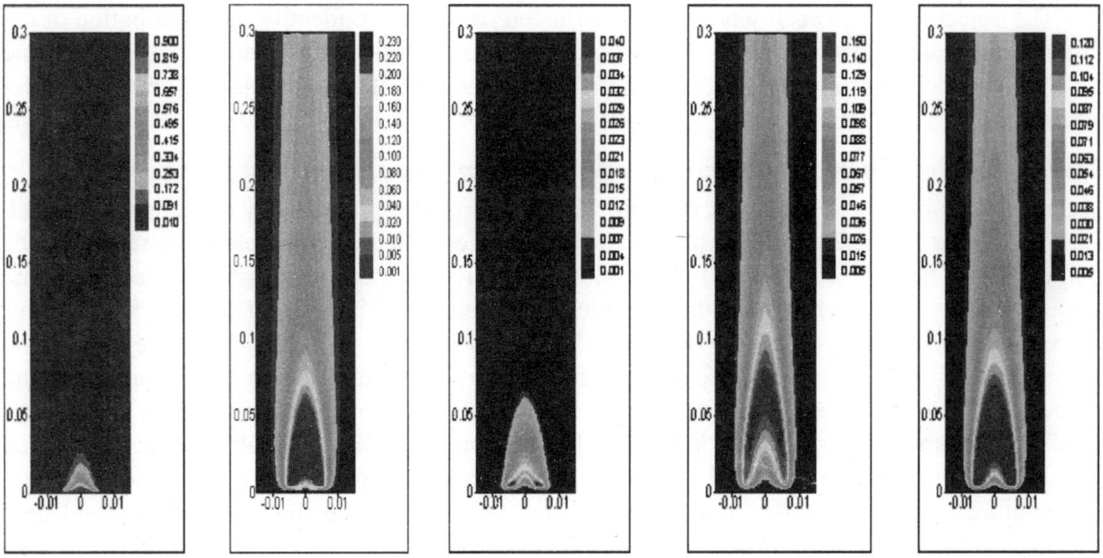


**Figure 7** Species distribution within part of the computational domain for diffusion flame at normal gravity. (a) $CH_4$, (b) $O_2$ (c) CO, (d) $CO_2$, (e) $H_2O$

flame region. The oxygen supplied from air enters as the co-flow and transports towards the flame from the peripheral side. It gets consumed at the flame surface and therefore within the flame region only a trace of it is located. CO is found inside the flame. It is formed on the inside surface of the flame, which burns in a rich atmosphere because of the shortfall of oxygen. The CO that comes towards the outer surface of the flame front gets readily consumed with oxygen. On the contrary, absence of oxygen inside the flame allows CO to diffuse inside showing its existence inside the flame. $CO_2$ and $H_2O$ are formed in the flame front and are transported in all directions based on the advective and diffusive rates. The transport of the product species continues downstream to the flame till the exit plane is reached. However, the major mass transport occurs in the flame region, where the sudden reduction in concentrations of the reactants and generation of products result in large rates of transport of the species.

Figure 8a-c illustrate the plots of the local volumetric entropy generation rate contours due to chemical reaction, heat dissipation and mass transfer respectively, for the diffusion flame at normal gravity. Incidentally, the entropy generation rates shown in the figures are calculated from a prediction, which has neglected the radiative exchange in the flame as a simplified approach. So the quantitative data of the variables corresponding to these are different from those in the earlier figures of 6 and 7. However, qualitatively they will be similar as both are the prediction for the same diffusion flame.

Quite naturally, it is evident from the figure that the entropy generation due to chemical reaction (Fig. 8a) takes place within the flame front. The volumetric entropy generation rate due to chemical reaction is higher near the base of the flame. It reveals that the flame is more intense near its base. At the tip, the flame is weak and the volumetric entropy generation rate is less than $1/20^{th}$ of the peak contour value shown in the figure. The concentration gradient of the fuel is more at

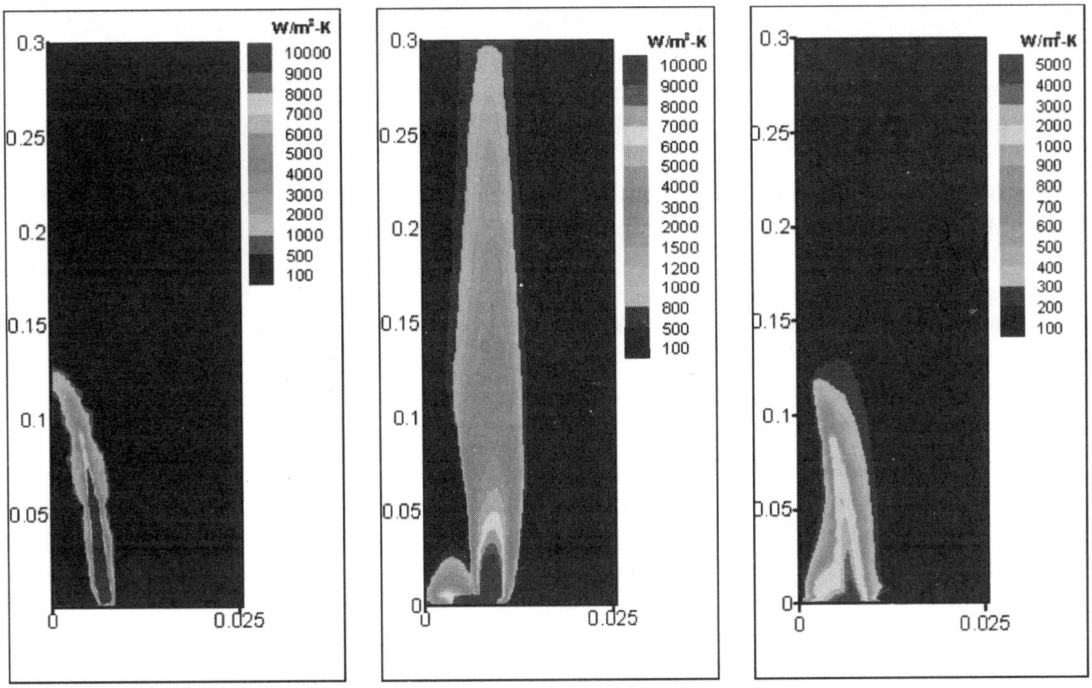


**Figure 8** Distribution of volumetric rate of entropy generation in diffusion flame at normal gravity for (a) chemical reaction, (b) heat transfer, (c) mass transfer

lower elevation and results in higher rate of diffusion of fuel towards the flame front. This makes the flame more intense there. With the consumption of fuel, the fuel concentration gradient reduces at higher elevation. This decreases the fuel transfer rate to the flame and the flame intensity also decreases.

Figure 8b shows the volumetric entropy generation due to heat dissipation along with the temperature distribution in the confined domain. The strong temperature gradient close to the burner rim, where the flame is stabilized, increases the entropy generation there. The peak volumetric entropy generation rate due to heat dissipation is less than that due to chemical reaction. However, the entropy generation due to heat dissipation occurs over a much greater volume compared to the volume of the flame, where entropy generates due to chemical reaction. In the flame zone, heat dissipation occurs from the high temperature flame. Moreover, at normal gravity, the buoyancy-induced acceleration increases the flow velocity at the core and results in a high entrainment of fluid towards the centre. The flow structure generates a thermal stratification with the high temperature gas near the core. Due to the axi-symmetric nature of the flow the gradient of temperature remains low near the axis. The ingress of atmospheric air through the outlet plane and formation of the recirculation zone around the peripheral wall maintains a low temperature shield at the periphery. There exists an annular region between the core and the wall, where the temperature falls rapidly causing a high rate of heat dissipation. The stratification of the flow and the formation of the recirculating zone due to atmospheric air ingress maintain the annular region of high temperature gradient till the exit plane. Therefore, high dissipation of heat continues over a much larger volume to cause a high entropy generation.

The entropy generation due to mass transfer is depicted in the figure 8c for the normal gravity condition. It is observed that the mass transfer results in entropy generation mainly around the flame. In a diffusion flame the reactants diffuse into the flame from the opposite sides of the flame surface and the products transport from the flame in all directions. Beyond the flame, though transport of heat continues, mass dissipation ceases to play any major role. The figure reveals that the magnitude of the rate of entropy generation per unit volume due to mass transfer is significantly lower compared to the other two major contributors even in the flame zone. Though generation and consumption of different species occur in the flame, the gradient of species concentrations never take a large value and the transport process always bring a uniformity in the distribution of the species around the flame. The only high rate of entropy generation due to mass transfer is observed just above the entry plane and around the interface between the fuel and air jets.

Figure 9a-c show the similar figures for zero gravity, viz. the distributions of the volumetric entropy generation rates due to chemical reaction, heat dissipation and mass transport, respectively. The entropy generation due to chemical reaction shows that the peak value of the volumetric rate of entropy generation due to chemical reaction is less at zero gravity than at normal gravity (Figure 9a). This indicates that the flame is less intense at zero gravity condition than that at normal gravity. The reason may be attributed to the fact that at normal gravity due to high rate of entrainment of flow, oxygen will be carried from the co-flow into the flame zone at a faster rate. In the absence of such high entrainment rate from the co-flow, the intensity of the flame reduces at zero gravity. However, the total rate of entropy generation due to chemical reaction is not much different at the two different gravity levels, which is shown later. This is because the volume of the flame front at zero gravity level is more due to the increased width of the flame. The higher volume of the flame front compensates the decrease in the volumetric entropy generation rate that takes place at zero gravity.

Figure 9b reports the plot for volumetric entropy generation rate due to heat dissipation at zero gravity along with the respective temperature distribution. Comparing the figure with the figure 8b for normal gravity, it can be seen that the maximum volumetric entropy generation rate is similar in both the cases. However, in absence of buoyancy, the flow structure is such that heat dissipation from the flame into the surroundings takes place more readily. The thermal stratification in the flow is not observed and there is no ingress of fluid from the exit. As a result, the radial temperature distribution is much uniform beyond a certain height above the burner and the entropy generation is low. Thus the volume over which high entropy generation takes place as a result of heat dissipation is diminished. This decreases the entropy generation rate due to heat dissipation at zero gravity.

The entropy generation due to mass transport takes place around the flame zone (Figure 9c). There is not much difference in the nature of this result from the corresponding result at normal gravity shown earlier. The magnitude of the rate of entropy generation due to mass transfer is low in most of the zone of interest. A high rate is again observed only above the entry plane and at the interface of the two jets.

Figure 10 shows a plot of the total rate of entropy generation in the confined domain and the rate of entropy generation due to individual processes in diffusion flame at various gravity levels.

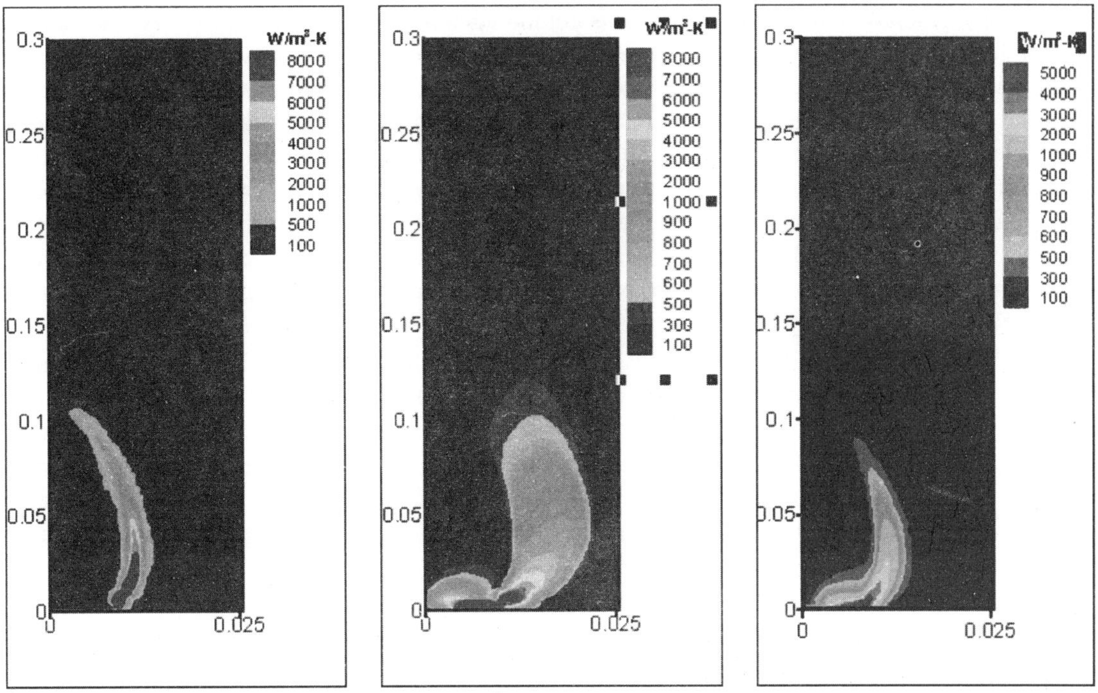


**Figure 9** Distribution of volumetric rate of entropy generation in diffusion flame at zero gravity for (a) chemical reaction, (b) heat transfer, (c) mass transfer

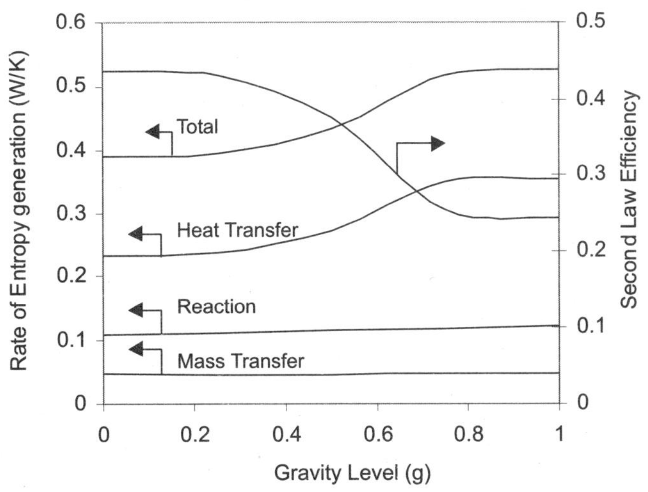


**Figure 10** Variation of the rate of entropy generation (total as well as for individual processes) and Second Law Efficiency in a confined laminar diffusion flame at different gravity levels

The entropy generation due to fluid friction is not included in the figure due to its very low value. It is seen that out of all the individual processes, the most dominant role towards entropy generation in diffusion flame is played by heat dissipation under all gravity conditions. Chemical reaction is the next significant contributor and mass transfer comes as third important. With the change in gravity level, the entropy generation due to reaction and mass transfer hardly changes in magnitude. However, the entropy generation due to heat transfer is grossly affected by gravity. It is already explained from the figures 8b and 9b that as the gravity level is reduced the temperature field becomes more uniform and the entropy generation due to heat dissipation decreases. This decrease in the entropy generation is also reflected in the total entropy generation rate and results in a decrease in the total entropy generation.

Therefore, it is evident that the gravity in earth's atmosphere has got a significant influence on the entropy generation and irreversibilities in diffusion flames within a confined domain. The influence is sensed by the change in the flow pattern caused by the buoyancy effect. Due to buoyancy a strong thermal stratification takes place in the flow, which generates high rate of heat dissipation. Buoyancy does not influence the contribution of chemical reaction towards entropy generation. The relatively higher rate of combustion and resulting higher volumetric rate of entropy generation is compensated by the thinner flame at normal gravity.

The incoming exergy into the confined domain is due to fuel only, as air is supplied at exergy reference temperature. The fuel flow rate and fuel temperature being the same for all the cases of different gravity, the incoming rate of exergy ($A_{in}$) remains always the same. The decrease in the entropy generation at micro-gravity condition also decreases the irreversibility of the reacting flow and increases the second law efficiency of the process. Figure 10 also shows a plot of second law efficiency of the diffusion flame in the confined domain for different gravity levels. It is seen that the second law efficiency decreases from 43.5% at zero gravity to 24.2% at normal gravity. The diffusion flame itself is highly irreversible in nature and causes a large loss of exergy. But this loss is much enhanced in presence of the buoyant motion at normal gravity because of the thermal stratification caused by the flow condition. The way to improve the second law efficiency of diffusion flame under earth's atmosphere is to go for conditions where the thermal stratification can be evaded and a uniform distribution of temperature can be obtained.

The following can be concluded from the above results.

- The rate of entropy generation due to heat transfer is found to be the maximum in diffusion flames. The second major contributor to the rate of entropy generation is the chemical reaction. Mass transfer is the third most important to this cause and fluid friction produces insignificant amount of entropy compared to the others.
- Gravity has no influence on the rate of entropy generation due to chemical reaction. Though, the flame intensity increases with the gravity level and the resulting volumetric entropy generation rate becomes higher, the decrease in the volume of the flame front compensates it.
- The rate of entropy generation due to heat transfer is very much influenced by gravity. The buoyant motion at normal gravity stratifies the high temperature gas and causes high gradient of temperature in the field. Therefore, the rate of entropy generation due to heat dissipation increases with the gravity level. As a result, the total rate of entropy generation in the confined domain also increases significantly with the gravity.

- The entropy generation due to mass transfer is only significant around the flame zone. However, the volumetric rate is low except just above the burner outlet and at the interface of the two jets. The contribution of mass transfer towards total entropy generation is low and remains almost same at all gravity levels.

Nomenclature

| | |
|---|---|
| $A$ | Preexponential factor |
| $C_j$ | Concentration (mass fraction) of the jth species |
| $c_p$ | Specific heat |
| $D_{jm}$ | Mass diffusivity of jth species in a binary mixture of j and $N_2$ |
| $E_a$ | Activation energy |
| $\dot{E}$ | Exergy flow rate |
| $e$ | specific exergy |
| $G$ | Generator output Power |
| g | Acceleration due to gravity, |
| $\bar{h}$ | specific molar enthalpy |
| $\dot{I}$ | Rate of exergy destruction / Irreversibility, |
| $M$ | Molecular weight |
| $\dot{m}$ | mass flow rate |
| $n$ | mass fraction |
| $P$ | Power |
| $p$ | Pressure, |
| $p_o$ | Exergy reference pressure |
| $Pr$ | Prandtl number |
| $q_{cv}$ | Heat transfer across the control volume |
| $\bar{s}$ | specific molar entropy |
| $\dot{s}_g'''$ | Volumetric entropy generation rate |
| $T$ | Temperature |
| $T_o$ | Exergy reference temperature (= 298 K) |
| t | Time, |
| v | Velocity, |
| $w_{cv}$ | Work transfer across the control volume |
| $\rho$ | Density, |
| $\mu$ | Viscosity, |
| $\lambda$ | Thermal conductivity, |
| $\dot{\omega}$ | Reaction rate, |
| $\eta_2$ | Second law efficiency |

Subscripts

| | |
|---|---|
| F | Fuel |
| o | oxygen |
| c | carbon dioxide |
| w | water vapour |

## References

1. Dunbar WR, Lior N. Sources of combustion irreversibility. Combust Sci Technol 1994;103:41–61.
2. Caton JA. A review of investigations using second law of thermodynamics to study internal combustion engines. SAE 2000-01-1081, 2000.
3. Lior N, Sarmiento-Darkin W, Al-Sharqawi HS. The exergy fields in transport processes: their calculation and use. Energy 2006; 31:553-578.
4. Sengupta S, Datta A, Duttagupta S. Exergy analysis of a coal based 210 MW thermal power plant. Int. J. Energy Research 2006 (in press).
5. Taniguchi H, Mouri K, Nakahara T, Arai N. Exergy analysis on combustion and energy conversion processes. Energy 2005; 30:111-117.
6. Hirschfelder J.O., Curtiss C.F., Bird R.B., Molecular theory of Gases and Liquids, John Wiley and Sons, New York, 1954.
7. Hiwase S., Datta A., Som S.K., Entropy balance and exergy analysis in the process of droplet combustion, J. of Physics D: Appl. Phys. 31 (1998) 1601-1610.
8. Datta A., Som S.K., Thermodynamic irreversibilities and second law analysis in a spray combustion process, Combustion Sci. Tech. 142 (1999) 29-54.
9. Datta A., Entropy Generation in a Confined Laminar Diffusion Flame. Combust. Sci. and Tech. 159 (2000) 39-56.
10. Nishida K., Takagi T., Kinoshita S., Analysis of entropy generation and exergy loss during combustion, Proc. of the 29th Symp. (Int.) of the Combustion Inst., 29 (2002) 869-874.
11. Datta A. Effect of gravity on structure and entropy generation of confined laminar diffusion flames. Int. J. Thermal Science 2005; 44:429-440.
12. Som SK, Sharma NY. Energy and exergy balance in the process of spray combustion in a gas turbine combustor. Journal of Heat Transfer ASME 2002; 124:828-836.
13. Stanciu D, Isvoranu D, Marinescu M, Gogus Y. Second law analysis of diffusion flames. Int J Appl Thermodyn 2001;4(1):1–18.
14. Datta A, Leipertz A. Thermodynamic optimization of a laminar diffusion flame. Proc. Of the 1st Int Conf. On Heat Transfer, Fluid Mechanics and Thermodynamics, Vol. 1, Part 1, pp. 111-116, Kruger National Park, South Africa, 8th-10th April, 2002.
15. Arpaci V, Selamet A. Entropy production in flames. Combust Flame 1988;73:251–9.
16. Yapici H, Kayatas N, Albayrak B, Basturk G. Numerical calculation of local entropy generation in methane air burner. Energy Conversion and Management 2004.
17. Yapici H, Kayatas N, Albayrak B, Basturk G. Numerical study on local entropy generation in a burner fueled with different fuels. Heat Mass Transfer 2005; 41:519-534.
18. Puri IK. Second law analysis of convective droplet burning. Int J Heat Mass Transfer 1992;35:2571–8.
19. Som SK, Mondal SS, Dash SK. Energy and exergy balance in the process of pulverized coal combustion in a tubular combustor. J Heat Transfer ASME 2005; 127:1322-1333.

20. DuPont V., Pourkashanian M., Williams A., Modeling of Process Heaters Fired by Natural Gas, J. of Inst. of Energy. 1993; 73:20-27.
21. Katta V.R., Goss L.P., Roquemore W.M., Effect of Nonunity Lewis Number and Finite Rate Chemsitry on the Dynamics of a Hydrogen-Air Jet Diffusion Flame, Combust. Flame. 1994; 96:60.
22. Daw S.C., Chakravarthy K., Conklin C.J., Graves R.L., Minimizing destruction of thermodynamic availability in hydrogen combustion. Int J Hydrogen Energy 2006;31:728–36.
23. Som S.K., Datta A., Thermodynamic irreversibilities and exergy balance in combustion processes. Progress in Energy & Combustion Science 2008; 34 (3): 351-376.

# GE Aviation Low Emissions Combustion Technology Evolution

**Hukam C. Mongia**
GE Aviation, Cincinnati, Ohio, U.S.A.

## I. INTRODUCTION

Many of the propulsion engine combustion system design requirements need to be optimized without adversely impacting any operational safety standards. It is well known that high-power NOx reduction potentials need to be balanced against low-power engine operation including ignition, acceleration to full power and deceleration back to ground idle power setting ("throttle burst-chop transients"), flame-out characteristics, total cost of ownership, weight, length, durability, combustion dynamics, acoustic signature, combustor exit temperature quality, combustion efficiency, smoke, CO and hydrocarbon emissions. Combustion system volume, length, pressure drop, the number of fuel nozzles and staging strategy can impact combustor exit temperature quality including pattern factor (PF) and profile which can adversely impact the hot-section durability and sometime the fuel-burn.

For a given maximum rated engine thrust level, the design and number of fuel nozzles and staging strategy may impact weight and cost of ownership. The cooling air requirement for durability needs to be balanced against combustion air for reducing NOx and/or smoke emissions, dilution and trim air for controlling profile and PF. The cooling technology selection process recognizes the importance of weight and total cost of ownership in addition to how it is impacted by the type of low-emissions technology.

Consequently, different low-emissions technology development approaches have been pursued, e.g., [1-4]. These approaches have continued to provide continuous improvements in the development of the low emissions technologies. Due mainly to the limited scope of this paper which is intended for presentation to the Long Term Technology Goal (LTTG) Review Panel for helping the ICAO/CAEP with the standard-setting process, we focus on the approach GE Aviation has taken during the last 30 years; the readers should refer to the extensive literature available as described in the following paragraph.

Propulsion gas turbine combustion R&D community has been very active in improving low- and high-power emissions during the last 30 years. A Google "Scholar" search for "low gaseous emissions propulsion engine technology" on February 15, 2006 gave a total of 1160 entries including the first relevant five references[5-9]. It is relatively easy to demonstrate potential feasibility of promising mixers to produce 70-90% reduction in the climb-out NOx emissions index (EINOx). However, demonstrating TRL[10] of 6 for propulsion engine combustors to achieve 50% reduction in EINOx is very challenging because the combustion system must meet all other design requirements including the ICAO[11] **L**anding **T**ake-**O**ff (LTO) CO, HC, smoke, sub-idle, idle and cruise combustion efficiency targets, design margin for flameout at critical operating conditions including fuel cutback to meet engine deceleration requirement and altitude descent, cold-fuel and cold day ground start and altitude restart to at least 25,000-30,000 feet, combustor exit temperature quality including peak pattern factor (PF), profile factor (PrF), gas temperatures near the hub and tip of the high pressure turbine nozzle, performance (pressure drop, turbine back-flow margin), operability (acceptable engine transient operation, water and hail ingestion), combustion noise (wide band and/ or at discrete frequencies) and dynamics, durability and field supportability (combustor, fuel nozzle including coking), combustion system cost and weight.

There is a general consensus among the combustion design community that combustion products based on recently demonstrated technology product are expected to continue evolving over a period of ~30 years. The first generation of low-emissions SAC's evolved out of the technology initiated in the middle 1960's, and the most popular of these for propulsion engines has been rich-dome single annular combustors[12]. The evolution of swirl-cup (of GE Aviation) based rich-dome combustion technology identified as **O**ptimized **R**ich-**Q**uench-**L**ean (ORQL) is described in Section 2. From this section we conclude that if one were to keep ~20% margin for NOx scaling, assuming 10-20% NOx reduction beyond ORQL, we will barely meet the ICAO single engine NOx standards at typical 40 Overall Engine Pressure Ratio (-1+2*EPR CAEP/6 for large engines) to having only 10% margin.

Bahr and Gleason[1] and Bahr[7], therefore, advocated for the development of a revolutionary low-emissions technology in the early 1970's, namely a lean **D**ual **A**nnular **C**ombustion (DAC) technology as discussed in Section 3. It took approximately 20 years to develop the DAC technology to achieve TRL9, which is consistent with industry's estimate on how long it takes to develop revolutionary low-emissions propulsion engine combustion technology and its transition into product. The DAC technology was developed for a nominal 30 Overall Engine Pressure Ratio (EPR). It was applied as a retrofit to the CFM56 engine with EPR in the range of 23 to 33. Its application in a new centerline engine (GE90 DAC1) with EPR in the range of 35 to 40 identified limitations of extrapolating emissions and heat transfer scaling design tools for promising advanced cooling technologies. Moreover the engine design cycle points drifted considerably between the conceptual and preliminary design phases. Consequently, some of the design requirements had to be compromised as a result of the combustion system tradeoffs, as summarized in Section 4.

The **T**win **A**nnular **P**remixing **S**wirler (TAPS), as described in Section 5 was initiated in 1995 as the next step in the development of revolutionary low-emissions combustion technology along with the objective to simultaneously meet or exceed all other attributes of ORQL. The SAC TAPS technology is TRL6 ready and DAC TAPS is TRL5. We are in concurrence with the industry's estimate that it takes approximately ten years to take TRL6 technology to TRL9.

As summarized in Section 6, the potential for the next generation of TAPS technology to achieve 30% and 15% of CAEP/2 NOx standard (32 + 1.6*EPR) is currently being investigated under government-sponsored programs, e.g., Ultra Efficient Engine Technology (UEET) and Propulsion 21.

## II. OPTIMIZED RICH-QUENCH LEAN COMBUSTION

In order to meet conflicting requirements of high idle efficiency, reliable operability (start, engine acceleration to rated thrust, flameout margin), low high-power NOx and smoke, acceptable combustion system durability, profile and PF, the technology evolution of single annular combustors, Figure 1a, started with rich-dome designs. An acceptable combustor design is achieved by optimizing fuel/air mixing processes within the mixer (swirl cup) and the resulting near-dome flow field which is affected also by the flare and splash plate design, the primary, intermediate and dilution zones which are identified as 1, 2 and 3 in Figure 1a. A number of configuration variables are optimized during the design processes including details of the swirl cup (e.g., air flow splits between the primary and secondary swirlers, swirl numbers, venturi shape and diameter, flare design, mixer exit streams velocities), equivalence ratio and residence times of the primary, intermediate and dilutions zones. Broadly speaking, the reacting flowfield in rich-dome diffusion flame combustor goes through rich, stoichiometric, quench and lean combustion regimes. For example at a typical takeoff conditions, the majority of the regions 1, 2 and 3 are fuel-rich, near stoichiometric and lean, respectively.

The average ICAO Landing Take-Off (LTO) NOx emissions, gm/kN of the first generation of these Rich-Quench-Lean (RQL) combustors (e.g., CF6-45, CF6-50, CF6-80A, CF6-80E and GE90 DACI, the latter has a near-stoichiometric primary zone) is shown in Figure 2 as a function of EPR at take-off.

Further optimization of this design approach has a produced several Optimized RQL (ORQL) combustors which include the CF34-3, CF34-8, CF34-10, CFM56TI, CF6-80C2, CF6-80E and GE90-115B. The average LTO NOx of ORQL family of combustors (Gen 2) as shown in Figure 2 is approximately 15% lower than that of Generation 1 RQL combustors. Many promising approaches are currently being pursued to achieve LTO NOx lower than that of ORQL. A successful approach will be able to deal with many challenges including rapid quench, smoke, cruise NOx, PF, cooling air requirement, durability and attendant cost of ownership. It is difficult to predict how much further NOx reduction can be realized with the next-generation RQL A reasonable entitlement of the next-generation (Gen 3) affordable RQL may be additional 10% reduction, and some designs might achieve up to 20% reduction, as presented in Figure 2. Accordingly, we assume that typical Gen 3 RQL could give average LTO NOx in the range of 50-57 g/kN at an overall Engine Pressure Ratio (EPR) of 40 with a high degree of probability of demonstrating the upper limit of 57. If one were to scale this technology for another engine application, it is advisable to use a 20% margin. Consequently, even though one can meet CAEP/4 single engine NOx characteristic value with 10% reduction beyond ORQL, it barely would meet the requirement for CAEP/6. On the other hand with a 20% reduction beyond ORQL, one can meet CAEP/6 requirement with 10% margin.

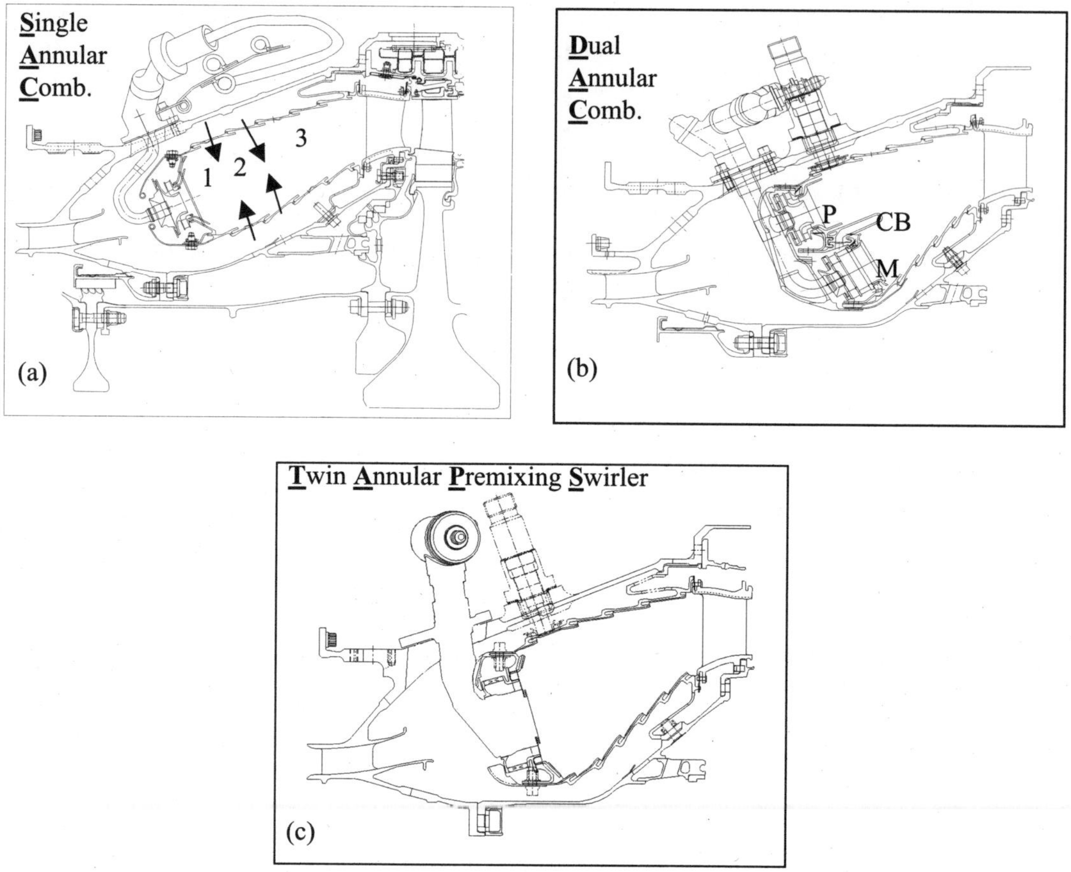


**Figure 1** Evolution of low emissions propulsion engines combustion technology

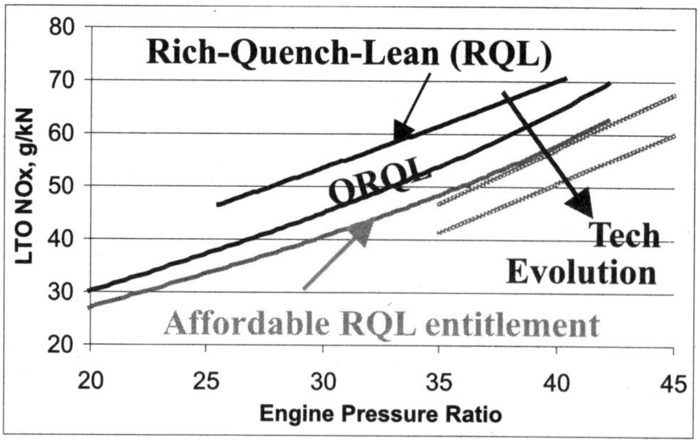


**Figure 2** RichQuench-Lean combustion NOx technology evolution

In summary, one needs to set approximately 20% NOx reduction objective for the next-generation RQL so that reasonable margin can be maintained from the CAEP/6 NOx standard in order to afford system level tradeoffs.

## III. DUAL ANNULAR COMBUSTION

Bahr and Gleason[1] introduced an innovative and practical combustion system, DAC in June 1975, and the first propulsion engine combustion product, the CFM56-5B1, B2 and B4 engine were emissions tested on June 1994, spanning a 20-year technology/product development effort. DAC consists of two annular combustors separated for a short axial distance by an annular center-body (CB) as shown in Figure 1b. The outer combustor (P) is optimized for low-power operation including ignition, flame propagation, engine acceleration to idle through some intermediate power before the inner annular combustor (M) is fueled. Several critical technologies development and maturation were successfully executed including CB cooling, DAC fuel nozzles and controls. The first generation of the lean DAC CFM56 included four engine models, namely the CFM56-5B1, B2, B4 and B6 covering the take-off EPR from 24.6 to 31.2. Further improvements, e.g. in radial profile for the pilot-only operation along with low-power efficiency and high-power durability were introduced in the second-generation DAC engines, CFM56-5Bx/2P and CFM56-7By/2. Here, $x = 1, 2, 3, 4$ and 6; and $y = 20, 22, 24, 26$ and 27. These engine models cover the take-off EPR range from 22.8 to 32.8.

Lean DAC combustion systems need to pay special consideration to the trade-offs between low-power combustion efficiency, cruise combustion efficiency, LTO and cruise NOx, average and peak temperature profiles, discrete tone noise; the latter three issues pertain specifically to highly-fueled pilot operation. Further design optimization of the lean DAC technology is feasible. However consistent with the intent of developing the next-generation of ultra-low emissions combustion technology, as described in Sections 5 and 6, only the evolutionary DAC technology development effort has been pursued for supporting the CFM DAC product.

The DAC went into service in the CFM56-5B engine in December 1994, and since that time has accumulated approximately 5 million flight hours and 3.3 million cycles on 375 CFM56-5B and –7B engines. The DAC met its primary goal of reducing NOx. The ICAO Emissions Data Bank shows that DAC NOx emissions are typically 30% lower than SAC emissions on the same engine model. Effects on other emissions are mixed. Smoke (related to particulate mass matter) is significantly reduced. CO and HC, while still meeting ICAO standards, are significantly increased.

Two engineering challenges that have been met with the more complex DAC fuel staging system are reliability and fuel injector durability. To optimize the combustion process, some of the fuel injectors are shut off during part power operation, both on the ground and in flight. This requires additional valves and control logic. Even with these additional features, recent DAC fleet statistics show that the reliability of DAC engines (as measured by in-flight shutdowns, aborted takeoffs and basic engine-caused delays and cancellations) is basically the same as that of SAC engines. The fuel injector durability challenge has been met with an actively cooled fuel injector tip to avoid formation of coke in the fuel injector passages when the injectors are shut off. There are noticeable changes in transient response with the fuel staged DAC, but response meets all airworthiness certification requirements and is similar enough that Airbus allows intermix of conventional and DAC engines on the A320 aircraft family.

There are tradeoffs with the DAC. Engine fuel consumption and maintenance cost are reportedly increased with the DAC[16]. Both of these effects are primarily related to non-uniform exit temperature profiles during part power operation, which affects turbine efficiency and thermal gradients in the combustor and turbine. Design and operating measures have been introduced to address the maintenance cost issues.

Using multi-passage diffusers and the dual-annular combustors, e.g., the GE90, can significantly reduce the overall combustion system length. The GE90 DAC combustor program illustrates a natural evolution of new technologies, e.g., higher EPR, cooling technology (optimized multi-hole with TBC, hybrid liner cooling), DAC fuel nozzle, optimized fuel staging, eliminating center-body, solving the profile issue along with and evolutionary reduction in LTO NOx.

## IV. COMBUSTION SYSTEM TRADEOFFS

No safety compromises can be allowed in introducing low-emissions propulsion engine combustion products. The combustion system pressure drop considerations are normally limited by the allowable prediffuser area ratio and the required turbine back-flow margin. It is very difficult to get any significant increase in pressure drop due solely to the introduction of a low-emissions combustion technology. It is therefore a good practice to develop the next-generation ultra-low emissions combustion systems with the objective of not adversely impacting pressure drop, engine fuel burn, ability to operate at the desired minimum thrust levels (e.g., ground idle and altitude descent) along with the required flameout margin, ground and altitude ignition, reliable operation under inclement weather conditions, engine acceleration and deceleration characteristics.

The tradeoff between LTO NOx and CO is well known. It is however worthwhile to note that when the same combustion system is used in different engine models, e.g., the CFM56-5A, CFM56-5B, CFM56-5B/P and CFM56-7B, there can still be a significant variation in the LTO NOx and CO characteristics, as presented in Figure 3a. Clearly, in this case the same build of material RQL combustion system is used. In an idealized situation the data should fall in a narrower band. The scatter in the data has not been fully explained. Nevertheless, it is now well recognized that a given RQL emissions technology cannot be represented by an average curve but rather a wide band as shown here. For example at LTO NOx value of 40, the LTO CO values vary between is 28 and 52, which might extend to ~70 if we had a bigger sample size. When LTO NOx and CO database of the two most recent lean DAC's, namely the CFM56-5B/P DAC and CFM56-7B DAC are added to those of the SAC's, the resulting Figure 2b indicates some increase in CO emissions with the introduction of the DAC technology. Considering the variability between the engine models with the same build of material SAC, a back-to-back comparison for CO emissions between SAC and DAC, e.g., for the CFM56-7B, should be made cautiously.

The variation of LTO HC versus LTO NOx is presented in Figure 3c. For example at LTO NOx value of ~40, the SAC's give LTO HC values in the range of 3.3-8.6 g/kN, the latter could be extended to ~11.5. On the other hand, the corresponding range for DAC's LTO HC is 6.8-11.2. In summary, there is some empirical relationship between LTO values of NOx, CO and HC, but it is

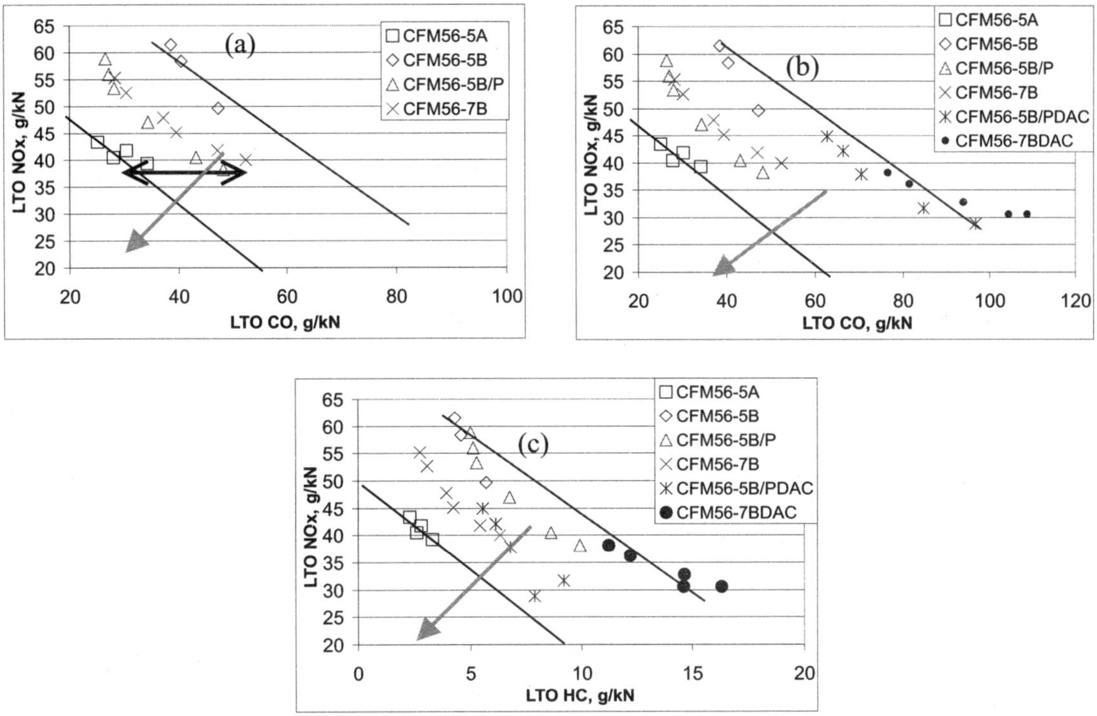


**Figure 3** Landing takeoff NOx, CO and hydrocarbon emissions are inter-dependent

quite complicated one. The conventional wisdom that improving emissions technology would move the NOx versus CO curve towards the origin is not quite obvious in Figure 3.

A complimentary approach for assessing the status of low-emissions technology is presented in Figure 4 where the lines represent emissions of ideal flames, namely a diffusion flame, partially-premix flames with 4 different scalar dissipation rates (c = 3, 4, 5 and 6 sec-1), and premix flames with equivalence ratios of 0.6, 0.65, 0.70, 0.75, 0.8 and 0.85, respectively. The NOx and CO emissions indices data are the typical values for a selected group of five different combustors along their respective engine operation lines. As expected, the degree of premixing and the effective equivalence ratio change as a function of the engine power setting. All the five combustors are reasonably well represented by a diffusion flamelet at the low-power operation. However each combustor behaves differently at intermediate through maximum power settings showing a combination of both partially-premix and premix flamelet behavior. Clearly for the simple flames, diffusion, partially-premix and perfectly premix, there is a unique relationship between EINOx and EICO as represented by the curves. However, this relationship is quite complicated for the propulsion engine combustors. Nevertheless, it is possible to get deeper insight of the developing low-emissions technology by utilizing both Figures 3 and 4.

The combustor exit temperature quality including average and peak radial temperature profiles need to be optimized to get the best overall turbine efficiency consistent with the hot-section durability objectives. The propulsion engine combustors have produced PF between 0.20 and 0.35.

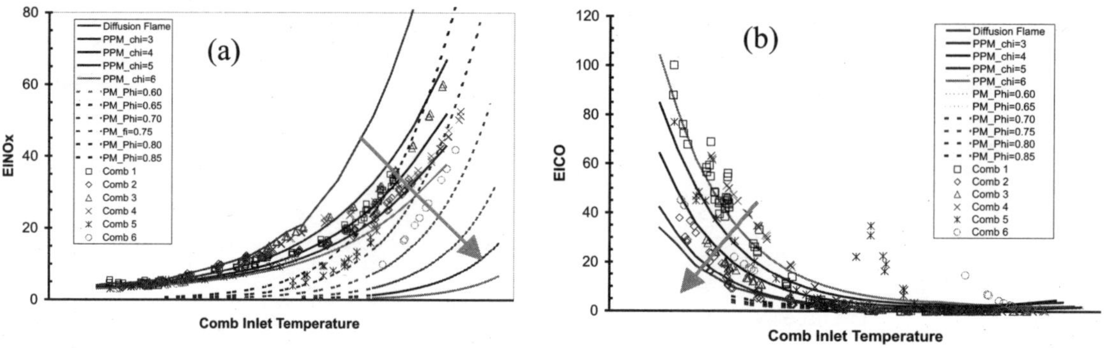


**Figure 4** Engine NOx and CO emissions indices compared with idealized flames, namely diffusion, partially-premix and premix flames

The radial location of the average peak profile lies around 50 ± 20% channel height as long a proper emphasis is placed on the near-hub gas temperature levels and gradients. However, excellent hot-section durability characteristics and turbine efficiency have been demonstrated with combustors having outer or inner peak profile locations, typical, shown in Figure 5 for two different engine combustors. The near-wall gas temperature characteristics are generally controlled by near-exit film cooling design features as long as the combustor has acceptable PF levels. It is obvious that if Engine A combustor is replaced by a low-emissions technology combustor giving profile similar to Engine B combustor, the Engine A turbine efficiency might be impacted. The specification of profile and PF requires a system-level trade-off and agreement with both the combustion and turbine system design communities is required.

The airplane flight envelope requested by a customer shown typically in Figure 6 is required to be met by current technology as well next generation combustion and control systems. Similarly, engine start, acceleration and snap deceleration, operating envelope, durability, cost, weight, maintainability need to be optimized at the system level.

## V. TWIN ANNULAR PREMIXING SWIRLER (TAPS)

It is a good commonsense engineering practice to critically assess evolution of the relevant technology of the years past (10 or 20 years!). The NASA initiative, Advanced Subsonic Technology (AST), lean- and near-stoichiometric DAC product introduction experience (for both the CFM56 and the GE90), Dry Low Emissions combustors[13] for industrial engines provided impetus for formulating an approach leading to the next-generation out of the box ultra-low emissions technologies. The first three years starting January 1995 were spent in exploring several promising approaches leading to the down select of a class of mixers based on the operating principles illustrated in Figure 7a.

It envisioned two independently controlled flames, notionally designed for low-power (Pilot) and high-power (Cyclone) operation along with an optimum flame-interaction region intended for transition and merging of the two (Twin) flames. The both flames are Swirl stabilized by a suitably selected arrangement of swirlers. The main flame, which is concentric with the low-power (pilot)

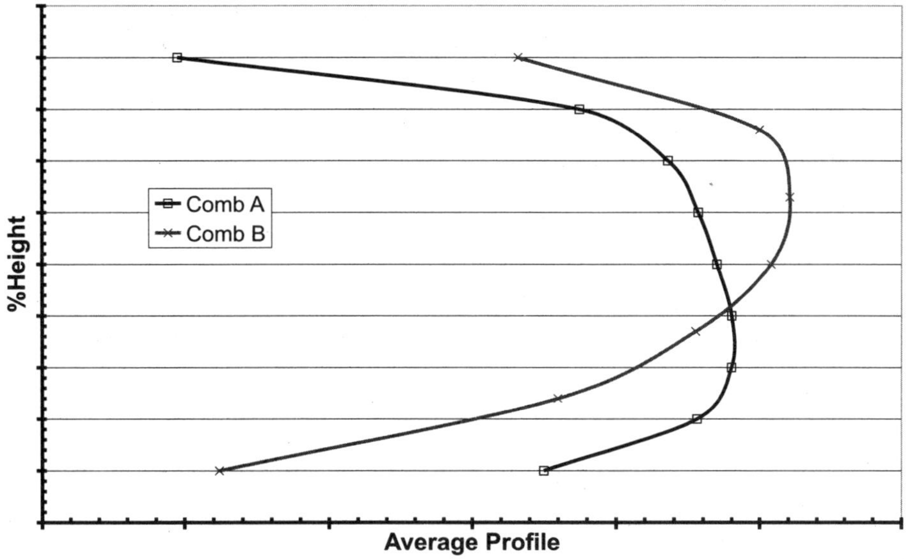


**Figure 5** Typical average combustor exit radial profiles of two different engine combustors

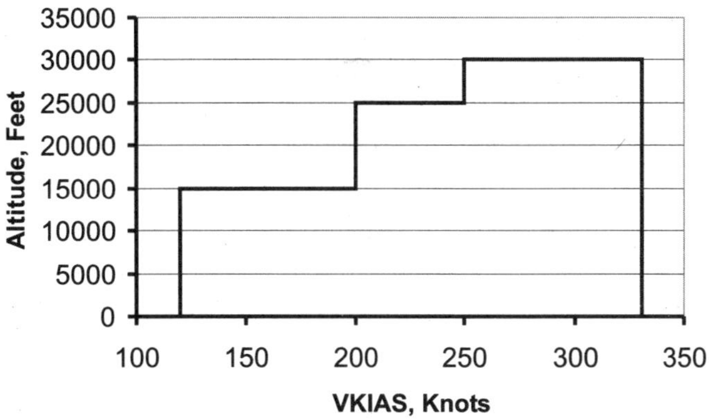


**Figure 6** Typical altitude relight envelope

flame, is therefore an <u>**A**</u>nnular-flame. The degree of <u>**P**</u>remixing of the both flames is controlled consistent with their functions. For example, if the main mixer's equivalence ratio is too high, being fully premix may not give the lowest EINOx for a given combustor inlet temperature ($T_3$) and $P_3$, as can be deduced from Figure 4a.

A number of independent design features can impact the characteristics of the twin flames including the following:

1. The number of swirlers for each flame (low-power (Pilot) and high-power (Main) flames), their relative orientation (co- or counter-flow), the air-flow splits when more than one swirler is used, swirl strength, the mixing passage geometry, the type of fuel injection

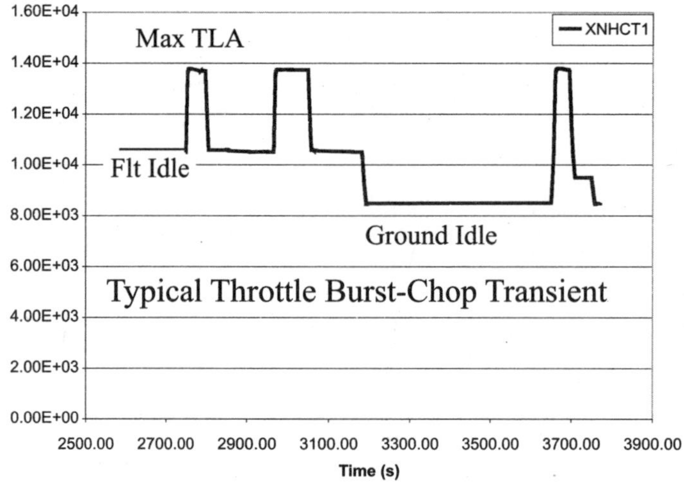


**Figure 7**

devices (pressure atomizer, prefilming, dual-fluid atomization), and the degree of unmixedness.

2. Airflow split between the pilot and main.
3. Type of fuel injection devices and their design features, macrolaminate, multi-point fuel injector along with airblast simplex for the pilot.
4. Details of the various mixing layers that impact operation of the pilot, main and the transition.
5. Fuel staging between the pilot and main.

In addition to a new set of design rules for the TAPS mixer, we used the design parameters relevant for the conventional combustors, e.g., the dome and reference velocities, cold and hot residence times, fuel nozzle spacing, combustor height (H) and length (L), L/H, aerodynamic loading parameter, volumetric heat release rate, cold and hot flow residence times.

Recognizing the limitations of the Reynolds averaged transport equations based on gradient diffusion approximation for handling complex two-phase swirling flows, limited comprehensive injection modeling capability, the TAPS development relied heavily on testing supplemented by CFD support when appropriate.

A number of test rigs were used during the development process including the flame tubes, single nozzle/mixer modules, low- and high-pressure sector rigs and full-scale annular combustor rigs.

The TAPS mixers' emissions measurements in flame tube rigs have been made for several engine operating conditions including the following most severe conditions:

Highest $T_3$ of 1200 F at 678 psi $P_3$ and 0.032 FAR

Highest $P_3$ of 778 psi at 1030 F $T_3$ and 0.032 FAR

Auto-ignition testing of the TAPS mixers has also been conducted for several engine operating conditions including the following most severe test points:

Highest mixer $\phi$ of 0.96 at 566 psi $P_3$, 1306 F $T_3$

Highest $P_3$ of 574 psia at 992 F $T_3$ with 0.79 mixer $\phi$

Highest $T_3$ of 1384 F at 527 psi $P_3$ with 0.75 mixer $\phi$

Very extensive lean blowout, ignition and flame propagation investigations have been conducted in low-pressure sector rigs as well as in full-annular rigs to simulate critical points along the flight envelope. Tests were run with ambient and cold air (down to –40 F) and fuel (down to –20 F air, –30 F fuel) and pressures down to 4.7 psia.

The TAPS technology development was conducted for potential application in both SAC and DAC configurations. A single annular combustor design was pursued for the CFM56 whereas a dual annular combustor design approach was adopted for the GE90. Both combustors went through a full suite of testing on the full-annular combustor rigs including the following:

- Pressure distribution around the combustor, and estimated airflow splits;
- Ignition, flame propagation and extinction at critical points simulated along the engine flight envelope;
- Combustor exit temperature quality including average and peak radial profiles as determined from multi-element thermocouple rake traverse at the different operating conditions including the simulated idle and take-off points;
- Assessment of thermal conditions of the combustion system structure including dome, liners and fuel nozzle as determined by surface-mounted and embedded thermocouples at several critical locations;
- Gaseous and smoke emissions at several engine operating conditions including measurements made to determine NOx, CO and HC emissions exponents for pressure and fuel/air ratio;
- Quantification of combustor annulus dynamics (frequency and amplitude) as detected by pressure transducers.

The SAC TAPS full-annular rigs have been used to collect data for an extensive range of operating conditions. The most severe conditions in regard to inlet temperature, pressure and fuel/air ratio (FAR) were:

Maximum T3 of 1232 F tested at 361 psia & 0.035 FAR

Maximum P3 of 486 psi tested at 1120 F & 0.031 FAR

Maximum FAR of 0.047 tested at 927 F and 195 psi.

Because of the higher design airflow rate required for the DAC TAP full-scale annular rig, it was tested to the maximum P3 of 350 psi (compared to the takeoff pressure of the GE90-94B being 596 psi) at 1175 F and 0.036 FAR.

Both SAC and DAC TAPS combustion systems (Figures 7b and 9) met or exceeded all the design requirements consistent with expectations for a TRL5 matured technology[10], including combustion efficiency at sub-idle, idle and cruise, design margin for flameout at critical operating conditions including fuel cutback to meet engine deceleration requirement and altitude descent, ground start and altitude restart to at least 25,000-30,000 feet, combustor exit temperature quality including peak pattern factor (PF), profile factor (PrF), performance, operability, combustion noise, estimated durability of combustor and fuel nozzle, combustion system cost and weight.

The CFM TAPS combustion system comprised of the combustor, fuel nozzles, valves, and the fuel injection control system was installed in a development CFM56-7B engine. Two engine tests were conducted accumulating a total of 909 hours of testing including more than 200 hours run time at the maximum rated thrust. The engine completed approximately 4,000 LCF cycles (~ 10 minute cycles) and 2,000 fuel nozzle purge cycles (~ 3 minute coking cycle) to validate the fuel injection system technologies including constant pressure fuel nozzle (CPFN) control system and main valve, macro-laminate fuel delivery passages, pressure augmented fuel purge, airblast simplex, liquid jet injection in confined swirling flow.

The first engine (FETT) shown in Figure 7c completed 174 test hours for documenting performance, pressure distribution, combustion dynamics, thermal and vibration surveys and engine transients, typical shown below:

FETT also included measurement of acoustics, gaseous and particulate emissions (following the ICAO procedures[11]) and documentation of engine operability including 14 rounds of hail ingestions (Figure 7d), which indicated that the TAPS combustion system would be certifiable for operation under inclement weather conditions.

The second engine test (SETT) completed a total of 735 hours of testing including 4,000 endurance cycles, 2,000 fuel nozzle purge cycles in addition to conducting operability, emissions and thermal surveys.

The combustion system hardware after the endurance test looked in excellent shape, shown typically in Figure 8. The hardware was subject to detailed teardown analysis including visual, mechanical and metallurgical analyses followed by in-depth thermal, stress and life prediction analytical evaluation. The post-test models for the combustion system including deflector, dome plate and liners indicated the basic design can be improved to meet the life, cost and weight objectives. The following issues can be addressed as part of the New Product Introduction (NPI) effort for transitioning TAPS TRL6 to TRL9.

- Cowls – contact observed on retention tabs, no wear.
- Cyclone – chatter on outer section of flange, contact area on outer and inner section of flange, contact on both anti-rotation tabs
- Deflector – crack in 7 of 20 deflectors at right side
- Dome plate – crack in 6 of 21 ribs at same locations as deflector cracks
- Outer liner – panel 1 spalled at 20 cups with crack at 2 cups, panel 2 spalled at 20 cups (uniform cooling can be replaced by preferential cooling)
- Inner liner – panel 1 spalled at 20 cups with overhang cracks at all 20 cups, panel 2 spalled at 20 cups (uniform cooling can be replaced by preferential cooling).

The GE90 DAC TAPS combustion system technology is the same as that of the CFM SAC TAPS except for the obvious mechanical design differences between SAC and DAC combustors. The main reason for selecting a DAC TAPS for the GE90 was to maintain configurationally similarity with the current engine model in regard to the number of DAC fuel nozzles, length and

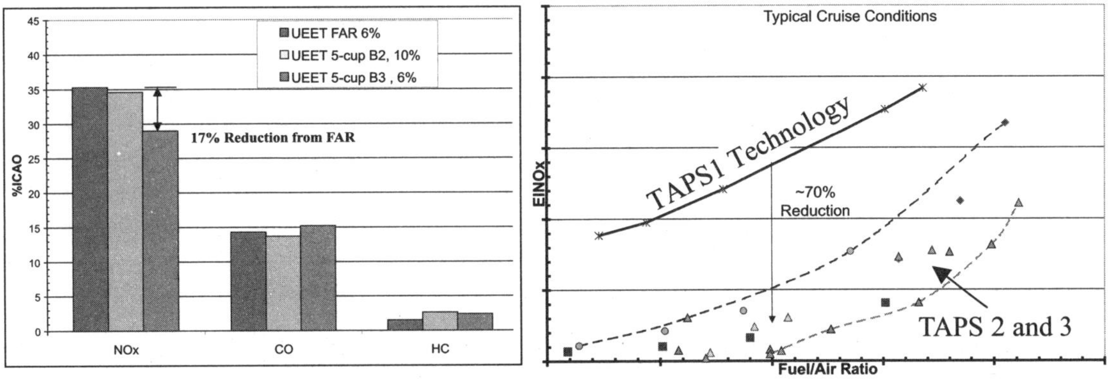


**Figure 8**

**Figure 9** GE90 DAC TAPS combustor and fuel nozzle

cross-sectional dimensions. We conducted an in-depth evaluation of the DAC TAPS combustion system similar to the SAC TAPS for performance, ignition, flame propagation and extinction, emissions, combustor exit temperature quality, thermal and dynamics surveys with a full-scale combustor and fuel nozzles in a full annular combustor rig. This brings DAC TAPS to TRL5 maturity.

Since functionally both SAC and DAC TAPS combustors have the common technology elements of the TAPS mixer and fuel injection system, and the resulting test data (performance, operability, emissions, temperature levels and gradients of the structure at critical locations) can be represented with the same set of correlations and analytical design tools, one could claim that DAC TAPS is also TRL6 proven technology similar to SAC TAPS.

It is generally assumed that cruise NOx emissions can be estimated from the sea-level engine emissions measurements along the engine operating line normalized by a suitable scaling function, e.g., $P_3^n$ $*FAR^m$, here FAR is combustor fuel/air ratio. Figure 10 shows typical EINOx/$P_3$^0.5 versus combustor inlet temperature of an optimized Rich-Quench-Lean engine combustor. Typical TAPS emissions curves are shown in Figure 10 for engines of 30 and 40 EPR, respectively. Fuel

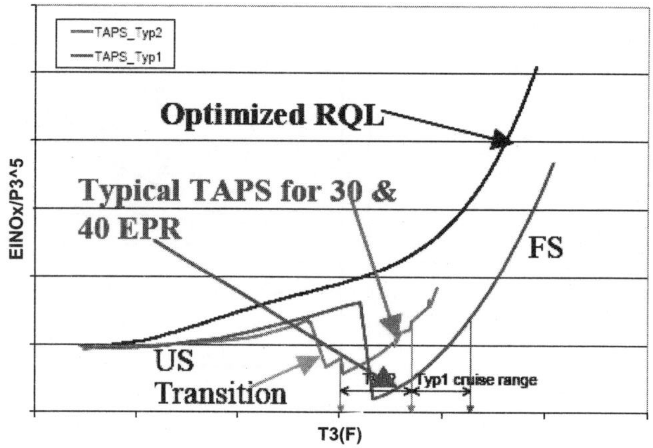


**Figure 10** Cruise NOx reduction typical from TAPS

flows through the pilot only during start and low-power operation, identified as un-staged (US) in Figure 10. Depending on the selected TAPS design features, the fully staged operation (identified as FS) is a function of FAR, T3 and P3. In between the US and FS operation is the transitional phase where the fuel staging strategy is adopted consistent with the system requirements. Also identified in Figure 10 are the typical ranges of $T_3$ that occur during the cruise operation from the minimum to the maximum thrust levels. Based on TAPS database collected to-date, it is a reasonable assumption that estimated reduction in cruise EINOx is ³ the LTO NOx reduction.

LTO NOx, CO and HC and maximum SAE smoke number of the CFM TAPS engine data were calculated using the standard ICAO procedure[11]. The calculations were done for the various engine models of the CFM56-5Bx/P and CFM56-7By, here x=1, 2, 3, 4, 5, 6; y= 20, 22, 24, 26 and 27. The TAPS emissions are compared with those of the CFM56-5B/P and CFM56-7B SAC and DAC combustors in Figure 11. Also shown is the calculated LTO NOx for the NASA/GE Energy Efficient Engine (E^3).

As shown in Figure 11a for the highest thrust rated CFM56-5B, DAC and TAPS give 24% and 39% reductions, respectively, in LTO NOx relative to the current product. The corresponding NOx reductions with DAC and TAPS for the CFM56-7B27 are 31% and 46%. On the other for the lowest thrust model of the CFM56-7B, i.e., CFM56-7B20 certified with DAC, which achieved NOx reduction of 27% compared to 49% reduction with TAPS.

From LTO HC data shown in Figure 11b, one can conclude the following. For the current product SAC combustors at the same EPR, there is an average variation of ~3.5. The CFM56-7B DAC gave the highest HC emissions, which ranges from 11.2 to 16.3. On the other hand, the average range for the SAC is 4.4 to 7.7. In summary, DAC's HC emissions are approximately twice that of the current product compared to no increase with TAPS.

LTO CO data of Figure 11c show approximately 63, 48 and 26 for the highest EPR engine emitted by DAC, TAPS and current technology SAC product, respectively. The corresponding LTO values for the 20K thrust CFM56-7B20 are approximately 100, 80 and 40 for the DAC, TAPS and current product.

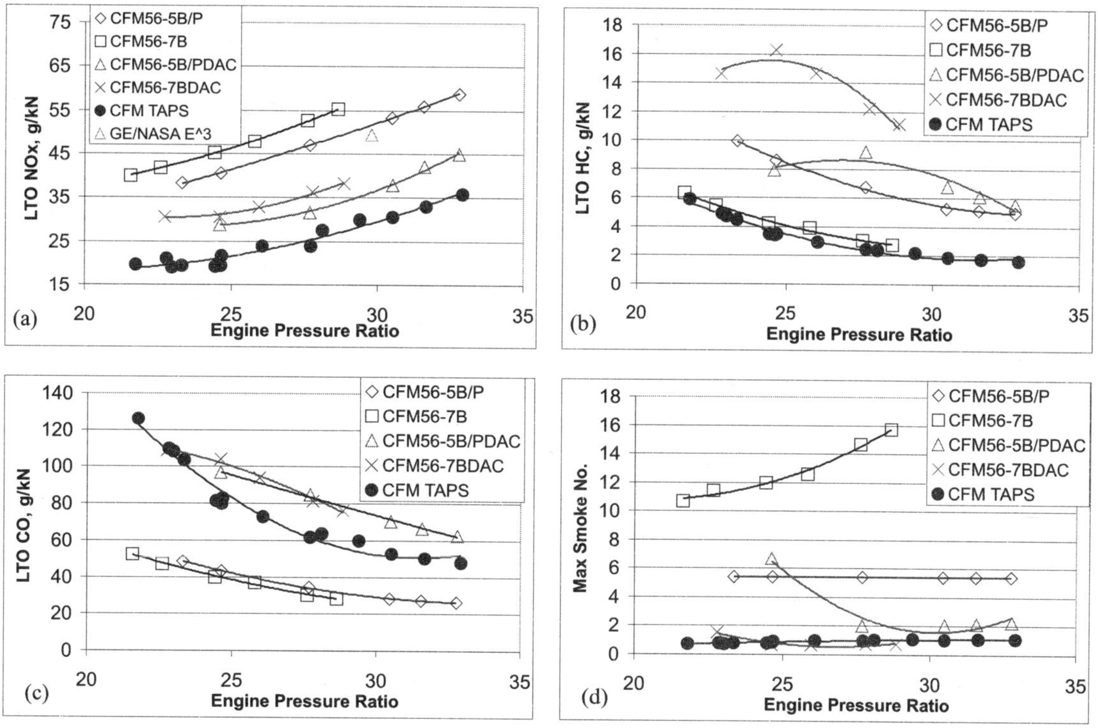


**Figure 11** Landing-takeoff emissions of NOx, HC and CO, and maximum SAE smoke number of nominal 30 pressure ratio engines

The CFM56-7B's maximum smoke number is ~5, compared to the range being 11 to 16 for the CFM56-5B/P. In other words, the average maximum smoke number of the current product is ~10 compared to ~1 emitted by the CFM TAPS, as shown in Figure 11d. Based on limited data on particulate mass emissions of propulsion engines it is approximately proportional to smoke emissions[14, 15]. Based on back to back comparison for the CFM56 engine, TAPS gives approximately 40 to 50% reduction in LTO NOx and cruise EINOx. TAPS' HC values are comparable. However, LTO CO emissions are doubled compared to ORQL. Maximum smoke number is lower by a factor 4 to 10. TAPS gives improved performance in regard to combustor exit temperature quality, liner wall temperature levels and gradients with attendant beneficial impacts on the hot-section durability. The fuel injection system with more required functionalities has adversely impacted the overall system specifically in regard to the design complexities, cost and weight. Some of these will be addressed as a part of NPI activities in addition to the natural simplifications that occur as a result of future evolutionary design improvement efforts. Further reductions in NOx and CO emissions are envisioned as explained in the next section.

TAPS technology development effort was impacted strongly by the direction future energy efficient engines might take in regard to the overall Overall Engine Pressure Ratio, fan bypass ratio and fuel/air ratio. With the higher operating pressure ratios, the current technology RQL combustors give ~100% increase in cruise NOx compared to their counterparts with nominal 30

EPR. TAPS technology is expected to give at least 50% reduction in cruise NOx based on back to back rig data. In order to establish DAC TAPS capability for lowering LTO emissions, we have to extrapolate the GE90 DAC emissions measured in a full-scale, full-annular rig at comparable $T_3$, $T_4$, pressure drop but $P_3$ lower than the corresponding engine climb out and takeoff pressures. Since the pressure extrapolation was less than a factor of 2, and the confidence in the NOx pressure exponent is relatively high, the calculated LTO emissions values can be considered accurate enough (within 3% error for the takeoff EINOx) to establish potential reduction with the first-generation TAPS (TAPS1) technology.

Figure 12a shows LTO NOx of GE90 DAC I (w/ near-stoichiometric primary zones), GE90 DAC II, GE90-115B, and DAC TAPS, which was calculated from the correlations based on the full-annular rig data. It should be noted that both DAC II and GE90-115B have rich-dome design. Compared to DAC I, GE90-115B produces approximately 15% lower NOx. DAC TAPS gives additional 44% reduction for the GE90-115B engine model. On the other hand, TAPS reduces NOx for the engine model GE90-77B DAC II by 58%.

LTO HC for the current technology GE90 combustors fall well below the ICAO standard as shown in Figure 12b. The GE90 TAPS gives LTO HC close to 1.0. Compared to current products with LTO CO falling in the range of 15 to 50, the DAC TAPS data fall between 31 and 42, see Figure 12c. The maximum smoke, Figure 12d, number of the DAC TAPS is ~1.0 compared to less than 4 with the current products which is quite small.

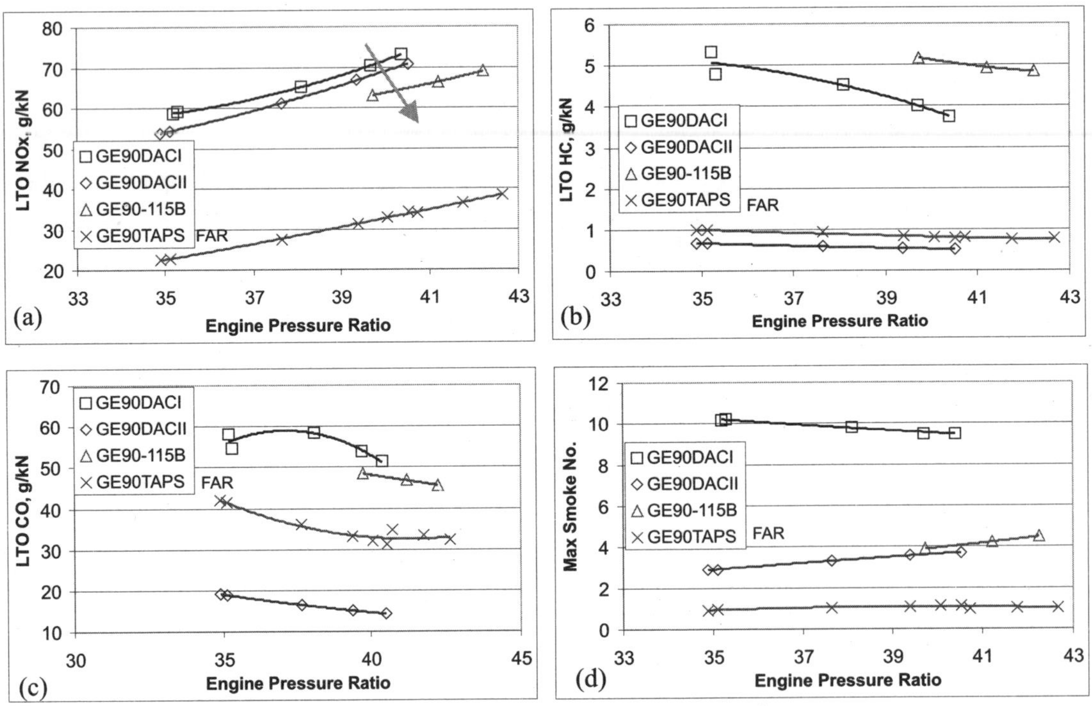


**Figure 12** Landing takeoff emissions of NOx, HC and CO, and maximum SAE smoke number of nominal 40 pressure ratio engines; TAPS data from full annular rig

As discussed in Section 4, the common practice of comparing the emissions technology advances by plotting data shown in Figure 3 may not be very useful any more. Nevertheless in keeping up with the tradition, we have plotted LTO NOx versus LTO CO data for the CFM56-5B/P, CFM56-7B and CFM TAPS in Figure 13a. A similar comparison for the GE90 DAC II, GE90-115B and DAC TAPS is presented in Figure 13b. LTO NOx versus LTO HC technology curves are summarized in Figures 13c and 13d, respectively. Perhaps we need data on more engines before making any definite conclusions. However tentatively we may conclude that DAC TAPS technology developed specifically for 40 EPR engines has made progress on all the three emissions, namely NOx, CO and HC. On the other hand, SAC TAPS technology for the 30 EPR-class of engines has achieved reduction in both NOx and HC.

In summary, the TAPS technology has demonstrated significant reduction (40-60%) in LTO NOx from the current propulsion engine combustors with rich-dome followed by primary and dilution holes to complete combustion and achieve rapid quenching to minimize NOx. The SAC TAPS technology has been validated to TRL6 by having successfully demonstrated in a CFM56 engine. The same mixer and fuel nozzle design features have been adopted in the DAC TAPS combustion system, which has been demonstrated in a full-scale full-annular test rig and thereby achieving TRL5. Because of the commonality of the SAC and DAC operational principles, one could claim TRL6 status for the DAC TAPS technology also.

Maturing TAPS technology from TRL6 to TRL9 involves the following three steps:

- TRL7: System prototype demonstrated in flight environment.
- TRL8: Actual system completed and "flight qualified" through test and demonstration.
- TRL9: Actual system "flight proven" on operational flight.

Depending on the relevance of the technology and business needs, TRL7 activities can add 6 months to 2 years to an on-going TRL1 through TRL6 effort. TRL8 and TRL9 activities are similar to TG6 through TG9 of the GE Aviation NPI plus airplane flight certification, which can take four to six years to complete. Depending on the business needs, the TRL7 activities can run concurrently with the TRL8 and TRL9. Therefore, TRL7-TRL9 effort can be completed in 4 to 8 years in concurrence with all the stakeholders. One could easily spend two years getting all the stakeholders on-board for applying "research program technology" to "products". In summary, it is reasonable to assume that it takes approximately ten years to go from TRL6 to TRL9.

## VI. LONG-TERM GOAL

The combustion technology community has been working hard consistently for the last forty years in executing the fundamental operating principles of flame stabilization and propagation by complex interacting swirlers, controlling spatial distribution of fuel/air along with the desired degree of mixedness, and mixing layers to design optimum propulsion engine combustors. We all have made significant progress in this area, and "all of us" believe in the statement: "We will meet all the low-emissions goals after the next test". We also know how hard it is to achieve 10-20% reduction in LTO NOx emissions along with meeting all other design requirements.

Recognizing the generic description of TAPS, Figure 7a, we have on our roadmap several promising concepts to achieve the following NOx reduction goals:

- 30% CAEP/2 NOx level for EPR>40 (TAPS2 Tech)
- 15% CAEP/2 NOx level for EPR>40 (TAPS3 Tech)

Assuming one can achieve these goals for 40 EPR on an average basis, applies 20% design margin for scaling the technology, the corresponding single characteristic LTO NOx values are 40 and 20 gm/kN, respectively, compared to the CAEP/6 value of 79. If TAPS2 and TAPS3 technologies come to fruition, it would give 50% and 75% margin from the CAEP/6 standard at 40 EPR. However, future engines may be operating at nominal design pressure ratio of approximately 50 resulting in even more challenging design trade studies.

Some promising results have been achieved under modestly funded UEET and Propulsion 21 programs as summarized in the following paragraph. But both of these programs have been prematurely terminated. The TAPS2 is between TRL3 and TRL4 and the combustion system is still long way off on meeting several of the key design requirements. Perhaps, it will take approximately another ten years to claim TAPS2 to TRL6 for 50 EPR engine applications. The current status of TAPS3 technology is estimated to be the beginning of TRL3.

It is hard to schedule inventions and estimate the resources required for maturing TAPS2 and TAPS3 to TRL6. A rough resource estimate for TAPS2 is within a factor of 2 of TAPS1 technology investment, and that of TAPS3 also within a factor of 2 of TAPS2 effort. Considering the scope of the undertaking, it can therefore be recommended only as a nationally funded program.

Under UEET funding we modified the AST TAPS mixer design along with an advanced liner cooling and tried several configurations on a flame tube rig before finishing three sector tests. The best mixer with simulated advanced liner cooling was tested in a 5-cup sector rig at the four ICAO points of the UEET cycle. However, the takeoff and climb out emissions were calculated from the data taken at the following test conditions because of the facility limitations:

$$T3 = 1050 \text{ F}, P3 = 391 \text{ psi}, FAR = 0.029.$$

The calculated LTO values for the UEET full-annular (FAR1), sector builds 2 and 3 (B2 and B3) are shown in the following figure:

The calculated LTO NOx can have an error of the order of 5%. However, the values shown are conservative indicating that the design would meet 70% reduction from the CAEP/2 standard. However, it should be remembered that although LTO NOx, CO and HC are very low, considerable more technology development effort is needed to meet other design requirements including the operability.

All the TAPS technology development configurations for use in the CFM TAPS, GE90 TAPS and UEET TAPS are identified as TAPS1 mixers. Conceptual concept evaluations continue to explore several co-annular flames in regard to the degree of mixedness, flame stabilization, low-power characteristics. These mixers are classified into TAPS2 and TAPS3 categories. Several of the TAPS2 and TAPS3 mixers have been evaluated in single cup flame tube at the simulated cruise conditions, P3 and T3. NOx emissions indices of these mixers as presented in the following figure indicate potential EINOx reduction of ~70% from the TAPS1 technology. However, these mixers have to be evaluated in regard to low-power combustion efficiency, ignition and flame propagation, combustion dynamics and durability.

## SUMMARY

A brief overview has been given on the combustion system tradeoffs, the evolution of the rich-quench-lean (RQL) of Single and Dual Annular (SAC, DAC) and the lean DAC Combustors leading to discussion on the medium terms goals and development of a first generation Twin Annular Premixing Swirler (TAPS1) for both SAC and DAC applications. Extensive rig testing was done to demonstrate TRL5 capability of SAC and DAC TAPS1. TAPS1 mixers have been tested to extreme operating conditions as described in the paper including the maximum $P_3$, $T_3$ and mixer equivalence ratio of 778 psi, 1384 F and 0.96, respectively. Even though neither SAC nor DAC TAPS1 have been tested in a flight test bed, very thorough flame ignition and propagation, and flameout investigations were undertaken in the sector and full annular combustor rigs under simulated conditions along the relight envelope. The severe conditions tested included $P_3$, $T_3$ and fuel temperature down to 4.7 psi, –40 F, and –30 F, respectively ensuring good ground start and altitude relight capability.

The CFM SAC TAPS1 combustion system comprised of the combustor, fuel nozzles, valves, and the fuel injection control system was installed in a development CFM56-7B engine. Two engine tests were conducted accumulating a total of 909 hours of testing including more than 200 hours run time at the maximum rated thrust. The engine successfully completed approximately 4,000 LCF cycles and 2,000 fuel nozzle purge cycles to validate critical technologies. The engine testing documented the emissions, performance, operability, noise, combustion dynamics, transients, thermal and vibration characteristics. As summarized in Table 1, the SAC TAPS is judged to be at TRL6; we conservatively call the restart capability demonstration being at TRL5 because we have not flown this system on a flying test bed. The CFM TAPS has demonstrated single engine NOx characteristics value of 44% CAEP2 standard with CO being at 62% for the CFM56-7B27; the corresponding values for the CFM56-7B22 (22 k thrust rating) are 35% and 86%. The SAC TAPS technology met all other requirements, as listed in Table 1.

Even though DAC TAPS1 technology development on the test rigs has been done over a broad range of operating conditions, and it uses the same SAC TAPS1 mixer technology, we judge it conservatively at TRL5, as indicated in Table 1 because the DAC TAPS has not been subject to engine testing. The LTO values calculated from the full-annular rig data that covering the range of T3 are considered reliable within 3% in regard to NOx. It is predicted to have single engine characteristic LTO NOx values of 44% and 30% CAEP2, respectively for the GE90-115B and GE90-76B. DAC TAPS is expected to meet all other design requirements.

The exploratory research for long-term goals (achieving 30% and 15% CAEP2 NOx standard) looks promising. However, the level of effort required to achieve TRL6 technology maturity is too great for justifying its commercial payback. If further reduction in the propulsion engine NOx emissions including low- and high-altitude NOx is desirable, a nationally funded R&D effort is recommended requiring 2 to 4 times the funding spent to-date in developing the TAPS1 technology.

**Table 1** Summary

| | *TAPS Medium Term Technology Status* | | | |
|---|---|---|---|---|
| | *SAC TAPS TRL6* | | *DAC TAPS TRL5* | |
| Combustion Attributes | Goal | Status[1,5] | Goal | Status[2,6] |
| Emissions, % CAEP2 | Goals | | Goals | |
| Single Engine NOx | Consistent | 44.0% | Consistent | 44.3% |
| Single Engine Smoke | with customers | 6.2% | with customers | 8.8% |
| Single Engine HC | & Environmental | 17.4% | & Environmental | 6.0% |
| Single Engine CO | Needs | 62.2% | Needs | 33.7% |
| Combustion Noise | None | P | None | P |
| Combustion Eff Idle | >98% | P | >98% | P |
| Combustion Eff Cruise | ~100% | P | ~100% | P |
| Pressure Drop | Comparable[3] | P | Comparable[3] | P |
| Altitude Restart | Relight Envelope[3] | P/TRL5 | Relight Envelope[3] | P |
| Descent Weak Extinction | No flameout | P | No flameout | P |
| Transient Operation | Comparable[3] | P | Comparable[4] | P |
| Hail/Water Ingestion | Comparable[3] | P | Comparable[4] | P |
| Peak Pattern Factor | | P | | P |
| Peak Profile Factor | Turbine Durability[3] | P | Turbine Durability[4] | P |
| Root & Tip Temperatures | | P | | P |
| Combustion Durability | Comparable[3] | P | Comparable[4] | P |
| Fuel Injector Coking | Comparable[3] | P | Comparable[4] | P |
| Combustion System Weight | Lower[4] | P | Comparable[4] | P |

1: CFM56-7B27 ICAO cycle points 2: GE90-115B ICAO cycle points

3: Compared w/current SAC's 4: Compared w/current DAC's

5. For CFM56-7B22 LTO NOx, Smoke, HC and CO: 35, 5, 28 & 86%, respectively.

6. For GE90-76B LTO NOx, Smoke, HC and CO: 30, 7, 8 & 43%, respectively.

## References


1. Bahr, D. W., and Gleason, C. C., "Experimental Clean Combustor Program Phase I Final Report", NASA CR-134732, June 1975.
2. Roberts, R., Peduzzi, A., and Vitti, G. E., "Experimental Clean Combustor Program Phase I Final Report", NASA CR-134736, October 1975.
3. Anderson, R. D. et al, "Pollution Reduction Technology Program Turboprop Engines Phase I Final Report", NASA CR-135040, March 1976.
4. Bruce, T. W., Davis, F. G., Kuhn, T. E., and Mongia, H. C., 1977, "Pollution Reduction Technology Program Small Jet Engines Phase I Final Report", NASA CR-135214.
5. Ballal, D., Zelina, J., "Progress in Aero-Engine Technology, 1939-2003", AIAA Paper 2003-4412.
6. Tacina, R., Mao, C., Wey, C., "Experimental Investigation of a Multiplex Fuel Injector Module for Low Emission Combustors," AIAA Paper 2003-827
7. Bahr, D. W., "Technology for the reduction of aircraft turbine engine exhaust emissions", AGARD Atmospheric Pollution by Aircraft Engines, p. 13, 1973

8. Corporan, Edwin; Roquemore, Mel; Harrison, William; Jacobson, Andy; Phelps, Donald, "Air Force programs to reduce particulate matter emissions from aircraft", 38th AIAA/ASME/SAE/ASEE Joint Propulsion Conference and Exhibit, Indianapolis, IN; 7-10 July 2002.
9. Mularz, Edward J., Gleason, Clifford C., and Dodds, Willard J., "Combustor Concepts for Aircraft Gas Turbine Low-Power Emissions Reduction", Journal of Energy, 1979.
10. Holsclaw, C, (Presenter) A Consideration of Technology Goal Demonstration Within The Standard-Setting Process", U.S. FAA/ICAO Committee on Aviation Environmental Protection Working Group 3, Emissions – Technical Issues, LTTG Meeting, Stockholm, Sweden.
11. International Civil Aviation Organization, Annex 16, "Environmental Protection", Volume II, "Aircraft Engine Emissions", Second Edition (1993) including Amendment 5 (November 11, 2005).
12. Lefebvre, Arthur H., "Gas Turbine Combustion Second Edition", Taylor and Francis, 1999.
13. Joshi, N. D., Mongia, H. C., Leonard, G., Steggmaier, J. W., and Vickers, E. C., 1998, "Dry Low Emissions combustor development", *ASME Paper 98-GT-310*
14. Iovinelli, Ralph, "Use of the first order approximation (FOA) to estimate aircraft engine particulate matter (PM) emissions in NEPA documents and Clean Air Act general conformity analyses", FAA Memorandum to Director of Environment and Energy (AEE-1), May 24, 2005.
15. Roger L. Wayson, Gregg G. Fleming and Ralph Iovinelli, "DRAFT - PARTICULATE MATTER EMISSIONS FROM AIRCRAFT: CONTINUED DEVELOPMENT OF THE FIRST-ORDER APPROXIMATION: FOA 3.0 BETA", FAA DTS-34-FA05-LR1, November 2005. Presented as part of ICAO CAEP Working Group 3, Alternative Emissions Methodology Task Group Working Paper 5/9, November 2005.
16. ICAO Forecasting and Economic Analysis Support Group, "ECONOMIC ANALYSIS OF NOx EMISSIONS STRINGENCY OPTIONS", CAEP/6 Information Paper 13, February 2004.

# Utilizing Combustion Imaging to Enable the Use of Alternative Fuels in Engines

**Stephen Ciatti**

Center for Transportation Research, Argonne National Laboratory, Argonne, IL USA

## 1. INTRODUCTION

*Hydrogen as fuel:* Hydrogen ($H_2$) has been garnering significant attention as a transportation fuel for a variety of reasons. Volatility in the price of petroleum, stringent emissions regulations, and the desirability of renewable energy are all factors leading to the interest in hydrogen. Hydrogen has primarily been presented as a fuel for fuel-cell (FC) applications, as proposed by the U.S. Department of Energy in the FreedomCAR initiative of 2002 [1]. However, hydrogen has also been studied as an automotive-type fuel for many years — most notably by BMW Group, Ford Motor Company, and others — for internal combustion engine (ICE) use, potentially leading to FC applications.

There are several advantages to using hydrogen as a fuel for ICE applications. First, ICE technology has undergone continuous refinement for over a century, with considerable resources invested in infrastructure to support ICE production, operation, and maintenance — resulting in a product that is relatively low cost and very reliable. Second, ICEs possess high power density — meaning that very little space onboard a vehicle is consumed by the device that provides the motive power. Finally, hydrogen ICEs are rather fuel efficient, providing brake thermal efficiencies on the order of 38% or better in many applications [2, 3].

Operating an ICE by using hydrogen fuel presents some very intriguing possibilities. When operated under stoichiometric mixture conditions with conventional port fuel injection (PFI), hydrogen-fueled engines are capable of producing roughly 85% power density and performance comparable to that of gasoline engines. Conventional PFI hydrogen engines tend to suffer a power density loss because the injection of hydrogen into the intake manifold tends to displace intake air. Direct injection (DI) can provide up to a 40% power density increase (roughly 17% better power density than gasoline) [4].

Under stoichiometric conditions, the after-treatment can also be similar to that for conventional gasoline engines, i.e., using a three-way catalyst to remove the oxides of nitrogen (NOx) produced

[5]. A disadvantage to stoichiometric operation is the propensity for hydrogen to pre-ignite and knock inside the combustion chamber [2, 3]. However, other than a negligible carbon-based emissions signature, no other tangible benefit is obtained by stoichiometric operation versus gasoline operation.

Significant differences between the two fuels are noticed when the $H_2ICE$ is operated at lean fuel conditions. The wide flammability limits of hydrogen allow for operation of the engine without a throttle under all operating conditions. This characteristic also provides the opportunity to operate the engine using direct injection, because fuel stratification does not greatly influence the mixture ignitability, as it does in gasoline [4, 6]. However, for the purposes of this study, the engine was operated using port fuel injection (PFI) but still without the throttle. The injector was installed on the intake manifold near the intake port.

Hydrogen engines offer significant promise for high efficiency, high power density and low emissions signature. However, the nature of hydrogen combustion as applied in ICEs has not been well documented. Combustion anomalies such as knock and pre-ignition are significant challenges in $H_2ICE$ operation. In addition, the significant $NO_x$ dependence upon fuel/air ratio is very pronounced in $H_2ICE$. Knock, pre-ignition and $NO_x$ production are all somewhat thermally dependent – therefore the ability to non-intrusively measure $H_2ICE$ combustion temperature would be highly desirable.

Since knock and pre-ignition in particular tend to be associated with engines operating at high speed and high load, the ability to make combustion temperature measurements under these conditions is also desirable. Fully accessible optical engines, with quartz pistons and liners, possess neither the proper operating range (they are low speed, low load) nor the proper heat transfer characteristics to effectively study these conditions.

The main focus of this study was to characterize the ability to utilize a fiber-optic access system to obtain ultraviolet (UV) spectral emission signals that can be processed to extract combustion temperature information. Optical emission due to excited hydroxyl (OH*) radicals was used as the primary signal to determine hydrogen combustion temperature.

Spectral signals were acquired via a fiber-optic pathway and recorded using an ICCD (512 X 512 pixels) equipped spectrograph. These signals were then compared to a signal generated by a computer simulation that predicts the energy distribution spectra for OH* as a function of temperature. The simulation code then iterates temperature until the simulated signal most closely matches the experimental signal, based upon a least squares fit.

*Biomass-to-Liquid Fuel:* Rising costs and dependency on foreign oil has renewed an interest in synthetically produced fuels for transportation needs. Stringent emissions regulations provide additional incentive to produce fuels which lower emission production, thus reducing the need for sophisticated, expensive after treatment devices. Biodiesel production and research is not new and the potential benefits/drawbacks are generally well understood. Various feed stocks have been successfully utilized to produce biodiesel fuels, including soybean and vegetable oil. SunDiesel is unique in that it is derived from wood pulp through the Fischer Tropsch process and is not oxygenated. The resulting fuel has significantly lower sulfur and aromatic contents, while maintaining a similar carbon to hydrogen ratio to that of conventional diesel fuel.

This type of fuel has the potential to provide two key benefits to industry and society. First, fuel derived from a renewable resource directly addresses the issue of a finite oil source and second the ability to tailor the constituents to improve combustion efficiency and reduce emissions while maintaining performance is quite promising.

## 2. EXPERIMENTAL SETUP

The engine used for the hydrogen study was a single-cylinder, four-valve, pent-roof, spark-ignition (SI) engine made by the Ford Motor Company. The relevant engine parameters and measurements are listed in Table 1.

**Table 1** Ford engine specifications

| *Engine Manufacturer* | *Ford Motor Company* |
|---|---|
| Displacement (L) | 0.5 |
| Compression Ratio | 11.4:1 |
| Bore/Stroke (mm) | 89 / 79.5 |
| Valve Train | 4V Dual Overhead Cam (DOHC) |
| Intake Valve Timing | IVO = 15 BTDC, IVC = 35 ABDC |
| Exhaust Valve Timing | EVO = 45 BBDC, EVC = 10 ATDC |
| Max Torque (N-m) | 30 |
| Speed Range (RPM) | 800-6000 |
| Injector | (6 bar) Quantum PFI |
| Hydrogen Measurement | Micro-Motion Coriolis Meter |

Figure 1 shows the schematic of the hydrogen engine test cell. The hydrogen system is designed to minimize the amount of hydrogen stored in the lines at all times. If, for some reason, a leak develops in the hydrogen supply system, the encasement will contain any leaked hydrogen. Because the leaked hydrogen is lighter than air, it will float toward the top of the case, where a hydrogen sensor is located. If the hydrogen sensor detects 2% hydrogen by volume, the system will shut down and de-pressurize. A completely separate hydrogen vent to the atmosphere is used to depressurize the hydrogen system. This vent system is plumbed so that the highest point of the system is outside the building, with no place in between for hydrogen to accumulate.

The target pressure in the intake surge tank was maintained at 1 bar. Intake air temperature was 25°C for each test run.

The purpose of this study was to evaluate the use of spectrographic signal characterization as applied to combustion temperature measurement, so the engine was operated under conditions that avoided combustion anomalies and provided intense OH* spectral emission signals.

Two different operating speeds were chosen: 1500 RPM and 3000 RPM. At each speed, a sweep of SI timings was performed at each equivalence ratio. Spark timing varied from 30° before top dead center (BTDC) to 20° after top dead center (ATDC). These numbers were adjusted on the basis of engine operation — if knocking or backfiring occurred, the spark timing was not advanced or retarded any further.

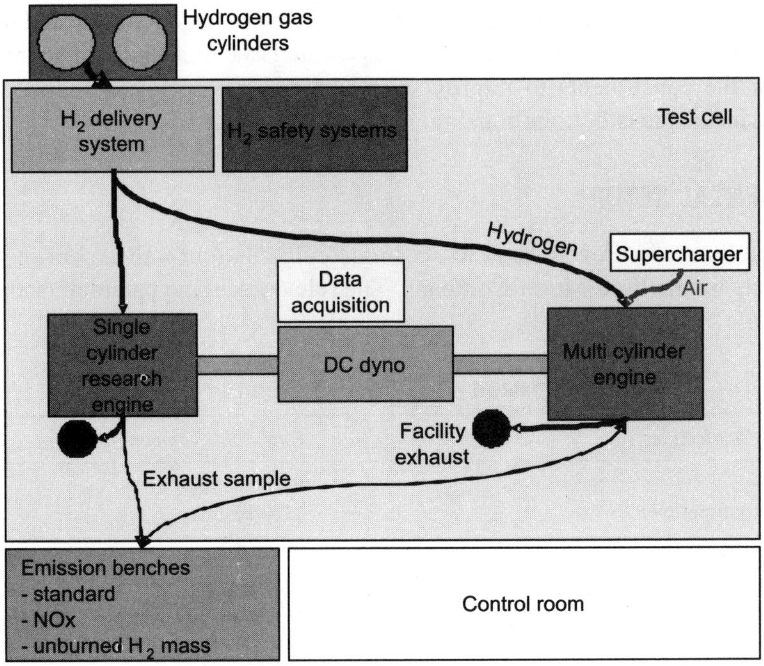


**Figure 1** Schematic of hydrogen engine test cell

In Figure 2a, the schematic of the optical access to the engine is shown. The fiber optic schematic is shown in Figure 2b. The optical fiber used was a 1mm core, step index silica fiber. The fiber was inserted into the access port and a sapphire window was used to seal combustion gases from the optical access.

*BTL Diesel Experimental Setup:* The experimental setup was comprised of a small, high speed diesel engine, AC dynamometer, various emissions measuring equipment, cylinder pressure sensor and analysis system and in-cylinder combustion visualization hardware.

The diesel engine used for this experiment was a 1.7L, four cylinder, Mercedes A class (MY 1999) with a four valve head, intercooled turbocharger and a Bosch common rail injection system. The fuel injection schedule utilizes both a pilot injection at approximately 35 CAD before top dead center and a main injection event. The engine was rated at 85 hp @ 4000 rpm. One exhaust valve from the number 4 cylinder was removed to gain optical access for the combustion visualization. Capability also exists to illuminate the combustion chamber to study spray preparation. A significant change in performance, efficiency or combustion characteristic was not expected, due to this modification.

The engine was controlled using Southwest Research Institute's (SwRI) Rapid Prototyping and Engine Control Software (RPECS) system. It provided complete control of the fuel injection system, EGR rate, and boost pressure. The amount of injected fuel was not modified during the experiments; however a fuel injection timing sweep was conducted, from 12 degrees BTDC to 3 degrees BDTC.

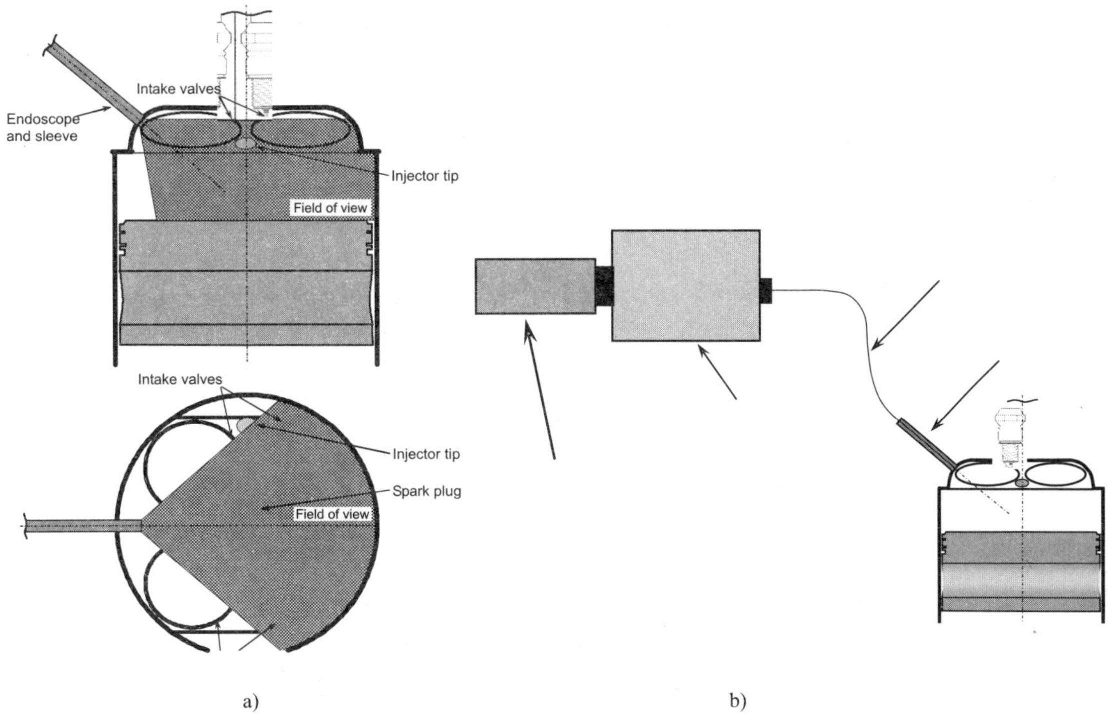


**Figure 2** a) Field of view of the endoscope – top: cross-section, bottom: view from top b) Fiber optic schematic for gas temperature measurement

## 3. EXPERIMENTAL DATA AND RESULTS

### 3.1 Temperature Measurements Using OH* Emission Spectroscopy

The spectrograph used for measuring OH* spectra is equipped with three gratings having 300, 600 and 3600 grooves/mm. The 300 groove/mm grating was used to record the spectrum covering entire UV-VIS-NIR (200-1100 nm) wavelengths. The 600 and 3600 grooves/mm gratings were used to record spectra for temperature measurements. Looking back at Figure 4, distinct, strong emission peaks are shown at about 281 nm and 309 nm for OH* chemiluminescence. The broad peak in the visible range (~500 to 700 nm) is due to water vapor and other radiating gases in the combustion chamber.

Figure 7 shows the more resolved, specific UV based OH* peaks in the spectral region 306-310 nm recorded with 600 and 3600 nm gratings. This spectral region was used to estimate the temperature using the formulation described in the previous section. For an engine running at 3000 RPM, 1° CA is roughly 55.5 ìs. Ideally the exposure time of the camera should be less than 1° CA to ensure data quality and maintain time resolution. The entire combustion process typically will last for about 1.39ms (~25 crank angles) at the studied equivalence ratio. Therefore it is reasonable to utilize “d” 50 ms gate-width for meaningful data acquisition. The intent is to use ~50 ìs gate times (approx. 1°CA) in increments throughout the entire combustion event, triggered at different timesteps throughout the combustion event, such that a thermal profile can be developed as a function of time.

The observed signal strength depends on several factors that include engine operating parameters such as equivalence ratio, spark timing, optical transmissivity of the system (window, optical fiber, lenses, etc.), and data acquisition settings such as ICCD gate width, gain, spectral resolution, etc. The cycle-to-cycle variations of the in-cylinder processes make it difficult to apply customary data treatments to improve signal to noise ratio in conventional spectroscopy, such as averaging over several cycles for temperature measurements. Also the temperatures during a single cycle are not constant for the entire cycle; therefore crank angle resolved data is desirable.

During the course of the experiments, it was observed that signal intensity increases exponentially with increasing $\Phi$, until stoichiometry has been reached. Richer than $\Phi = 1$, the signal intensity begins to decline again [8]. This dependence upon $\Phi$ has also been observed in other experiments [12]. It will be shown later that the OH* chemiluminescence peak intensity is a function of equivalence ratio. At higher $\Phi$, the emission intensity is higher and shorter gate widths along with lower camera gain can be used to collect data - for the purposes of this discussion, $\Phi \leq 1.0$ in all cases.

However, at leaner operation the signal to noise ratio degrades significantly, resulting in tradeoffs regarding data acquisition. Maximizing the light intensity to the camera is critical – using the most transmissive optics improves the signal quality significantly. Once the optical equipment is chosen, several tradeoffs need to be made with regard to acquiring the emission signal. The choice of diffraction grating will influence the light intensity at the CCD detector; high resolution gratings provide easily discerned spectral peaks but the light is much more dispersed and very little reaches the detector. Lower resolution gratings degrade the ability to resolve peaks in the 4 nm bandwidth but much more light reaches the detector.

The camera settings influence the data as well. It is desirable to collect the spectral data with as small gate widths as possible, such that the temperature information deduced from the spectrum represent the instantaneous temperature as opposed to the cycle average. Short gate widths limit the light intensity at the detector. Therefore, to maximize the light intensity and keep the gate width reasonably small, the lower-resolution grating of 600 g/mm was chosen for the temperature measurements as the best compromise between light intensity, speed of exposure, and spectral resolution.

The first experiments were run with the engine operating at 1000 RPM, with 3 different loads or equivalence ratios. The peak temperature data was generated by averaging 10 combustion cycles for each data point. Under throttled conditions (MAP = 0.83 bar) at $\Phi = 0.8$ and 2.2 bar BMEP, the temperature estimates were ~ 2600K to 2900K. Running unthrottled (MAP = 1 bar), $\Phi = 0.55$ and 3.0 bar BMEP, the temperature was in the range of 1500-1800K. The third point with $\Phi = 0.4$ and 1.8 bar BMEP the signal was very weak and it was necessary to use maximum gain setting of ICCD to observed the signal. It was very difficult to infer any temperature value with this set of spectra. The equivalence ratio measurements, shown in Figure 11, display a similar trend – light intensity or the peak combustion gas temperature increases as the load, or equivalence ratio, increased.

The next experiment was run to investigate the ability to perform crank-angle resolved temperature measurements. Figure 8 shows OH* light intensity as a function of crank angle. Plotted on the same graph is the Gross Apparent Heat Release Rate (GAHRR), derived from in-cylinder pressure, as a function of crank angle. The GAHRR is generated by performing a differential upon the pressure trace as a function of crank angle, producing an estimate of how rapidly the combustion event is producing work in the cylinder. The two measurements display similar characteristics, especially the slope of signal rise and location of peak signal. Since OH* intensity is related to parameters such as temperature, pressure, and equivalence ratio, the correlation of OH* intensity to heat release rate is especially appropriate.

Of note is the divergence of the two signals after the peaks are reached at around 4 degrees ATDC. The OH* intensity signal diminishes much more slowly than the heat release signal. This is likely due to thermally excited OH* radicals still emitting photons even beyond the chemical reactions taking place [12]. GAHRR is a function of pressure and temperature, while OH* intensity is a function of pressure, temperature, Φ, and field of view (FOV). The engine's Φ is constant (homogeneous mixture within the combustion chamber) since hydrogen is injected into the port during the intake stroke, so the primary difference between the two measurements is then FOV.

Looking at Figure 2a, the endoscopic field of view (which is roughly 80°) shows the ability to see the upper portions of the combustion chamber. However, when the fiber optic is used, the field of view becomes roughly 25°, meaning that the several portions of the combustion chamber may not be seen by the fiber. Since the combustion event is initiated at the spark plug in the hidden portion of the combustion chamber, it is possible that the delay seen in Figure 8 in the OH* signal is due to this reduced FOV. It is also possible that the OH* signal at the end of the combustion event is a local phenomenon that does not contribute significantly to the GAHRR. There may be other contributing factors to this discrepancy that need to be explored. However, during the flame initiation and peak portions of the OH* intensity signal, the temperature measurements should be rather close to combustion temperature.

Figure 9 shows a typical spectrum recorded at 1000-rpm engine speed, Φ = 0.82, BMEP = 2.25 bar with ignition timing at 0° CA BTDC. The gate width in this case was 100 ms. The gate delay was kept minimum ~85 ns and trigger was provided at ~ 12 ATDC. At this time in cycle the chemiluminescence signal is the maximum value. The temperature estimated from this spectrum is ~3000 K (± 100K). The simulated spectrum is also shown in this figure.

To see how temperature measurements are affected when data is collected at different times during a combustion event, the ICCD trigger position was varied in step of 2 degrees CA. Figure 10 show the temperature as a function of CA. The range of temperatures during the duration of the combustion event (1000 RPM, Φ = 0.82, 2.25 bar BMEP, MAP = 0.83 bar), was measured to be between 2300K and 2900K, with an error of roughly ± 100K. This error primarily results from uncertainties in the match between the simulated spectra and the experimental spectra. The best estimate, given the limited data set, is roughly ± 100K. The experimental spectral measurement error is reduced by calibrating the spectrograph and grating together with a mercury vapor lamp. This error appears to be roughly constant regardless of temperature measured, although more extensive data sets will assist in quantifying this assessment.

Figure 3 shows the corrected power output of the diesel engine over a range of injection timings, operated on SunDiesel and conventional diesel fuel. As expected, power output increased for both fuels as main injection timing was advanced. However, at each injection timing, power

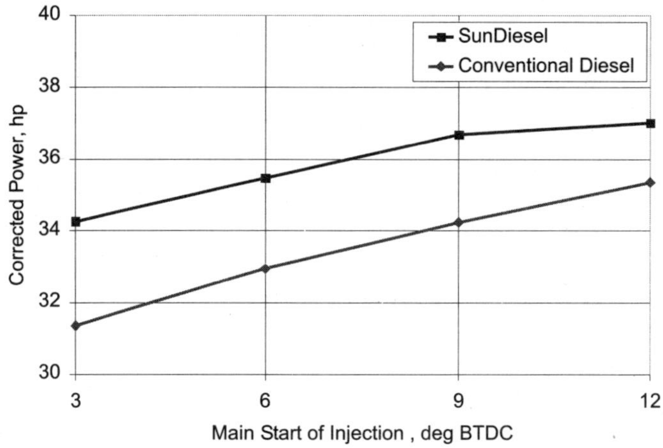


**Figure 3** Corrected power output of engine for sunDiesel and conventional diesel

output was improved with the SunDiesel, due to the increase in specific energy content. The resulting implication could be an engine operated on SunDiesel can produce a similar power level at retarded timings. Retarded timings phase the combustion further into the expansion stroke, thus reducing NOx production.

The conventional diesel fuel apparent heat release rates (AHRR) for four injection timings are shown in Figure 4. As main injection timing is retarded, the peak of pre-mixed combustion is delayed, as expected. Additionally, the diffusion burn duration decreases as well.

When the SunDiesel fuel AHRR curves are analyzed as shown in Figure 5, the same shift in the peak of the pre-mixed flame as main injection timing is retarded is observed. For the same injection timing, the peak of the pre-mixed combustion for SunDiesel occurs earlier in the cycle compared to conventional diesel fuel. This is a result of a shorter ignition delay caused by the increased volatility

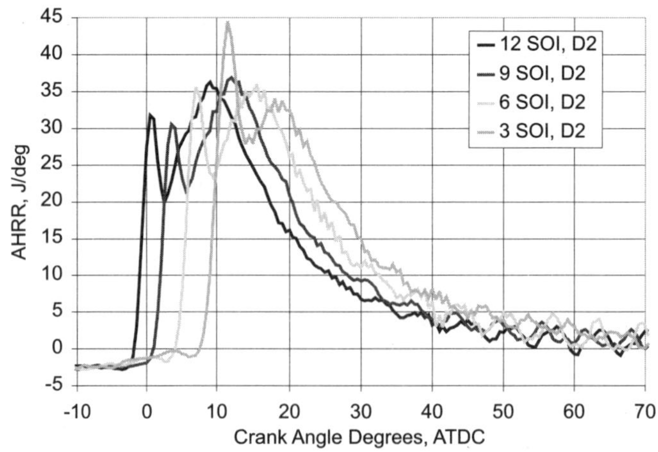


**Figure 4** Crank angle resolved apparent heat release rate (AHRR) for conventional diesel fuel (D2)

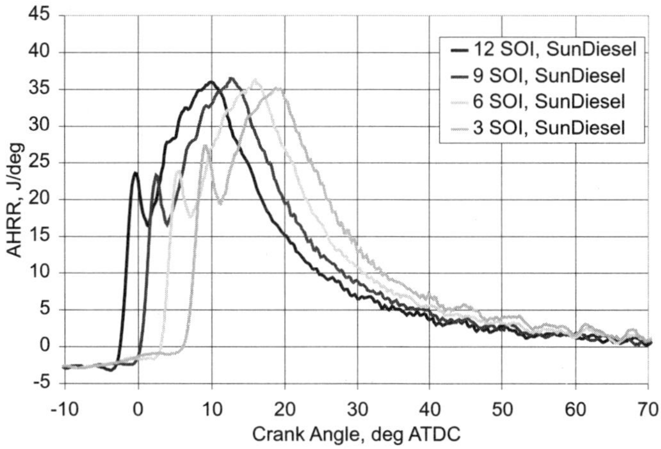


**Figure 5** Crank angle resolved apparent heat release rate (AHRR) for SunDiesel

of the fuel. The chemical composition of the SunDiesel fuel makes it easier to evaporate, as shown in Figure 4 and thus leads to shorter ignition delays. The peak of the pre-mixed combustion is lower than that for conventional diesel fuel because the fuel ignites earlier after delivery into the cylinder. The net effect is a smoother running engine with improved power output for the same fuel consumption rate.

## EMISSIONS ANALYSIS

Brake specific fuel consumption verses NOx trade-off curves are shown in Figure 6. As injection timing is advanced, BSFC decreases and NOx production increases, as expected. At each specific injection timing however, there is a reduction in NOx production with the SunDiesel fuel. This reduction is related to the increased specific power output and slightly reduced combustion temperatures.

The energy density of SunDiesel fuel is higher than conventional diesel fuel. Therefore, a BSFC comparison is not unbiased, due to the energy content discrepancy. Brake specific energy consumption takes into account the differences in energy density between the two fuels and provides an unbiased comparison. Figure 7 shows the effect of injection timing on brake specific energy consumption and NOx for both conventional and SunDiesel fuel. The same trends appear compared to the BSFC analysis and there is still a measurable reduction in energy consumption for the SunDiesel.

The CO verses NOx trade-off curve shown in Figure 8 reveals similar trends as the two previous figures. The NOx reduction from SunDiesel is apparent along with a reduction in CO emissions, at each test condition. The reduction in CO is counter-intuitive based on the work of Siebers, et.al. (2002) where reduced ignition delays resulted in increased sooting. An explanation for the overall reduction in soot production is provided by the combustion visualization results.

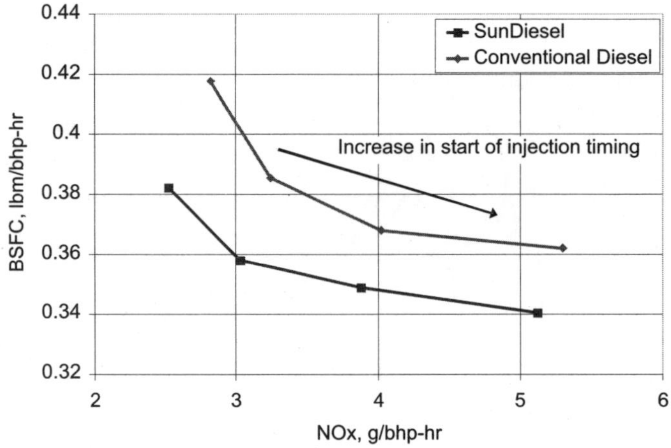


**Figure 6** Brake specific fuel consumption verses NOx production for conventional diesel and SunDiesel, at four main injection timings

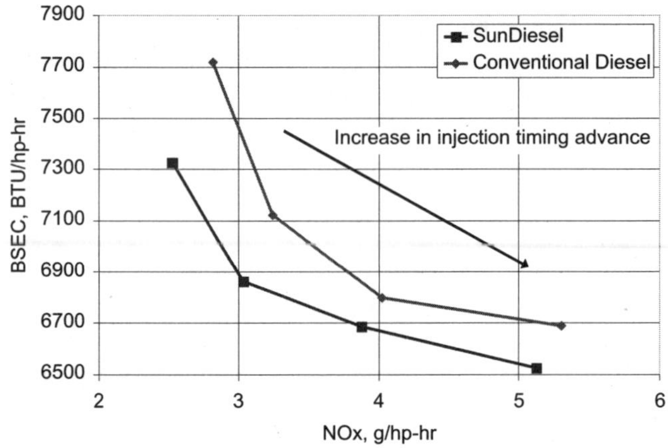


**Figure 7** BSEC vs. NOx for Mercedes engine

Endoscopic imaging provides the opportunity to study the soot volume fraction, inside the cylinder, on a crank angle basis. Gravimetric soot data cannot be analyzed on a crank angle basis. This provides the opportunity to determine the critical parts of the cycle for soot production and oxidation. As shown in Figure 9, soot production rate is relatively constant over the four injection timings and as expected, peak soot volume fraction reduces as timing is retarded.

Also notice the amount of soot in the late part of the cycle (30-40 degrees ATDC), for the later injection conditions (3 SOI and 6 SOI). The oxidation rate of the soot is mixing controlled for the conventional diesel fuel and at these late timings, the in-cylinder pressure and turbulence has decreased to the point to significantly reduce soot oxidation.

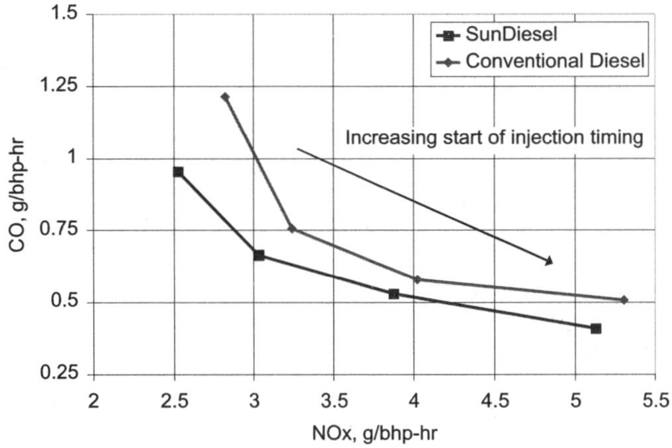


**Figure 8** CO verse NOx production for conventional diesel and sunDiesel, at four main injection timings

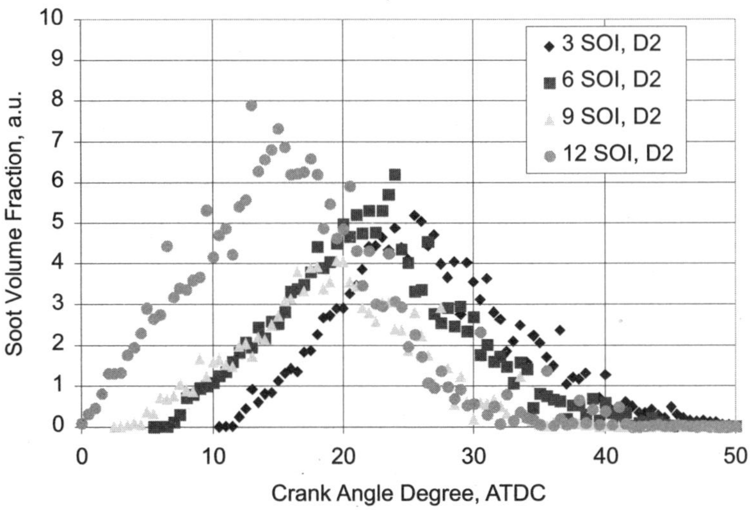


**Figure 9** Soot volume fraction vs. crank angle for D2 using endoscopic measurements

The soot volume fraction for SunDiesel, shown in Figure 10 is noticeably different compared to conventional diesel fuel. The rate of soot production is slightly greater for SunDiesel as well as the peak value of soot volume fraction. This is an expected result, based on the short ignition delay and low viscosity of the fuel. Both tend to produce higher soot due to decreased air entrainment of the fuel spray, as noted by Siebers, Higgins and Picket (2002) and Siebers and Higgins (2001).

The late cycle soot volume fraction is also significantly less for the SunDiesel fuel. This tends to support the theory that the chemistry of the fuel is the controlling factor in late cycle soot oxidation, compared to fluid dynamics, for this particular fuel.

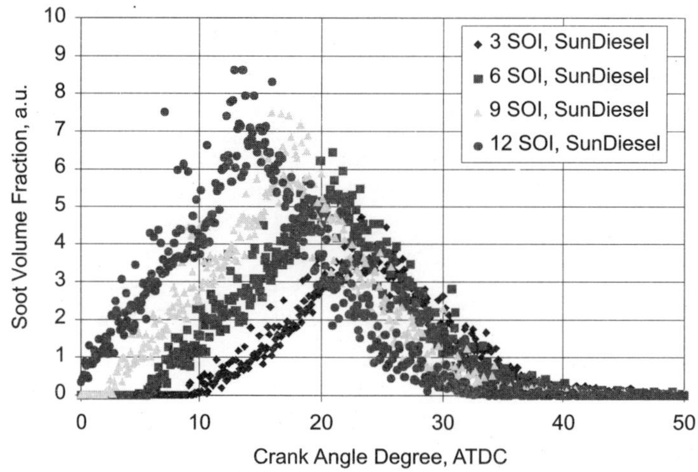


**Figure 10** Soot volume fraction verses crank angle position ATDC for sunDiesel at four start of injection timings

## Summary and Conclusions

The premise of this study was to discover whether combustion temperature in a practical, operating hydrogen powered engine is measurable utilizing spectrographic techniques. A single cylinder, hydrogen fueled engine made by Ford Motor Company was operated at realistic speeds and loads with a non-intrusive fiber optic system. This fiber optic system allowed for the UV emission of the OH* radical to be captured, quantified and analyzed to allow for the calculation of combustion gas temperature based upon the spectrographic signal.

A computer simulation was generated that predicted the spectra of OH* radical at various gas temperatures. This simulated spectrum was compared to the measured spectrum, iterating temperature in the simulated spectra until the simulated and measured spectra matched as closely as possible based upon a least squares error type curve fit. This approach provides for a consistent means to assess gas temperature in a hydrogen powered engine with a minimum of error and interference. Hydrogen powered engines do not produce any soot radiation to mask the signal. In addition, the potential to utilize OH* emission signals to determine equivalence ratio (Φ) appears possible for hydrogen engines. Expanding the measurements beyond hydrogen, with proper application, the temperature measurement approach may be potentially adaptable to low temperature combustion systems like HCCI.

# Undesirable Combustion Control Techniques of a Hydrogen Engine

**L.M. Das**

Centre for Energy Studies, IIT Delhi, Hauz Khas, New Delhi – 110016

Hydrogen, as a fuel, exhibits an entirely different spectrum of combustion characteristics as compared to the conventional petroleum fuels, more so in respect of engine operation. It is therefore essential that various combustion parameters related to engine operation must be identified and optimised to do away with undesirable combustion phenomena such as backfire, pre ignition, and rapid rate of pressure rise resulting in detonation.

Exhaustive tests show that fuel induction techniques play a critical role in achieving situations of controlled normal combustion in a hydrogen engine. Tests involving Carburetion, Continuous Manifold Injection (CMI), Timed Manifold Injection (TMI), and Low Pressure Direct Injection (LPDI) as fuelling modes have shown Timed manifold Injection (TMI) to be most pragmatic mode in view of good performance and low emission characteristics without any symptoms of undesirable combustion such as backfire and pre-ignition.

Flashback happens to be the most conspicuous of all the problems associated with hydrogen combustion in a Spark Ignition engine. Combustion problems such as backfire were experienced even by the earliest researchers, but the causes were not identified. Extensive research has revealed that the suspected causes of flashback are (i) Hot spots in the combustion chamber (ii) Particulate matter in the residual from oil (iii) Extreme lean operation (iv) Communication of the fresh charge with burning exhaust gases. The TMI system was specifically designed by the authors to eliminate all such possible causes.

Apart from circumventing these combustion problems, conditions of lean misfire limit (LMI), rate of pressure rise, maximum cylinder pressure, combustion duration and cyclic pressure variations with a TMI operated hydrogen engine have also been experimentally investigated. The problems of knocking in a hydrogen-fuelled Compression Ignition engine have been identified and curative steps have been recommended on the basis of tests.

# Performance and Emission Test of Biodiesel Fuelled In-direct Injection Diesel Transportation Engine System

**Avinash Kumar Agarwal**
Department of Mechanical Engineering, Indian Institute of Technology Kanpur, Kanpur-208016, India
E-mail: akag@iitk.ac.in

## INTRODUCTION

The world is presently confronted with the twin crises of fossil fuel depletion and environmental degradation. Indiscriminate extraction and lavish consumption of fossil fuels have led to reduction in underground-based carbon resources. The search for alternative fuels, which promise a harmonious correlation with sustainable development, energy conservation, efficiency and environmental preservation, has become highly pronounced in the present context.

Vegetable oils may provide one such alternative fuel and their potential has been examined in the past years by several researchers [1-7]. Vegetable oils have their own advantages: first of all, they are available everywhere in the world. Secondly, they are renewable as the vegetables which produce oil seeds can be planted year after year. Thirdly, they are "greener" to the environment, as they seldom contain sulphur element in them. This makes vegetable fuel studies become current issue among the various popular investigations. In the present investigation Jatropha curcas oil, a non-edible vegetable oil which has been considered as a potential alternative fuel for compressed ignition (CI) engines has been chosen to find out its suitability for use as fuel. Jatropha curcas is a large shrub or tree native to the American tropics but commonly found and utilized throughout most of the tropical and subtropical regions of the world. Several properties of the plant, including its hardness, rapid growth, easy propagation and wide ranging usefulness have resulted in its spread far beyond its original distribution. The Jatropha oil is slow-drying oil which is odorless and colorless when fresh but becomes yellow on standing. The oil content of Jatropha seed ranges from 30 to 50% by weight and the kernel itself ranges from 45 to 60%. The fatty acid composition of Jatropha classifies it as a linoleic or oleic acid type, which are unsaturated fatty acids. The fatty acid composition of Jatropha oil consists of myristic, palmitic, stearic, arachidic, oleic and linoleic acids.

**Table 1** Fatty acid composition of Jatropha oil [6]

| *Fatty acid* | *Composition (%)* |
|---|---|
| Palmitic acid | 12.8 |
| Stearic acid | 7.3 |
| Oleic acid | 44.8 |
| Linoleic acid | 34.0 |
| Other acids | 1.1 |

However, from the properties of this oil, it can be predicted that the oil would be suitable as fuel. The oil compares well against other vegetable oils and more importantly to mineral diesel itself in terms of its calorific value etc. But the greatest difference between Jatropha oil and mineral diesel is viscosity. A number of investigations have been made, and the test results have proved that vegetable oils are feasible substitutes for diesel fuel [1-9]. But there is still a lot of work that needs to be done to make vegetable oil applicable to diesel engines as a partial replacement of mineral diesel. The important chemical and physical properties of Jatropha curcas oil compared with diesel are shown in Table 2. Vegetable oils possess almost the same heating values as that of diesel. But a major disadvantage of vegetable oils is their inherent high viscosity. Modern diesel engines have fuel-injection systems that are sensitive to fuel viscosity changes. High viscosity may lead to poor fuel atomization, incomplete combustion, coking of the fuel injectors, ring carbonization, and accumulation of fuel in the lubricating fuels [2-15].

Caterpillar (Brazil) in 1980 used pre-combustion chamber engines with a mixture of 10% vegetable oil to maintain total power without any alterations or adjustments to the engine. At that point, it was not practical to substitute 100% vegetable oil for diesel fuel, but a blend of 20% vegetable oil and 80% mineral diesel was successful. Some short-term experiments used up to a 50/50 ratio [7].

**Table 2** Physical and chemical properties of diesel and Jatropha oil vis-à-vis mineral diesel [7]

| *Properties* | *Diesel* | *Jatropha curcas oil* |
|---|---|---|
| Density (gm/cc), 30°C | 0.836–0.850 | 0.93292 |
| Kinematic viscosity (cSt), 30°C | 4–8 | 52.76 |
| Cetane No. | 40–55 | 38.00 |
| Flash point °C | 45–60 | 210.00 |
| Calorific value (MJ/Kg) | 42–46 | 38.20 |
| Saponification value | - | 198.00 |
| Iodine No. | - | 94.00 |

Direct use of vegetable oils and/or the use of blends of the oils have generally been considered to be not satisfactory and impractical for both direct and indirect diesel engines. The high viscosity, acid composition, free fatty acid content, as well as gum formation due to oxidation, polymerization during storage and combustion, carbon deposits and lubricating oil thickening are obvious problems.

One of the ways to avoid these problems is to reduce the viscosity of vegetable oil in order to improve its engine performance. There are several methods to reduce the viscosity of vegetable oils. Fuel blending, transesterification, pyrolysis are some of these methods. The present research investigates the performance and emission characteristics of diesel/ Jatropha (SVO and Biodiesel) blends in a diesel engine. Three blends were obtained by mixing diesel and Jatropha oil in the following proportions by volume: 95% diesel/5% Jatropha oil; 90% diesel/10% Jatropha oil and 80% diesel/ 20% Jatropha oil. The experiments were also conducted using B100 (100% Jatropha Oil Methyl Ester). Performance and emission parameters for the fuels and the blends were evaluated under a range of varying torque applied to the engine at a constant speed.

## EXPERIMENTAL DETAILS

The experimental setup used in this work consists of a Mahindra & Mahindra (model XD-3P) four-cylinder engine coupled with an eddy current dynamometer. This is water-cooled, indirect-injection, four-stroke, four cylinder diesel engine. The eddy-current dynamometer (Model: ASE-50, Schenck Avery, India) is equipped with a dynamometer controller capable of loading the engine at the desired speed/ load combination. Detailed specifications of the engine are given in table 3. Smoke opacity meter (AVL- 437) and exhaust emission analyzer (AVL- 444) were used to measure exhaust gaseous emissions. Several temperature sensors was installed to measure exhaust gas temperature, inlet and outlet temperature of water, lube oil temperature etc.

**Table 3** Technical Specifications of four cylinder IDI diesel engine

| | |
|---|---|
| Model | Mahindra XD-3P |
| Type | water- cooled, four-stroke |
| Number of cylinders | 4 |
| Compression ratio | 23:1 |
| Bore diameter | 94 mm |
| Stroke length | 90 mm |
| Power output | 72.5 hp @4000 rpm without oil cooler |
| Max. torque | 150 N-m @ 2000 rpm |
| Loading device | Eddy current dynamometer |

Three blends of Jatropha oil by volume, biodiesel and Diesel were tested (V5, B10, V20, B100, B0). The tests were repeated three times for every fuel samples and then average readings are reported in this paper. For each fuel, the engine was operated at six different torque values (0, 20, 40, 60, 80, and 100 Nm) at constant speed of 2000 rpm. Several parameters were recorded such as torque, fuel consumption, engine exhaust temperature, smoke opacity and emissions etc.

## RESULTS AND DISCUSSION

The following engine performance parameters were calculated and presented in the form of graphs:

(a) Brake specific fuel consumption (bsfc)

(b) Brake thermal efficiency

(c) Exhaust gas temperature

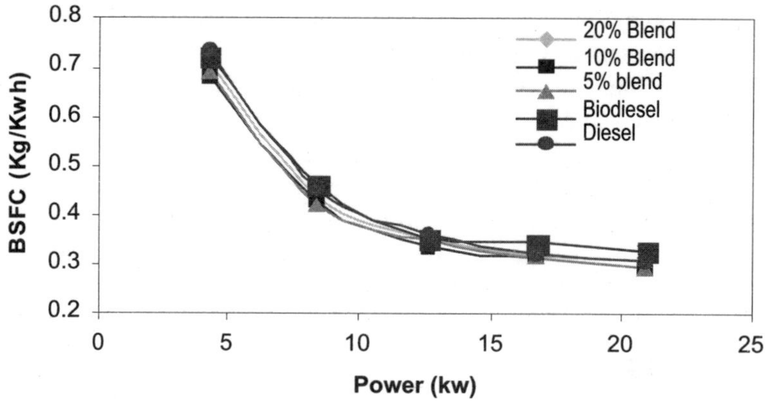


**Figure 1** Brake specific fuel consumptions for different fuel samples

Figure 1 shows brake specific fuel consumptions (bsfc) vs. Power curve for different fuels. The bsfc decreases with increase in engine power/ load. The differences of bsfc are very small at different engine torque, when using different fuels. At 17 kW, there is a reduction in bsfc for other fuels compared to mineral diesel; V05, it is 7.8%; V10 5.7%, V20, it is 5.7% however for biodiesel it is similar to mineral diesel.

Figure 2 shows the trend for brake thermal efficiency vs. engine power for different fuel samples. Brake thermal efficiency increases with engine power output for all fuels. It can be seen that the brake thermal efficiency of all blends of Jatropha and is higher in comparison with mineral diesel.

Figure 3 shows the exhaust temperatures versus the engine power. The exhaust temperatures rise with the increase of engine power for all of the fuels. It can be seen that the exhaust temperatures at different engine power, when using different fuels, are nearly similar.

## EMISSION RESULTS

The following engine emissions were observed and presented in the form of graphs:

(a) CO

(b) $CO_2$

(c) NO

(d) Smoke Opacity

Fig.4 shows the comparison of the CO emissions of different fuels at different engine power. Within the experimental range, the CO emission from the vegetable oil and vegetable oil/diesel blends are nearly all higher than that from mineral diesel at initial stage. This is possibly because of the high viscosity of vegetable oil and its blends, which has slightly lower mixing with the air leading to higher CO formation during combustion. The higher the viscosity, the more difficult it is to atomize for the vegetable oil and its blends. This may have resulted in locally rich mixture regions in the engine leading to more carbon monoxide generation.

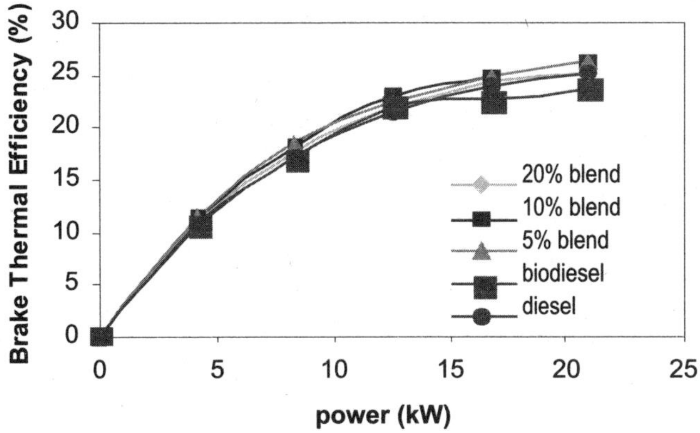


**Figure 2** Brake thermal efficiency of different fuel samples

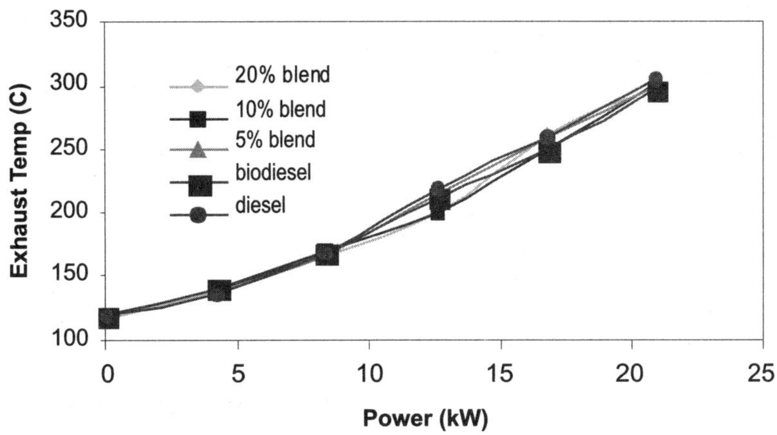


**Figure 3** Exhaust gas temperatures of different fuel samples

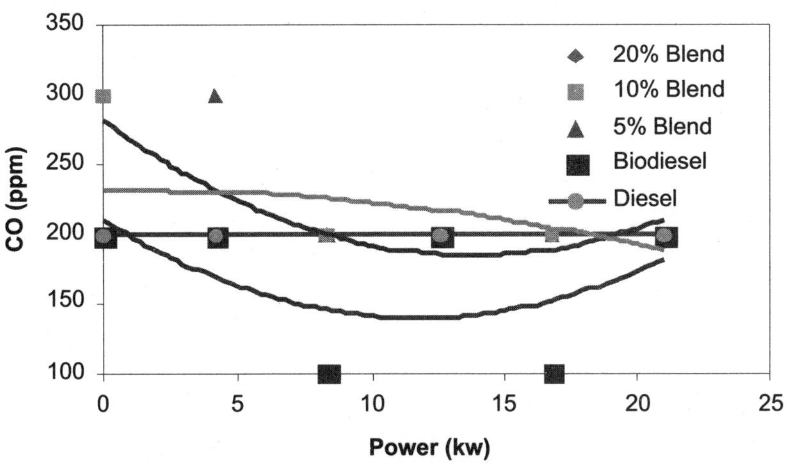


**Figure 4** Carbon Monoxide emissions using different fuel samples

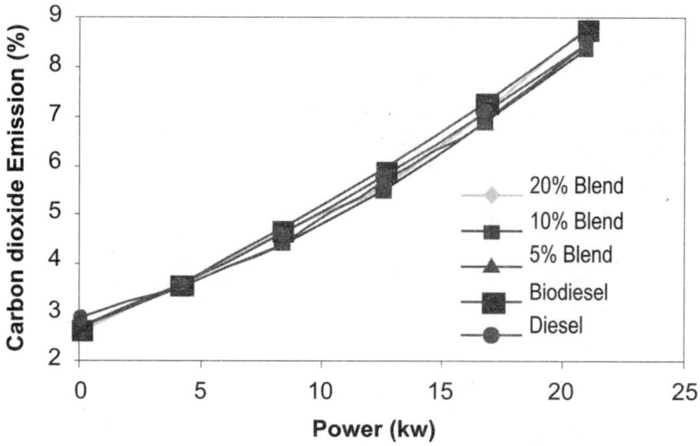


**Figure 5** $CO_2$ emissions from different fuel samples

$CO_2$ emissions of different fuels from the engine are shown in Fig. 5. In the range of whole engine power, the $CO_2$ emissions of biodiesel and diesel fuel are all higher than that of the other fuels.

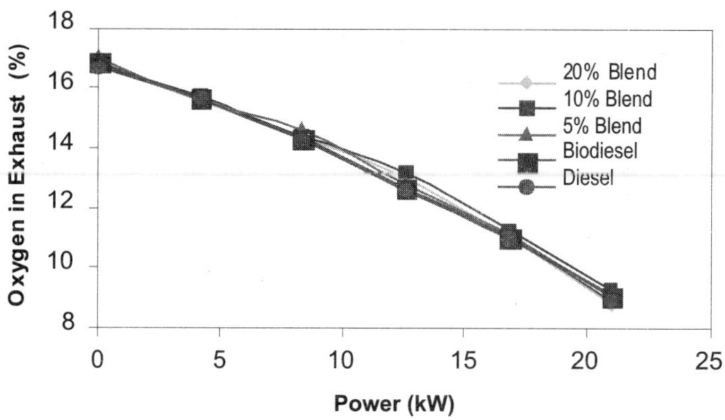


**Figure 6** Oxygen content in engine exhaust for different fuel samples

Fig. 6. shows the oxygen in exhaust vs. engine power trend. Oxygen content in exhaust decreases with increasing engine power for all fuel samples.

As shown in fig. 7, that all the biofuels emit relatively higher NO compared to mineral diesel however at higher engine load, there is a reduction in NO emissions observed and the gap between the NO emission from diesel and other biofuels also reduces.

Smoke opacity percentages in figure 8, during each of the vegetable oil blend operations were greater than that of diesel fuel except for biodiesel, which consistently showed lower smoke opacity. 5% and 10% blends were relatively closer to diesel. The greater smoke opacity percentages of vegetable oil blends are mainly due to the heavier molecules of hydrocarbons in the fuel, which

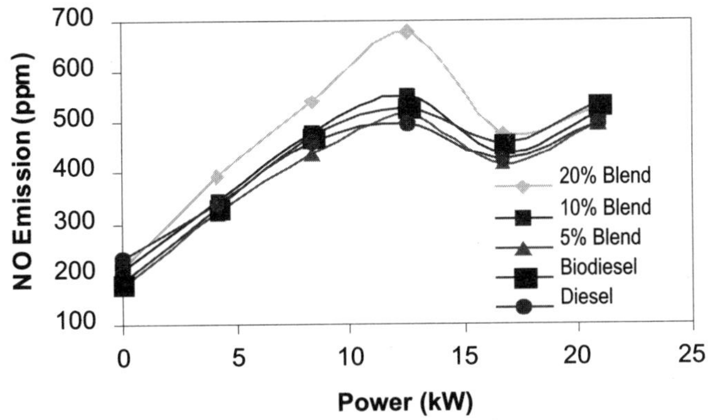


**Figure 7** NO emissions from different fuel samples

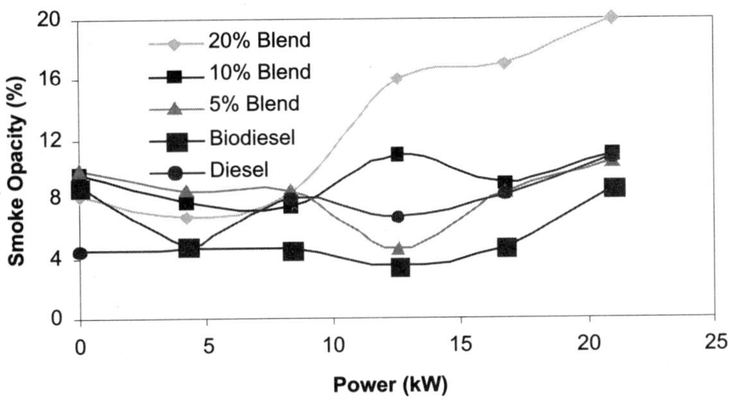


**Figure 8** Smoke opacity (%) from different fuel sample

do not burn completely and there is larger formation of particulates. On the other hand, biodiesel undergoes more complete combustion and shows lower smoke opacity.

The unburnt hydrocarbon (HC) emissions are presented in Fig. 9. The HC emissions of all fuels are lower at partial engine load, but they increase at higher engine load/ power. This is due to relatively less oxygen availability for the combustion when more fuel is injected into the engine cylinder to meet higher load requirement. Fig. 9 also shows that the HC emissions for 5% blend lower than that of diesel fuel.

## Conclusion

The main objective of the present investigation was to reduce the viscosity of Jatropha curcas oil and bring it closer to that of conventional diesel to make it suitable for use in an unmodified transportation diesel engine and to evaluate the performance of the engine with the modified fuels as well as with pure biodiesel in comparison to baseline data of mineral diesel. The experimental results show that the engine performance of vegetable oil blends is comparable to the performance

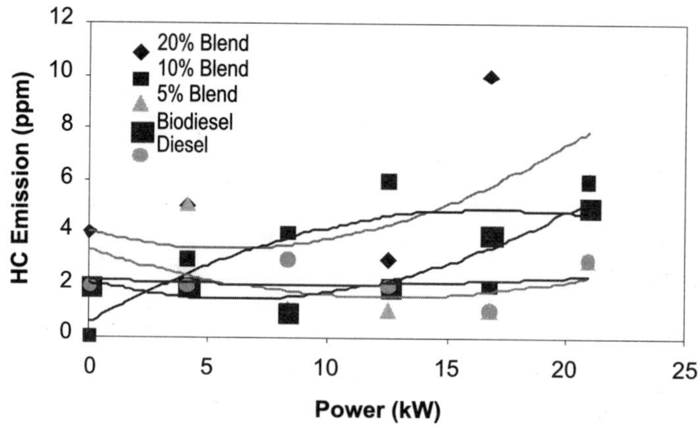


**Figure 9** Hydrocarbon emissions from different fuel samples

of biodiesel and mineral diesel. The engine power output and the fuel consumption of the engine are almost similar when the engine is fueled with vegetable oil blends compared with biodiesel and diesel fuel. The carbon monoxide (CO) emission from the biodiesel is lower than that of the vegetable oil blends and diesel fuel at all engine loads except lower loads. The emission of nitrogen oxide (NO) from vegetable oil blends and biodiesel however higher than that of mineral diesel at all loads. The results from this experiment prove that vegetable oil blends and biodiesel are potential partial substitute fuels for transport diesel engines in the near future, when availability of petroleum based fuels become dearer.

## References


1. A.S. Ramadhas, Use of vegetable oils as I.C. engine fuels—A Review, Renewable Energy, 29 (5): 727–742, 2004.
2. F. Karaosmanoglu, Vegetable oil fuels: a review, Energy Sources, 21 (3): 1999.
3. Recep Altan, Selim Cetinkay, Huseyin Serdar Yucesu. The potential of using vegetable oil fuels as fuel for diesel engines, Energy Conversion and Management, 2001.
4. Masjuki H, Salit. Biofuel as diesel fuel alternative: an overview. J. Energy Heat Mass Transfer 15:293–304, 1993.
5. J. B. Kandpal and Mira Madan, Jatropha curcus: a renewable source of energy for meeting future energy needs, Renewable Energy, 6(2):159-160, 1995.
6. S. Sinha, N.C. Misra, Diesel fuel alternative from vegetable oils, CEW, Chemical Engineering World, 32 (10): 77–80, 1997.
7. K. Pramanik, Properties and use of jatropha curcas oil and diesel fuel blends in compression ignition engine, Renewable Energy, 28: 239–248, 2003.
8. Masjuki H, Sohif M. Performance evaluation of palm oil diesel blends on small engine. J. Energy Heat Mass Transfer, 13:125–33, 1991.
9. C. D. Rakopoulos, Comparative performance and emission studies when using olive oil as a fuel supplement in DI and IDI diesel engines, Renewable energy, 2(3): pp. 327 331, 1992.

10. A.K. Agarwal, Vegetable oils verses diesel fuel: development and use of biodiesel in a compression ignition engine. TIDE, 8(3):191–204, 1998.
11. Nag A, de Bhattacharya KB. New utilizations of vegetable oils. J. Am. Oil Chem. Soc, 72(12):1391–3. 1995.
12. Demirbas, Biodiesel from vegetable oils via transesterification in supercritical methanol, Energy Conversion and Management, 43: 2349–2356, 2002.
13. M.S. Graboski, R.L. McCormick, Combustion of fat and vegetable oil derived fuels in diesel engines, Progress in Energy and Combustion Science, 24 (2):125–164, 1998.
14. Peterson CL, Wagner GL, Auld DL. Performance testing of vegetable oil substitutes for diesel fuel, ASAE, 1981.
15. Y.D. Wang et al, An experimental investigation of the performance and gaseous exhaust emissions of a diesel engine using blends of a vegetable oil, Applied Thermal Engineering, 26: 1684–1691, 2006.

# Control of Combustion Processes

**Sumanta Acharya**
Mechanical Engineering, Louisiana State University, Baton Rouge, LA 70803, USA

Key performance metrics for gas turbine combustion include: combustion dynamics, emissions, volumetric heat release and the uniformity of temperature distribution at the combustor exit. Many of these metrics can be controlled through passive and/or active control concepts. In active control, the combustion process is altered through an external perturbation of the air or fuel feeds. The control can be closed-loop, based on a feedback signal, or open-loop based on heuristics and/or an understanding of the flow physics. The primary goal of this control is to alter the spatio-temporal distribution of the heat release distributions. In this talk, we will discuss closed loop and open loop control strategies for controlling combustion performance. We will focus on dynamics, emissions and the volumetric heat release as the key metrics of interest.

We will examine a strategy where we have utilized a high-momentum air-jet as the control actuator, and demonstrated effective combustion control (dynamics, emissions and heat release) through either a closed-loop or an open-loop strategy. The advantages and disadvantages of the two strategies will be discussed. We will also examine how the control authority can be enhanced through the addition of catalytic particles, and make a case for using energetic fuels for combustion control purposes. Measurements presented will primarily include: pressure, temperature, heat release, and NO emissions in a swirl-stabilized combustor. All measurements presented will be for a laboratory scale combustor operating under atmospheric pressure conditions.

# Thermoacoustic Instability without Eigenvalues

**Koushik Balasubramanian and R.I. Sujith**[a]
Department of Aerospace Engineering, Indian Institute of Technology Madras, Chennai 600036, India

## I. INTRODUCTION

The occurrence of combustion instabilities has been a plaguing problem in the development of combustors for rockets, jet engines and power generating gas turbines[1]. Predicting and controlling combustion instability requires an understanding of the combustion-acoustic interactions.

Combustion instabilities are self-sustained large amplitude oscillations of pressure and velocity in combustors with the flame acting as an acoustic actuator and the combustion chamber as an acoustic resonator. If the heat released in the system depends on the pressure and velocity fluctuations in the system, a feedback loop is established that can destabilize the system. The occurrence of combustion instability depends on the phase between heat release fluctuations and pressure fluctuations at the flame[2] (Rayleigh 1878). Amplification of the pressure oscillations by the heat addition process will take place if the maximum and minimum of the heat addition occur during the compression and rarefaction phases of the pressure oscillation, respectively. In contrast, the pressure oscillations will be attenuated if the maximum and minimum of the heat addition occur during the rarefaction and compression phases of the pressure oscillation, respectively.

When pulsations start spontaneously, as they do in a Rijke tube with a heater at ¼ length from the bottom of the tube, the system is said to be linearly unstable; i.e., the system is unstable with respect to any small amplitude disturbance that may be present in the combustor. This scenario has been successfully modeled in various systems by using classical linear stability analysis of the normal modes that model the system as a network model, in which each element is modeled using a linear transfer function. The stability of the system can then be determined easily by examining the eigenvalues of the system.

It is also possible that a linearly stable combustor (i.e., one that does not pulse spontaneously) could be "triggered" into pulsating operation by the introduction of a finite amplitude disturbance such as might be caused by a spark plug ignition or a small explosion. Such a system will be stable

[a]Author to whom correspondence should be addressed. Electronic mail: sujith@iitm.ac.in

with respect to all disturbances whose amplitudes are below a certain threshold value, but transition into pulsating operation will occur when the amplitude of the disturbance exceeds this threshold value. Furthermore, there are instances of "bootstrapping" where a mode that decays initially can grow later and ultimately become unstable[3]. A linear stability analysis using normal modes would always indicate stability in the case of bootstrapping and miss the ultimate behavior. A comprehensive prediction of the conditions for the onset of instabilities is a difficult task, which is not yet mastered. In particular, predicting the conditions under which finite amplitude disturbances destabilize a linearly stable system and predicting the limit cycle amplitude of the instability remain a key challenge, as little is known, even in a qualitative sense about the key parameters controlling nonlinear flame dynamics even in simple laminar flames[4]. The pressing need to control this phenomenon in combustion chambers compounded by a lack of understanding of combustion instability due to the complex nature of combustion acoustic interaction in flames led to interest in simpler thermoacoustic systems such as Rijke tubes. The Rijke tube serves as a convenient prototypical system for the study of thermoacoustic phenomena.

Nonlinear effects in a Rijke tube were investigated in detail by several authors. A Rijke tube is a relatively simple system: it is a duct with a heat source (often electrically heated wires) at quarter length from the bottom (if located vertically). Heckl[5] studied the nonlinear acoustic effects leading to limit-cycles in unstable oscillations experimentally and theoretically for the case of a Rijke tube. She showed that the important nonlinear effects are (1) the reduction of the rate of heat transfer when the velocity amplitudes are of the same order as the mean velocity and (2) increased losses at the ends of the tube at very high amplitudes. Hantschk and Vortmeyer[6] showed that the limit-cycle amplitude in a Rijke tube is determined by nonlinearities in the heat flux from the heating element to the flow. Yoon et al.[3] proposed a nonlinear response model of a generalized Rijke tube. Their oscillatory heat release model was not derived from physical principles. They derived both closed form and numerical solutions for the acoustic field by an approximate modal analysis using a two mode formulation. The two-mode nonlinear model is capable of predicting the bootstrapping effect which characterizes nonlinear velocity sensitive combustion response in rocket motors. However, they have neglected nonlinear convection in the present model. In order to explain the nonlinear effects of a Rijke tube, Matveev[7,8] constructed a simple theory by using an energy approach. The equilibrium states of the system are found by balancing thermo-acoustic energy input and acoustic losses. It is again confirmed that the nonlinearity of the unsteady heat transfer is a dominant factor in limit-cycle saturation. Further, he demonstrated the necessity of accurately modeling the effects of temperature gradient on the mode shapes to obtain accurate results for stability[9].

Although considerable research has been performed on the nonlinear nature of thermoacoustic oscillations, their non-normal nature is an aspect that has not received any attention. Non-normality can lead to transient growth of a system even when the eigenvalues indicate linear stability. When the transient growth leads to high enough amplitudes, it can trigger nonlinearities in the system which can cause "nonlinear driving", thereby driving a system which was thought to be linearly stable under the framework of classical linear stability. Further, non-normality leads to coupling between the modes, although this effect has been often attributed to nonlinearity. The role of non-normality in the context of thermoacoustic oscillations has been shown in the context of ducted diffusion flames by Balasubramaniam and Sujith[10] using a Galerkin type analysis. Nicaud et al.[11]

has shown that the eigenvectors of a thermoacoustic system are non-orthogonal in the presence of heat release, or in the presence of general complex impedance boundary conditions.

Thus there are two distinct aspects of combustion acoustic interactions: non-normality of the modes and nonlinearities in the system.

Non-normality has been found to play an important role in several fields such as turbulence[12, 13, 14], instability of magnetic plasmas[15], formation of cyclones[16] etc. Recently the authors have shown that the combined role of non-normality and nonlinearity can lead to the triggering of thermoacoustic oscillations in a linearly stable system, in the context of ducted diffusion flames[10]. This is a general feature of all thermoacoustic systems and need not be restricted to diffusion flames. In the case of ducted diffusion flames, the combustion modes themselves are non-normal[10].

The objective of this paper is to examine the role non-normality in a simpler model for a thermoacoustic system, constructed for a Rijke tube. The rest of this paper is organized as follows. In section II, the characteristics of a non-normal operator and its consequences are discussed in the context of thermoacoustic instabilities. The schematic of a horizontal Rijke tube considered in this study is described in section III. The equations governing the evolution of the acoustic field in this horizontal Rijke tube are derived in section IV. In section V, examples highlighting the effects of non-normality and nonlinearity are discussed. The results and inferences are summarized in section VI.

## II. NON-NORMALITY AND TRANSIENT GROWTH OF THERMOACOUSTIC OSCILLATIONS

A system is said to be non-normal if its eigenvectors are not orthogonal. The evolution of such a system is governed by an operator which does not commute with its adjoint. The non-orthogonal nature of eigenvectors can lead to transient growth of oscillations before they eventually decay. This property is illustrated in Figures 1a and 1b. The vectors $e_1$ and $e_2$ represent the direction of the eigenvectors and $\Phi$ is a vector which is expressed as a linear combination of the eigenvectors. Figure 1a shows that for a normal system, $\Phi$ decreases monotonically if the amplitudes of the individual eigenvectors themselves decay. On the contrary, for the non-normal system shown in figure 1b, $\Phi$ increases even when the amplitudes of individual eigenvectors decay. However $\Phi$ decays after a sufficiently long time if nonlinear effects do not become significant during the transient growth. There could be situations where, the short-term growth of fluctuations can lead to significant amplitudes where nonlinear effects could cause "nonlinear driving" as illustrated in figure 2. Such a scenario arises in the evolution of thermoacoustic oscillations.

The acoustic equations[b] in the presence of a heat source can be written in as:

$$gM\frac{\partial u'}{\partial t} + \frac{\partial p'}{\partial x} = 0 \qquad \text{...(1)}$$

$$\frac{\partial p'}{\partial t} + \gamma M\frac{\partial u'}{\partial x} = (\gamma - 1)\,\frac{L_a}{c_0}\,\frac{\dot{Q}'}{\rho_0 c_0^2} \qquad \text{...(2)}$$

[b]The assumptions in deriving these equations and the details of non-dimensionalization are given in section 4.

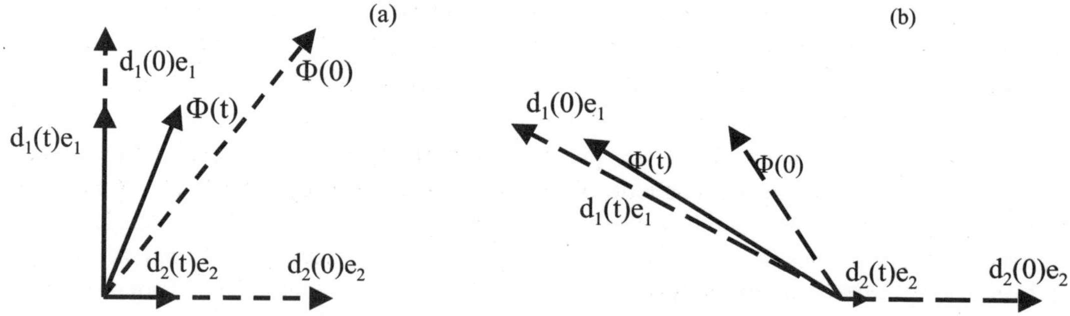


**Figure 1** Shows (a) monotone decay of a normal system, (b) transient growth of a non-normal system. The initial state is $\Phi(0) = d_1(0)e_1 + d_2(0)e_2$, and the final state $\Phi(t) = d_1(t)e_1 + d_2(t)e_2$. The dashed lines denote the vectors at time t = 0 and the solid lines denote the vector at some time t

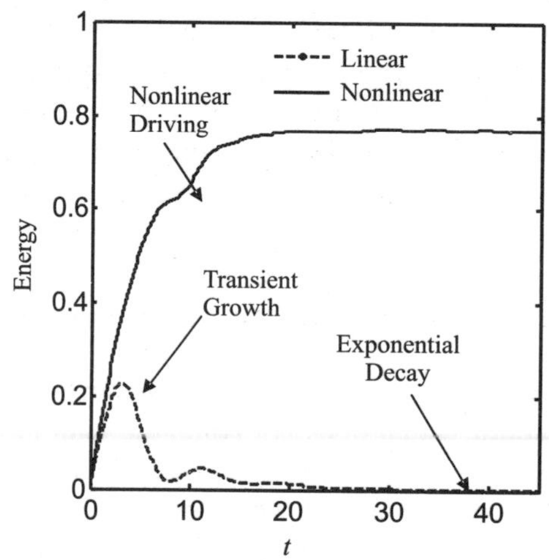


**Figure 2** Shows the evolution of the non-dimensional acoustic energy $\left(\frac{1}{2}p'^2 + \frac{1}{2}(\gamma Mu')\right)^2$ after locally weighted regression smoothing (LOWESS) from linear and nonlinear simulations. $\eta_1(0) = 0.17$. $\tau = 0.002$, $x_f = 0.3$. The constants used for damping are 0.135 and 0.015. $c_0 = 399.6$ m/s, $\bar{\rho} = 1.205$ kg/m$^3$, $L_w = 1.5$ m, $\lambda = 0.0328$ W/mK, $T_w = 1000$ K, $\bar{u} = 0.05$ m/s, $C_v = 719$ kJ/kgK

The heat release rate in the above equation is calculated from a model for the heat source. The linearized oscillatory heat release rate, can be written as $\dot{Q}' = R(x, \varepsilon_i)\,\gamma Mu' + S(x, \mu_i)p'$, where $R$ and $S$ can be treated as a continuous function of $x$ (which could even be sharply peaked at the flame location as in the case of a compact heat source), $\varepsilon_i$ and $\mu_i$ are parameters which affect heat release rate. The heat release rate could have an explicit dependence on time as well. Equations (1) & (2) can be recast in the matrix form as:

$$\begin{bmatrix} \dfrac{\partial}{\partial t} & \dfrac{\partial}{\partial x} \\ \dfrac{\partial}{\partial x} - \dfrac{RL_a(\gamma-1)}{\rho_0 c_0^3} & \dfrac{\partial}{\partial t} - \dfrac{SL_a(\gamma-1)}{\rho_0 c_0^3} \end{bmatrix} \begin{bmatrix} \gamma Mu' \\ p' \end{bmatrix} \qquad \text{...(3)}$$

The matrix in Eq. (3) is a matrix of operators. The above operator does not commute with its adjoint[3] for non-zero $R$ and $S$. Therefore, it is clear that the thermoacoustic interaction is non-normal. In the absence of heat release, the matrix is symmetric and hence normal.

The non-normal behavior of the system causes large transient growth which can potentially trigger nonlinearities in the system. Classical linear stability analysis then becomes a poor indicator of system stability due to this algebraic growth[13]. This phenomenon has been studied in detail in the context of turbulence by Baggett *et al.*[12]. They explain that in the non-normal evolution, the input and output structures (such as streamwise vortices, streaks etc.) are different and nonlinearity closes the feedback loop by converting some of the output into input. In a similar manner, the interplay between transient linear growth resulting from non-normality and "nonlinear mixing" can indeed lead to the growth of the thermoacoustic oscillations.

Equation (3) is analyzed by reducing the infinite dimensional operator to a finite dimensional operator, such as a finite dimensional matrix. In this paper, this is achieved by reducing the partial differential equations governing the thermoacoustic interaction to a set of ordinary differential equations (ODE) using the Galerkin technique. ODEs in time domain are obtained by decomposing the spatial variation using basis functions. This is similar to decomposing a vector along some basis. The basis functions used in this study are not the eigenmodes of the linearized system, but they are the eigenmodes of the self adjoint part of the linearized system. Such an approach has been used in solving partial differential equations; see for e.g., Henningson and Schmid[17]. These evolution equations are solved numerically using the $4^{th}$ order Runge-Kutta scheme. The complete evolution equations are linearized and the linearized equations are found to be non-normal. It must be emphasized that the eigenvalues of the linearized equations are not the wave numbers of the basis functions used in Galerkin technique.

## III. THE HORIZONTAL RIJKE TUBE

A horizontal Rijke tube with an electric heat source is a system convenient for studying the fundamental principles of thermoacoustic instabilities both experimentally and theoretically. In such a set-up, the mean flow is provided by a blower, which sucks air in the tube. This enables us to control the heater power and the mean flow independently. If the tube were oriented vertically, as in the classical Rijke tube, the effect of mean flow component caused by natural convection will have to be accounted for in the stability analysis. The horizontal orientation of the Rijke tube is implemented to exclude the influence of natural convention on the mean flow rate. Such a setup has been used by Matveev[7, 8], Heckl[5], and Kopitz and Polifke[18]. A schematic of the horizontal Rijke tube setup is shown in figure (3). The current study models such a set-up.

## IV. GOVERNING EQUATIONS

Neglecting the effect of mean flow and mean temperature gradient in the duct, the governing equations for the one-dimensional acoustic field are:

[c]The adjoint of a real matrix is simply the transpose of the matrix. The adjoint of a differential operator is its negative.

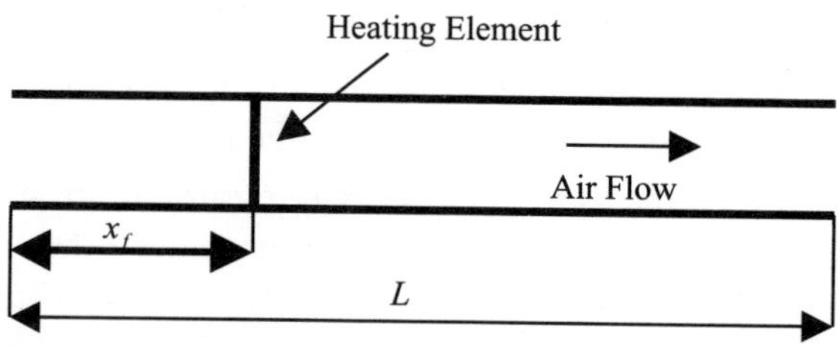


**Figure 3** Shows a schematic of the horizontal Rijke tube setup

Acoustic Momentum:
$$\bar{\rho}\frac{\partial \tilde{u}'}{\partial \tilde{t}} + \frac{\partial \tilde{p}'}{\partial \tilde{x}} = 0 \quad ...(4)$$

Acoustic Energy:
$$\frac{\partial \tilde{p}'}{\partial \tilde{t}} + \frac{\partial \tilde{u}'}{\partial \tilde{x}} = (\gamma - 1)\dot{\tilde{Q}}' \quad ...(5)$$

A modified form of King's law is used to model the heat release rate. Since King's law exhibits nonlinearity only for velocity perturbations greater than the mean fluctuations, Heckl[5] suggested the following empirical model.

$$\dot{Q}' = \frac{2Lw(T_w - \bar{T})}{S\sqrt{3}}\sqrt{\pi\lambda C_v \bar{p}\frac{d_w}{2}}\left[\left|\frac{u_0}{3} + u_f'(t-\tau)\right| - \sqrt{\frac{1}{3}}\right]\delta(\tilde{x} - \tilde{x}_f) \quad ...(6)$$

In the above expression, $L_w$ is the equivalent length of the wire, $\lambda$ is the heat conductivity of air, $C_v$ is the specific heat of air, $\tau$ is the time lag, $\bar{\rho}$ is the mean density of air, $d_w$ is the diameter of the wire, $(T_w - \bar{T})$ is the temperature difference, and $S$ is the cross-sectional area of the duct. The above equations can be non-dimensionalized as follows;

$$\tilde{x} = L_a x; \tilde{t} = \frac{L_a}{c_0}t; \tilde{u}' = u_0 u'; \tilde{p}' = \bar{p}p'; M = \frac{u_0}{c_0}; \tilde{\rho} = \bar{\rho}\rho$$

where $c_0$ is the speed of sound, $L_a$ is the duct length and $\bar{p}$ is the pressure of the undisturbed medium and $u_0$ is the mean flow velocity. The acoustic equations in the non-dimensional form can be written as follows:

$$\gamma M\frac{\partial u'}{\partial t} + \frac{\partial p'}{\partial x} = 0 \quad ...(7)$$

$$\frac{\partial p'}{\partial t} + \gamma M\frac{\partial u'}{\partial x} = k\left[\sqrt{\frac{1}{3} + u'(t-\tau)} - \sqrt{\frac{1}{3}}\right]\delta\left(x - x_f\right) \quad ...(8)$$

$$\text{where, } k = (\gamma - 1)\frac{2L_w(T_w - T_0)}{Sc_0\bar{p}\sqrt{3}}\sqrt{\pi\lambda C_v\bar{\rho}\frac{d_w}{2}u_0} \quad ...(9)$$

The above set of partial differential equation can be reduced to ordinary differential equations using the Galerkin technique[19]. The velocity and pressure field can be written in terms of the duct's natural modes as follows[20]:

$$u' = \sum_{j=1}^{N} \eta_j \cos(j\pi x) \quad \text{and} \quad p' = -\sum_{j=1}^{N} \frac{\gamma M}{j\pi} \dot{\eta}_j \sin(j\pi x) \qquad \text{...(10)}$$

The Galerkin technique makes use of the fact that any function in a domain can be expressed as a superposition of expansion functions which form a complete basis in that domain. The basis functions are chosen such that they satisfy the boundary conditions. However the choice of the basis functions is not unique. The basis functions chosen here are just an arbitrary basis and not the eigenfunctions of the system. They are the eigenfunctions of the self adjoint part of the linearized system. Clearly, the expansion functions chosen here satisfy the boundary conditions and they form a complete basis.

Substituting the above expansions into Eq. (7) and (8), and projecting along the basis functions[d], the following evolution equations are obtained,

$$\frac{d\eta_j}{dt} = \dot{\eta}_j \qquad \text{...(11)}$$

$$\frac{d\dot{\eta}_j}{dt} + k_j^2 \eta_j = -\frac{2k}{\gamma M} j\pi \left[ \sqrt{\frac{1}{3} + u_f'(t-\tau)} - \sqrt{\frac{1}{3}} \right] \sin(j\pi x_f) \qquad \text{...(12)}$$

In the presence of damping the above set of equations can be modified as follows[8]

$$\frac{d\eta_j}{dt} = \dot{\eta}_j \qquad \text{...(13)}$$

$$\frac{d\dot{\eta}_j}{dt} + 2\xi_j \omega_j \dot{\eta}_j + k_j^2 \eta_j = -\frac{2k}{\gamma M} j\pi \left[ \sqrt{\frac{1}{3} + u_f'(t-\tau)} - \sqrt{\frac{1}{3}} \right] \sin(j\pi x_f) \qquad \text{...(14)}$$

where the damping constant is given by:

$$\xi_j = \frac{1}{2\pi} \left[ c_1 \omega_j / \omega_1 + c_2 \sqrt{\omega_1 / \omega_j} \right] \qquad \text{...(15)}$$

In this paper, the effect of damping is studied by varying the constants $c_1$ and $c_2$ in the above expression. In experiments, the changes in these constants in Eq. (15) can be effected by adjusting the end conditions and other experimental conditions. Equations (13) & (14) can be expanded to 2nd order for low amplitudes to yield the following matrix differential equation,

[d]The component or the projection of a function $f$ along a basis function $\psi_n$ is given by their inner product $\langle f | \psi_n \rangle$ which is defined as $\int_{domain} f(x)\psi_n(x)\,dx$.

$$\frac{d\chi}{dt} + B_{NL}(\chi^T, \chi) + B_{NN}\chi = 0 \qquad \text{...(17)}$$

where, $\chi = [\eta_1 \quad \dot{\eta}_1 \quad \eta_2 \quad \dot{\eta}_2 \quad \text{...........}\eta_N \quad \dot{\eta}_N]^T$

$$B_{NN} = \begin{bmatrix} 0 & -1 & 0 & 0 & \cdots\cdots & 0 & 0 \\ \omega_1^2 + \beta_1\cos(\pi x_f) & 2\omega\xi_1 & \beta_1\cos(2\pi x_f) & 0 & \cdots\cdots & \beta_1\cos(N\pi x_f) & 0 \\ 0 & 0 & 0 & -1 & \cdots\cdots & 0 & 0 \\ \beta_2\cos(\pi x_f) & 0 & \omega_2^2 + \beta_2\cos(2\pi x_f) & 2\omega_2\xi_2 & \cdots\cdots & \beta_1\cos(N\pi x_f) & 0 \\ \cdots & \cdots & \cdots & \cdots & \cdots\cdots & \cdots & \cdots \\ \cdots & \cdots & \cdots & \cdots & \cdots\cdots & \cdots & \cdots \\ 0 & 0 & \cdots & \cdots & \cdots\cdots & 0 & -1 \\ \beta_N\cos(\pi x_f) & 0 & \beta_N\cos(2\pi x_f) & 0 & \cdots\cdots & \omega_N^2 + \beta_N\cos(N\pi x_f) & 2\omega_N\xi_N \end{bmatrix}$$

$$B_{NL} = u'(t-\tau) = \begin{bmatrix} 0 & 0 & 0 & 0 & .. & 0 & 0 \\ \alpha_1\cos(\pi x_f) & -\alpha_1\tau\cos(\pi x_f) & \alpha_1\cos(2\pi x_f) & \alpha_1\cos(N\pi x_f) & .. & \alpha_1\cos(N\pi x_f) & -\alpha_1\tau\cos(2\pi x_f) \\ 0 & 0 & 0 & 0 & .. & 0 & 0 \\ \alpha_2\cos(\pi x_f) & -\alpha_2\tau\cos(\pi x_f) & \alpha_2\cos(2\pi x_f) & \alpha_2\cos(N\pi x_f) & .. & \alpha_2\cos(N\pi x_f) & -\alpha_2\tau\cos(2\pi x_f) \\ \cdots & \cdots & \cdots & \cdots & .... & .... & \cdots \\ \cdots & \cdots & \cdots & \cdots & .. & .... & \cdots \\ 0 & \cdots & \cdots & 0 & .. & 0 & 0 \\ \alpha_N\cos(\pi x_f) & -\alpha_N\tau\cos(\pi x_f) & \alpha_n\cos(2\pi x_f) & \alpha_N\cos(N\pi x_f) & .. & \alpha_N\cos(N\pi x_f) & -\alpha_N\tau\cos(N\pi x_f) \end{bmatrix}$$

where, $\beta_j = (k/\gamma M)\sqrt{3}\, j\pi \sin(j\pi x_f), \alpha_j = \beta_j/2$.

It can be seen that $B_{NN}$ does not commute with its adjoint and hence it is non-normal. In the above equation, $B_{NL}$ is the nonlinear matrix. Various cases highlighting these features and their consequences are discussed using examples in the following section.

## V. RESULTS AND DISCUSSIONS

The duct acoustic modes are normal in the absence of heat addition. However, heat addition makes the system non-normal as can be seen from the following physical arguments. The heat transfer is a function of the velocity as given by Eq. (6). A small disturbance in the velocity can alter the heat release rate which in turn acts as the source of disturbance for the acoustic oscillations. The heat addition drives the acoustic oscillations in the duct when it is in phase with acoustic pressure. The phase lag itself evolves with time and hence the phase difference between the heat release fluctuation and acoustic pressure oscillations at the heater at some instant depends on their phase difference at an earlier instant. The acoustic modes that are driven by the heat release oscillations at these two instants are different. Since the phase difference between the heat release oscillations and a particular mode of the acoustic field depends on the phase difference at an earlier time, the interaction would depend on that mode of the acoustic field which was in phase with the thermal process at an earlier time. Hence the mode which is driven at a particular instant of time depends on which mode got driven at an earlier time. Hence, the energy in a mode at some instant will depend on the energy of another mode which was driven at an earlier instant of time. This indicates that there is a non-zero projection of a mode along another mode indicating non-normal behavior.

Another process which plays a crucial role in thermoacoustic interaction is damping. Damping tries to decay all the modes. Hence damping inhibits any kind of growth of oscillations. It can also be seen from Eq. (15) that modes with higher wave numbers decay more rapidly than the modes with smaller wave number. Damping causes different modes to have different phase relation with the heat release rate, as modes with larger wave numbers damp faster than the waves with smaller wave number. Hence, damping can make the non-normal effects significant, resulting in larger transient growth. However, when damping is too high, the oscillations get completely damped out. Further, there are situations where these two process counter balance each other and such situations can lead to a very complicated thermo-acoustic situation where the interaction becomes highly sensitive to small changes in the various parameters and amplitude. If non-normality overcomes the damping process, one can expect transient growth which can in turn trigger the nonlinearities in the system.

The time lag is an important factor which affects the dynamics of the system. As discussed earlier, it is one of the factors which affect the non-normal nature of the system. Apart from affecting the non-normal nature of the system, time lag also determines whether a system is conservative or non-conservative. In the following, it is shown analytically that the system behaves differently for different ranges of time-lag. Further, it is also shown that there is a critical time lag below which the energy of the system decreases. It is assumed that the time lag is small for analytical convenience. It is also seen that in many experimental situations the time lag is small compared to the acoustic time scale[18]. Hence, Eq. (13) and (14) can be written upto first order in $\tau$ as:

$$\frac{d\eta_j}{dt} = \dot{\eta}_j$$

$$\frac{d\dot{\eta}_j}{dt} = -2\xi_j\omega_j\dot{\eta}_j - \omega_j^2\eta_j - k\omega_j\sin(j\pi x_f)\left[\sqrt{1+u'(t)} - \frac{\tau}{2}\frac{\partial u'/\partial t}{(1+u')^{1/2}} - 1\right] = F_j \qquad \text{...(18)}$$

The above set of equations can be written in the following vector form

$$\frac{d\tilde{\eta}}{dt} = \tilde{F}(\tilde{\eta}); \text{ where } \tilde{\eta} = \left[\eta_1\ \dot{\eta}_1\ \eta_2\ \dot{\eta}_2 \ldots\ldots \eta_N\ \dot{\eta}_N\right]^T \text{ and } \tilde{F}(\tilde{\eta}) = \left[0\ F_1\ 0\ F_2 \ldots\ldots 0\ F_N\right]^T \qquad \text{...(19)}$$

The dissipative or conservative nature of the system can be studied by calculating the divergence of the forcing term. The dissipative nature of the system indicates that the energy of the system will decay. However, for a conservative system, the energy remains constant; i.e., the oscillations neither grow nor decay when the system is conservative. The divergence of the forcing function can be expressed as:

$$\nabla.\tilde{F} = -\sum_j\left[2\xi_j\omega_j - \frac{k\tau\cos(j\pi x_f)}{2\sqrt{1+u'(t)}}\omega_j\sin(j\pi x_f)\right]; \nabla_j \equiv \frac{\partial}{\partial\tilde{\eta}_j} \qquad \text{...(20)}$$

It is clear that when both time lag and damping are absent, $\nabla.\tilde{F} = 0$. Hence, the system is conservative when damping and time lag are absent. Since the system is conservative, the volume occupied by a set of points in the phase space remains invariant as the system evolves. Further, if a system is conservative, then the trajectory will lie a *2N* – dimensional hypersurface which is given by the following equation:

$$\sum_{j=1}^{N} \frac{\cos(j\pi x_f)}{2\beta_j}\left[\left(\frac{d\eta_j}{dt}\right)^2 + k_j^2\eta_j^2\right] + \frac{2}{3}\left(\frac{1}{3} + u_f'\right)^{3/2} - \frac{u_f'}{\sqrt{3}} = E_0 \qquad ...(21)$$

Further, during different parts of the evolution of the thermoacoustic oscillatins, $\nabla.\tilde{F}$ can have positive or negative sign depending upon the time-lag and the damping factor. This leads to a wide variety of dynamical behavior as $\nabla.\tilde{F}$ can change sign during the evolution process itself. When $\nabla.\tilde{F}$ is negative, the volume occupied in the phase space will reach zero after sufficiently long time. This feature can lead to saturation of the oscillations by reaching a limit cycle behavior if there is exponential separation of trajectories.

The energy growth rate is given by,

$$\frac{dE}{dt} = -\sum_j \beta_j \frac{2}{3}\left[\frac{1}{3} + u_f'\right]^{3/2} + \sum_j \beta_j \left[\frac{1}{3}\right]^{1/2} u_f' - \int \left[\frac{d\eta_j}{dt}\right]^2 dt + \tau \int \left[\frac{d\eta_j}{dt}\right]\left[\frac{du_f'}{dt}\right] dt \qquad ...(22)$$

where *E* is the acoustic energy. Hence the energy decreases if the time lag is less than the following critical value:

$$\tau_{cr} = \max_t \frac{\sum_j \beta_j \frac{2}{3}\left(\frac{1}{3} + u_f'\right)^{3/2} - \beta_j \sqrt{\frac{1}{3}} u_f' + \int \left(\frac{d\eta_j}{dt}\right)^2 dt}{\sum_j \int \left(\frac{du_f'}{dt}\right) \frac{d\dot{\eta}_j}{dt} dt}$$

The role played by non-normality and damping on the thermo-acoustic interaction can be studied through the following examples. The values used for various parameters such as time lag, wire temperature and base flow velocity used for the current study are adopted from Heckl[5]. The damping model used for the current analysis is similar to that of Matveev[8]. The Mach number of the steady base flow is chosen as $1.5 \times 10^{-4}$. The numerical simulations were performed by keeping some of the parameters fixed and varying others. The parameters that were kept fixed are $\lambda$ = 0.0328 W/mK, $C_v$ = 719 kJ/kgK, and $\bar{\rho}$ = 1.205 kg/m$^3$. The set of first order equations in (19) was integrated numerically using the 4$^{th}$ order adaptive Runge Kutta technique. The numerical simulations were performed with six acoustic modes so that the change in the solution with increase in number of modes is less than 5%.

## A. Triggering

If the system is non-normal as well as nonlinear, oscillations can grow even when the individual eigenvalues indicate linear stability. For such systems, there exists some initial condition for which the oscillations decay and some other initial conditions for which they grow.

Figures (4a) and (4b) show triggering in the absence of damping. Figure (4a) shows that the oscillations decay, whereas figure (4b) shows that the oscillations grow if the initial conditions are different. Further, figure (4b) shows that the amplitude of the oscillations saturates. Hence, this example shows that saturation can occur in the absence of damping. This interesting feature was discussed by Balasubramanian and Sujith[10] in the context of diffusion flames. The authors have explained that saturation can occur in the absence of damping if the phase difference between the acoustic oscillations and the heat release oscillations evolves to 90°[21]. The phase between pressure and heat release rate is the correlation between the acoustic pressure and heat release rate. This is confirmed from the evolution of phase to 90° as the system approaches limit cycle, as seen in figure 4(c). The evolution of the non-dimensional acoustic energy $\frac{1}{2}p'^2 + \frac{1}{2}(\gamma M u')^2$ after locally weighted regression smoothing (LOWESS) calculated from both linear (integration of Eq. 17 without nonlinear terms) and nonlinear simulations (with nonlinear terms included) are presented in Figure 4(d). The linear simulation shows that the acoustic energy eventually grows initially and eventually decays; however, the nonlinear simulation shows that the acoustic energy grows and saturates. This shows that transient growth causes nonlinear driving to occur which causes the growth of oscillations even though the system has stable eigenvalues.

The following example demonstrates triggering in the presence of damping. In this example, triggering is demonstrated in a Rijke tube with a heat element located at 0.3 duct length from the inlet. $L_w$ and $T_w$ were chosen as 1.5m and 1000K, respectively and $u_0 = 0.05$ m/s. Figure 5a shows the evolution of acoustic velocity at the heating element for the initial condition $\eta_1(0) = 0.05$, $\eta_{i\neq 1}(0) = 0$ and $\dot{\eta}_i(0) = 0$. The damping coefficents $c_1$ and $c_2$ were chosen as 0.135 and 0.05 respectively. It can be seen that the acoustic velocity grows for a short time and decays eventually. Figure 5b shows the variation of acoustic velocity for the system with a different initial condition; i.e., $\eta_1(0) = 0.09$, $\eta_{i\neq 1}(0) = 0$ and $\dot{\eta}_i(0) = 0$. The acoustic velocity in this case grows and saturates. This indicates that the amplitude grows to a value sufficient to trigger nonlinearity. It can be seen from the two figures that the acoustic velocity in figure (5a) grows to a value which is lesser than 1/3 of the base flow, whereas the acoustic velocity in figure (5b) grows to a value greater 1/3 of the base flow causing the nonlinearity to get triggered. The evolution of the non-dimensional acoustic energy calculated from both linear and nonlinear simulations are presented in figure 2. It shows that there is linear growth of the acoustic oscillations for a short period and it decays eventually. The fact that the nonlinear simulation matches the linear simulation initially indicates that the linear growth of oscillations caused the evolution of the amplitudes to a high enough value (1/3 of the base flow) where nonlinearities in the system to "pick up".

The phenomenon of triggering has been observed in experiments with Rijke tube[9], and also in other thermoacoustic devices such as solid rocket motors[22, 23]. This is usually attributed to nonlinearities. However, as shown in this example, this is not the complete picture. This example

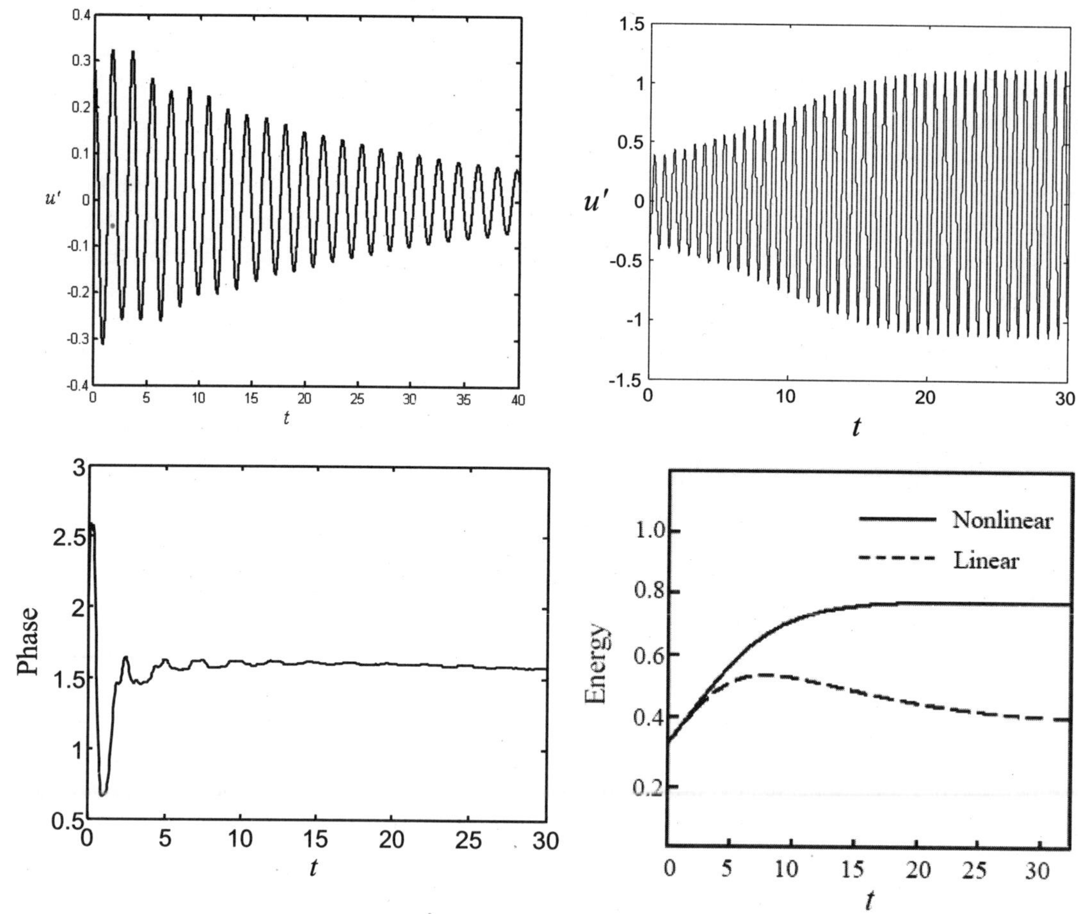


**Figure 4** Shows evolution of acoustic velocity at the wire gauze located at $x_f = 0.3$ for a case with the initial condition (a) $\eta_1(0) = 0.6$, $\eta_{i\neq1}(0) = 0$ and $\dot{\eta}_i(0) = 0$; (b) $\eta_1(0) = 0.7$, $\eta_{i\neq1}(0) = 0$ and $\dot{\eta}_i(0) = 0$. (c) shows that the phase corresponding to the initial condition in (b) evolves to 90 degrees. (d) shows the evolution of the non-dimensional acoustic energy $\left(\frac{1}{2}p'^2 + \frac{1}{2}(\gamma M u')\right)^2$ from linear and nonlinear simulations (for the initial conditions in (b). This simulation is performed in the absence of damping. $\bar{\rho} = 1.205$ kg/m$^3$, $\bar{u} = 0.05$ m/s, $L_w = 1.5$ m, $\lambda = 0.0328$ W/mK, $C_v = 719$ kJ/kgK

shows that non-normal nature of the evolution equations is indeed responsible for triggering the nonlinearity in the system.

## B. Growth of Oscillations in an Initially Decaying System

This section discusses bootstrapping in a thermoacoustic system which is stable according to classical linear stability analysis based on eigenvalues. In this example, the heater is located at ¼ duct length. The initial conditions chosen are $\eta_1(0) = 0.02$, $\eta_{i\neq1}(0) = 0$ and $\dot{\eta}_i(0) = 0$. $L_w$ and $T_w$

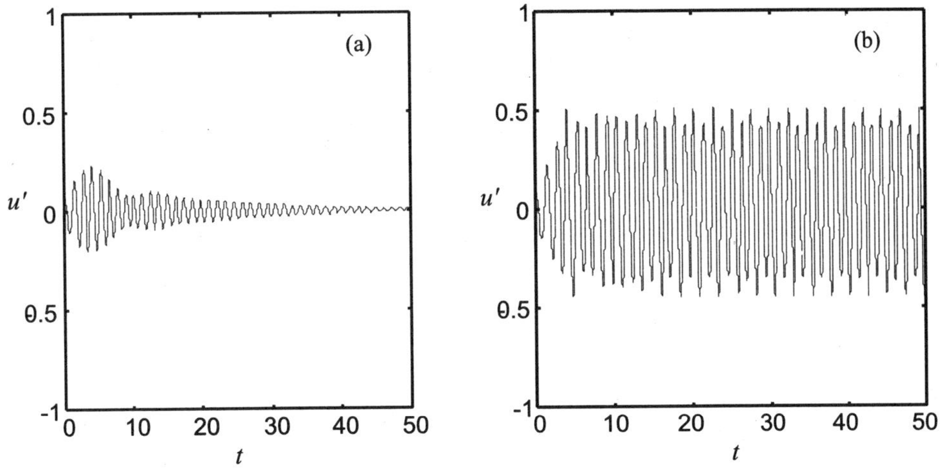


**Figure 5** Shows the evolution of non-dimensional acoustic velocity when the initial condition is a) $\eta_1(0) = 0.12$; b) $\eta_1(0) = 0.17$. $\tau = .002$, $x_f = 0.3$. The constants used for damping are $c_1 = 0/.135$ and $c_2 = 0.015$. $c_0 = 399.6$ m/s, $\lambda = 0.0328$ W/mK, $L_w = 1.5$ m, $\bar{u} = 0.05$ m/s, $\bar{\rho} = 1.205$ kg/m$^3$, $C_v = 719$ kJ/kgK, $T_w = 1000$ K

were chosen as $1.5m$ and $1000K$, respectively. The damping coefficients are $c_1 = 0.05$ and $c_2 = 0.005$ the heater is located at quarter duct.

Figure 6a shows the evolution of acoustic velocity at the heater location. It can be seen that low frequency oscillations that are initially present in the system decays and high frequency oscillations sets in after some time. Further, it can be seen that the oscillations eventually saturate after nonlinear growth. It must be emphasized that classical linear stability analysis based on the eigenvalues shows all eigenmodes of this coupled system to be stable.

Figures 6 b-d show the evolution of the acoustic velocity projected on the first three Galerkin expansion functions. It can be seen that while the acoustic velocity projected to the first expansion function decays, the projections on the second and third expansion functions grow. After sufficient energy is projected onto the second and third expansion functions, they project the energy back causing the energy projected on the first expansion function to grow. This bootstrapping results in a shift in frequency during the evolution. The net effect of all these energy transfer causes the acoustic velocity to grow and eventually saturate. This feature has been discussed in the context of turbulence and it is known as "bootstrapping"[13,14]. Yoon, et al.[3] has discussed "bootstrapping" in the context of Rijke tube using an ad hoc nonlinear model for the heat release rate.

There are two mechanisms by which energy can get redistributed in the system. The first mechanism is a nonlinear mechanism where two individual eigenmodes interact directly causing exchange of energy between the two modes. Another mechanism which causes redistribution of energy is the interaction of various modes with the base flow. At low amplitudes when the nonlinear effects are not significant, the redistribution of energy mainly occurs due to the interaction of various modes with the base flow. When the disturbance is caused by exciting only one eigenmode, it is expected that the oscillations will decay if the system is linearly stable. However, if the system is non-normal, then the oscillations can grow if there is a small amount of energy in the unexcited

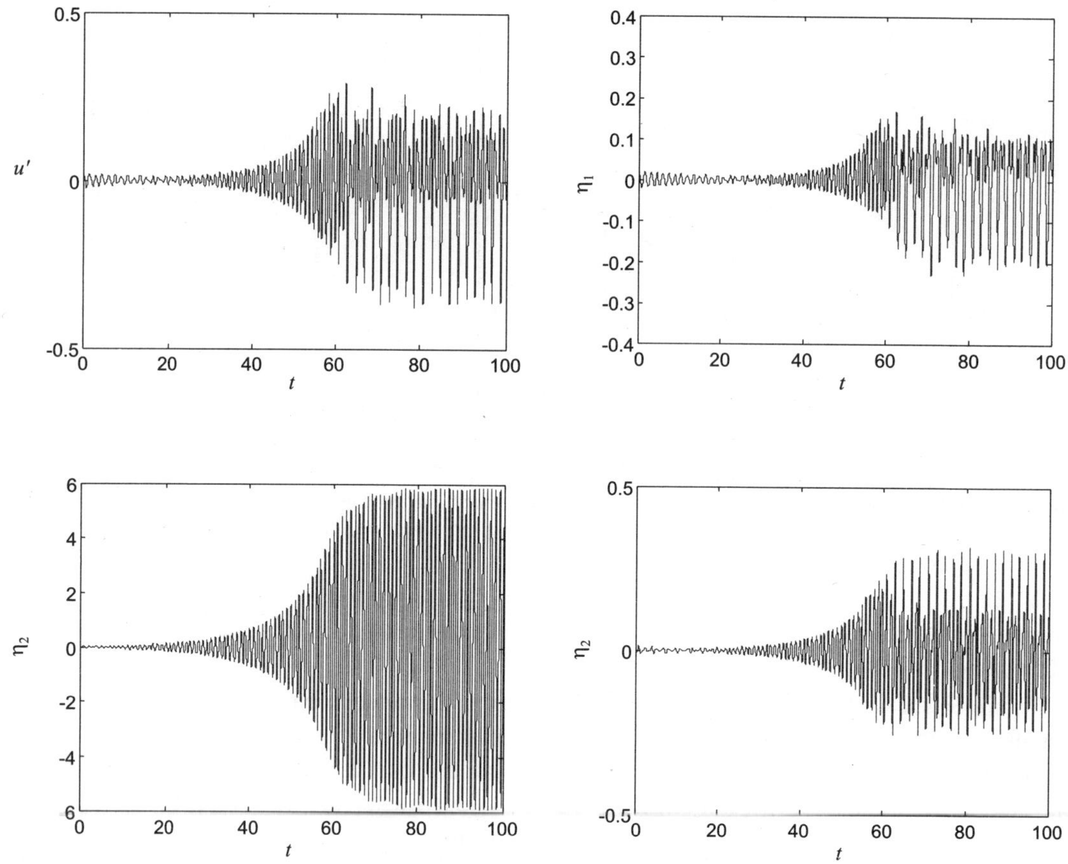


**Figure 6** a) shows the non-dimensional evolution of acoustic velocity when the initial condition is h1(0) = 0.02; (b)-(d) shows the evolution of the acoustic velocity projected onto the various Galerkin modes. $L_w$ = 1.5 m, $\lambda$ = 0.0328 W/mK, $C_v$ = 719 kJ.kgK, $\bar{\rho}$ = 1.205 kg/m$^3$, $\bar{u}$ = 0.05 m/s, $T_w$ = 1000 K, $\tau$ = 0.002, $x_f$ = 0.25, $c_0$ = 399.6 m/s

modes. This could be due to noise in the system or due to a mild nonlinearity. The authors would like to emphasize that the mild nonlinearity just transfers some energy from the excited modes to other modes. Though, the individual modes can decay there can be an overall growth of amplitude due to non-normality. This overall growth can cause nonlinear effects to become significant and more energy can get exchanged between the various modes. Such a situation cannot occur in a linearly stable normal system (classical linear stability) if the nonlinearity is initially mild. This is because the amplitudes of a normal system decay if the individual eigenmodes decay and hence nonlinearity becomes milder and milder. This explains how non-normality plays an important role in the exchange of energy between various eigenmodes.

## C. Transient Growth

As discussed earlier, the non-normal nature of a system can cause transient growth of oscillations which can trigger nonlinearities in the system. In this section, a method to analyze transient growth is discussed. Schmidt and Henningson[24] give a detailed discussion on the analysis of transient growth, in the context of transition to turbulence in shear flows. They analyzed the stability of shear flows by studying the energy growth of the system. This analysis is general and can be applied to thermoacoustic systems as well. Balasubramanian and Sujith[10] performed such an analysis in the context of diffusion flames. The evolution of the acoustic oscillations in a Rijke tube is also non-normal as $B_{NN}B_{NN}^{\dagger} \neq B_{NN}^{\dagger}B_{NN}$. The symbol $^{\dagger}$ denotes the adjoint (conjugate transpose) of an operator. The solution of the system of evolution equations in (17) can be written in the operator form as [24]:

$$\tilde{\eta}(t) = \exp(Lt)\tilde{\eta}(0) = S^{-1}\exp(L_D t)S\tilde{\eta}(0) \qquad ...(23)$$

where, $L = -B_{NN}$ is the stability operator, $S$ is the similarity transformation that diagonalizes L, $L_D$ is the diagonal form of $L$. Since $L$ is non-normal, $S$ is non-unitary indicating that it is not possible to make the eigenvectors perpendicular by a simple rotation. Transient growth is quantified by maximum growth factor which is defined as[24],

$$G(t) = \max_{\eta}\left(\|\tilde{\eta}(t)\|^2 / \|\tilde{\eta}(0)\|^2\right) = \|\exp(Lt)\|^2 \qquad ...(24)$$

where, 'max' indicates that the ratio of the norms is maximized over all initial conditions. Growth factor is a measure of maximum amplification of energy density at an instant of time. The expression in (24) is maximized for various instants of times over all possible initial conditions. This is achieved through singular value decomposition. The stability of a system can be studied using the maximum growth function in a particular time interval [0, $t$] which is defined as $G_{max} = \max_t G(t)$.

The above maximum value is infinite if $L$ has an eigenvalue with a positive real part. This corresponds to a linearly unstable system. In the present paper, $G_{max}$ values for various parameters such as time lag and flame location are calculated and the regions with large transient growth are identified. Transient growth cannot occur for those parameters which have $G_{max} = 1$. When $G_{max} = 1$, the energy at any instant is less than the initial energy of the system indicating that the energy of the system decays. When $G_{max} > 1$, the system will exhibit transient growth. Hence, a system which is linear initially will behave like a linear system throughout its evolution as there is no amplification of the oscillations to trigger nonlinear effects, when $G_{max} = 1$. This fact is used in the next section to obtain necessary and sufficiency conditions for the stability of the system.

Figure (7) shows the variation of growth factor with time lag for different locations of the heating element. The step-like behavior of the variation indicates that the growth factor is a non-monotonic function of time lag. This is due to the fact that the decay rates and phase are different for different eigenmodes, even when the time lag is same. However, it can be seen that when the heating element is located at 0.2 duct length, the growth factor drops linearly. Further the growth factor is of the order of 10. The authors would like to point out that in the present analysis a simple model for the heating element is used. This simplified approach was used to focus on the non-normal nature of the acoustic equations alone in the presence of heat source. However, the energy released at the heating element is governed by an advection diffusion equation. It has been shown that the advection-diffusion equation is non-normal[25, 26]. This will cause the growth factor to be much larger as the number of eigenmodes of the coupled system will be much larger.

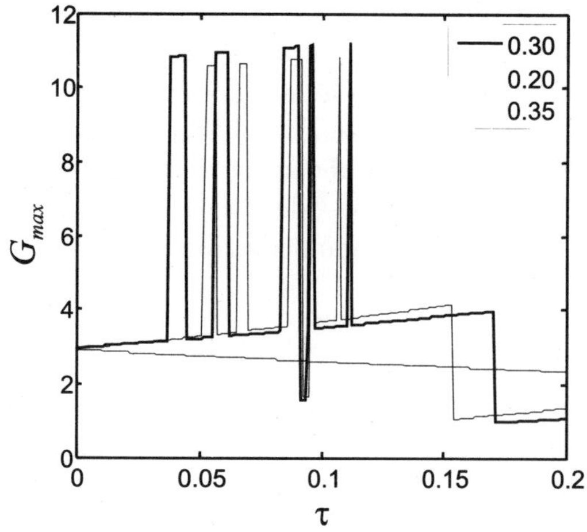


**Figure 7** Shows the variation of maximum growth factor with time lag for different locations of the heating element. $L_w$ = 2.0 m, $\lambda$ = 0.0328 W/mK, $C_v$ = 719 kJ/kgK, $\bar{\rho}$ = 1.205 kg/m$^3$, $c_0$ = 401.2 m/s, $\bar{u}$ = 0.6 m/s, $T_w$ = 900 K

## D. Pseudospectra

The degree of resonant amplification that may occur in a normal system in response to an input frequency is inversely proportional to the distance in the complex plane between the input frequency and nearest eigenvalues. However, in the case of a non-normal operator, the resonant amplifications may be orders of magnitude larger[13] and cannot be determined by the eigenvalues alone. Such a resonance of a non-normal system is known as pseudoresonance.

The concept of *e*-pseudoeigenvalues can be used to analyze the behavior of evolution governed by such non-normal operators[27]. $z$ is an $e$ pseudoeigenvalue of $A$ if it satisfies $\left\|\left(zI - A^{-1}\right)\right\| \geq \varepsilon^{-1}$. There are other equivalent definitions of pseudo eigenvalues and they have been discussed in great detail by Trefthen and Embree[24]. In the above expression, $e$ is a measure of the distance of the forcing frequency from the eigenvalues of the system. This can also be interpreted as the noise-level in a system when there is no external forcing. It should be noted that the system discussed in this paper is a self-excited system. In a linearly stable self-excited system, if the initial condition is along one of the eigenmodes, then there is no growth and the oscillations will decay throughout the evolution. However, in most situations, the initial conditions do not correspond to any particular eigenmode and hence transient growth can occur. In these situations, ε is a measure of the deviation of the initially excited mode from the eigenmodes of the system.

Transient growth and the non-normal nature of the operator can be studied using pseudospectra. The pseudospectra of normal operators are closed circles. When a contour corresponding to some ε value does not lie entirely in the left-half plane, the system exhibits transient growth causing the amplitudes to increase to high values[13]. Further, it is possible to obtain necessary and sufficiency conditions for an oscillation to be stable based on the geometry of pseudospectra.

Trefetehen *et al.*[13] have used the relation between the geometry of the pseudospectra and the lower bound on the transient growth factor to analyze hydrodynamic instability in Couette and Poiseuille flows. The lower bound on the transient growth factor is given by

$$\max_t \| e^{tA} \| \geq \max_{\varepsilon} \frac{\max\limits_{z\in\Gamma(\varepsilon)} \mathrm{Re}(z)}{\varepsilon} \qquad ...(24)$$

This inequality serves as a necessary condition. If a contour protrudes far into the right half plane, then the system will exhibit large transient growth. Hence, for a system not to grow (either transient or exponential), pseudospectra must lie entirely on the left half plane.

Similarly, a sufficiency condition for a system to be stable can be obtained by relating the upper bound on transient growth to the geometry of pseudospectra. The upper bound on the transient growth factor of a non-normal operator can be obtained from the pseudospectra as follows[27]. The exponential of an operator $A$, $e^{tA}$ can be defined by the Dunford-Taylor integral (opearator analogue of Cauchy integral),

$$e^{tA} = \frac{1}{2\pi i}\int_{\Gamma} (z-A)^{-1} e^{tz} dz \qquad ...(25)$$

where $\Gamma(\varepsilon)$ is the boundary of pseudospectra corresponding to some $\varepsilon$. Hence the norm of the evolution operator is bounded by the Cauchy integral of $|e^{tz}|\|(z-A)^{-1}\|$. When $\Gamma$ encloses the $\varepsilon$-pseudospectra then the upperbound for the transient growth factor can be written as:

$$\| \exp(tA) \| \leq \left[ \frac{L_\varepsilon}{2\pi\varepsilon} \max_{z\in\Gamma(\varepsilon)} |\exp(tz)| \right] \qquad ...(26)$$

where, $L_\varepsilon$ is the length of the contour (or convux hull) $\Gamma(\varepsilon)$.

The following sufficiency condition for a thermoacoustic system to be stable can be obtained by choosing $A$ as the stability operator $L$. The upper bound on the maximum growth rate is given by,

$$G_{\max} = \max_t \| \exp(tL) \|^2 \leq \max_t \left[ \frac{L_\varepsilon}{2\pi\varepsilon} \max_{z\in\Gamma(\varepsilon)} |\exp(tz)| \right]^2 \qquad ...(27)$$

Hence a system is stable if the real parts of all the eigenvalues are negative and if the right hand side of the above expression is 1. Hence, if the pseudospectra lies entirely in the left half plane and if $G_{\max} = 1$, then the system will not grow for any initial condition.

Rayleigh criterion gives the condition for acoustic driving to occur. However, the prediction of transient growth by Rayleigh criteria requires the precise knowledge of initial conditions. The ambiguity of initial conditions due to noise makes the identification of transient growth using Rayleigh criteria difficult. However, the necessary and sufficiency conditions obtained in this paper are conditions on the evolution operator and hence do not depend on the initial conditions.

Figure (8) shows the pseudospectra for a Rijke tube with length 1m and time lag $\tau = 0.05$. It is clear from the non-circular behavior that the system is non-normal. It is clear that the contour is not

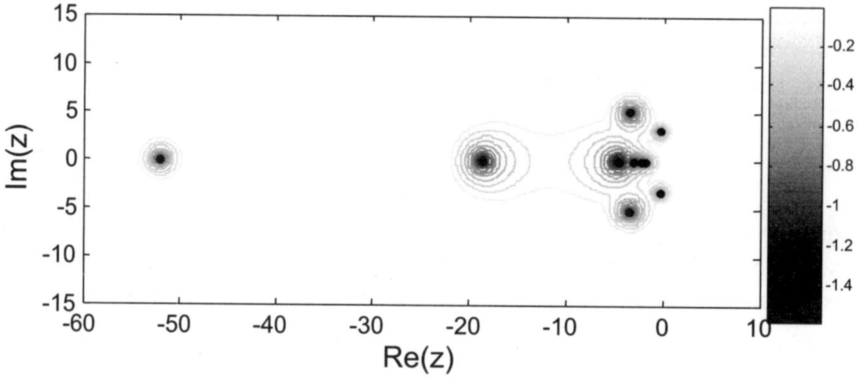


**Figure 8** Shows the pseudospectra of a Rijke tube in which the heater is located at quarter duct ($x_f = 0.3$). The time lag was chosen as $\tau = 0.05$, $L_w = 2.0$ m, $\lambda = 0.0328$ W/mk, $C_v = 719$ kJ/kgK, $c_0 = 401.2$ m/s, $\bar{\rho} = 1.205$ kg/m3, $\bar{u} = 0.6$ m/s, $T_w = 900$ K

entirely on the left half plane. Hence, it can be inferred from Eq. (24) that this system shows transient growth. This indicates that there is an "unstable" pseudo-eigenvalue for some ε. Even if a system behaves linear, the transient growth can cause the amplitudes to reach high values and trigger nonlinearities which can cause the oscillation to grow further.

## Conclusion

In this paper, the nature of thermoacoustic interaction is studied in the context of a simple model for a horizontal Rijke tube. The coupled thermoacoustic system is non-normal and hence the eigenvectors are non-orthogonal. Non-normality can lead to short time growth of oscillations before they eventually decay, even when the system is linearly stable. Transient growth can cause the amplitudes of the oscillations to trigger "nonlinear" driving. It has been observed that the various eigenmodes of the coupled thermoacoustic system interact resulting in the growth of oscillations, even when the eigenvalues indicate stability. The current methodology to study the onset of thermoacoustic oscillations involves looking for exponentially growing or decaying modes by calculating the individual eigenvalues of the linearized system. The role of non-normality has not been considered in predicting the stability margins of a system.

The stability of a system can be studied by calculating the transient growth factor. The system is conservative when both time lag and damping are absent, while it is non-conservative if time lag is present. The dissipative nature of the system changes as the system evolves, leading to a complex behavior. The growth factor shows a non-monotonic step-like variation with time lag for some locations of the heating element while it drops linearly for other locations. Further, a critical value for time lag was obtained below which the energy of the oscillations cannot grow.

For a non-normal system, "pseudoresonance" can occur at frequencies far from the spectrum and pseudospectra can be used to analyze such systems The pseudospectra of normal operators are disjoint circles. The pseudospectra of the thermoacoustic system considered in this study are non-circular implying a highly non-normal nature of the system. Estimates of transient growth factors can be obtained by studying the geometry of the pseudospectra. The relation between the bounds of transient growth factor and the geometry of the pseudospectra provide necessary and sufficient conditions for the stability of a system. If the pseudospectra are not entirely within the left-half plane, then the system can show transient growth. It is possible to identify systems which cannot show transient growth by analyzing the pseudospectra.

## Acknowledgement

This work was funded by the Department of Science and Technology. The authors wish to thank Prof. W. Polifke (Technishe Universität München), S. Muralidharan (Princeton University) Prof. R. Govindarajan (Jawaharlal Nehru Centre for Advanced Scientific Research, Bangalore) for their critical comments and suggestions.

## References

1. K. Mcmanus, T. Poinsot, and S. M. Candel "A Review of Active Control of Combustion Instabilities", *Progress in Energy and Combustion Sciences*, **19**, 1 (1993).
2. J. W. S. Rayleigh "The Explanation of Certain Acoustic Phenomena", *Nature*, **18**, 319 (1878).
3. H. G. Yoon, J. Peddieson and K. R. Purdy "Nonlinear Response of a Generalized Rijke Tube", *International Journal of Engineering Science*, **39**, 1707 (2001).
4. B. T. Zinn, and T. C. Lieuwen, Combustion Instabilities: Basic Concepts, Chapter 1, Combustion Instabilities in Gas Turbine Engines: Operational Experience, Fundamental Mechanisms, and Modeling, Edited by T. C. Lieuwen and V. Yang, **210** *Progress in Astronautics and Aeronautics*, AIAA, Inc (2005).
5. M. A. Heckl "Nonlinear Acoustic Effects in the Rijke Tube", *Acustica*, **72**, 63 (1990).
6. C. C. Hantschk and D. Vortmeyer "Numerical Simulation of Self-Excited Thermo-acoustic Instabilities in a Rijke Tube", *Journal of Sound and Vibration*, **277**, 511 (1999).
7. K. I. Matveev "Energy Consideration of the Nonlinear Effects in a Rijke Tube", *Journal of Fluids and Structures* **18**, 783 (2003).
8. K. I. Matveev "Thermo-acoustic Instabilities in the Rijke Tube: Experiments and Modeling", Ph. D. Thesis, California Institute of Technology (2003).
9. K. I. Matveev and F. E. C. Culick "A study of the Transition to the Instability in a Rijke Tube with Axial Temperature Gradient", *Journal of Sound and Vibration*, **264**, 689 (2003).
10. K. Balasubramanian and R. I. Sujith "The Role of Non-Normality in Combustion-Acoustic Interactions in Diffusion Flames", *AIAA-2007-0567*, $45^{th}$ Aerospace Sciences Meeting and Exhibit, Reno, Nevada, USA, Jan 8-11 (2007).
11. F. Nicoud, L. Benoit, C. Sensiau and T. Poinsot "Acoustic Modes in Combustors with Complex Impedances and Multidimensional Active Flames", AIAA J. **45**, 426 (2007)
12. J. S. Baggett, T. A. Driscoll and L. N. Trefethen "A Mostly Linear Model of Transition to Turbulence", *Physics of Fluids*, **7**, 833 (1995).
13. L. N. Trefethen, A. E. Trefethen, S. C. Reddy and T. A. Driscoll "Hydrodynamic Stability without Eigenvalues", *Science*, **261**, 578 (1993).
14. T. Gebgart and S. Grossmann "Chaos Transition Despite Linear Stability", *Physical Review E*, **50**, 3705 (1994).
15. W. Kerner "Large Scale Complex Eigenvalue Problems", *Journal of Computational Physics*, **85**, 1 (1989).

16. B. F. Farrell "Optimal Excitation of Baroclinic Waves", *Journal of Atmospheric Sciences*, **46**, 1193 (1989).
17. D. S. Henningson and P. J. Schmid, "Vector Eigenfunctions for Plane channel Flows", *Studies in Applied Mathematics*, **87**, 15 (1992)
18. J. Kopitz and W. Polifke, "CFD Based Analysis of Thermoacoustic Instabilities by Determination of Open-Loop-Gain", *Twelfth International Congress on Sound and Vibration*. Lisbon (2005).
19. L. Meirovitch, *Analytical Methods in Vibrations*, Macmillan Publishing Co., Inc., New York, 1967, Chapter 6.
20. A. P. Dowling "The Calculation of Thermoacoustic Oscillations", *Journal of Sound and Vibration* **180** (4) 557 (1995).
21. W. Polifke (Private Communication)
22. N. Ananthkrishnan, S. Deo and F. E. C. Culick "Reduced-Order Modeling and Dynamics of Nonlinear Acoustic Waves in a Combustion Chamber", *Combustion Science and Technology,* **177**, 221 (2005).
23. J. M. Wicker, W. D. Greene, S. I. Kim and V. Yang "Triggering of Longitudinal Combustion Instabilities in Rocket Motors: Nonlinear Combustion Response", *Journal of Propulsion and Power*, **12**, 1148 (1996).
24. P. J. Schmid and D. S. Henningson, Stability and Transition in Shear Flows. Springer (2001).
25. S. C. Reddy, and L. N. Trefethen, Pseudospectra of the Convection-Diffusion Operator, SIAM Journal of Applied Mathematics, **54**, 1634 (1994).
26. L. N. Trefethen, "Pseudospectra of Linear Operators", *SIAM Review*, 39, 383 (1997).
27. L. N. Trefthen and M. Embree, Spectra and Pseudospectra, Princeton University Press, Princeton and Oxford. (2005)

# Experimental and Numerical Investigation of Fundamental Aspects of Combustion Dynamics

**S.R. Chakravarthy**
Department of Aerospace Engineering, Indian Institute of Technology – Madras, Chennai – 600 036

## INTRODUCTION

Most of the work on combustion instability in the last decade or two has focussed on premixed flames [1-10]. This has been done in the context of stationary gas turbine engines for power generation, which are required to follow stringent emissions norms, since they are land-based. For this purpose, the combustors of these power-plants have been operated in the lean premixed pre-vaporized mode. The accompanying feature is one of combustion instability that plagues these combustors. However, flight gas turbine engines operate in a manner where the fuel is injected at the core of a swirl flow region located downstream of a sudden expansion region. The fuel mixes with the swirl flow of air as it gets atomized and vaporizes, eventually burning in a partially premixed flame. Partially premixed flames can be viewed as non-premixed flames with finite Damköhler number. Therefore, it is important to study the dynamics of non-premixed flames as well.

In non-premixed flame, an additional time-scale, namely that of fuel-air diffusion comes into play in causing the delay between the heat release rate fluctuations and prevalent acoustic velocity fluctuations. The fuel-air mixing follows the flow field fluctuations, similar to the way the flame area fluctuations do in the case of premixed flames. However, in relatively complex geometries such as gas turbine engine combustors, the recirculation of hot combustion products complicates the fuel-air mixing process. Hence, it is important to understand first the mechanics of the oscillatory flow field before examining the flame response in such geometries.

To study these aspects with clarity, it is important to reduce the complexity of the combustor geometry to a simpler one, but retaining the essential features of the flow field. Two levels of simplification of the typical gas turbine combustor geometry are shown in Figure 1(a) and (b). In general, combinations of one or more of the following three constructs are employed in combustors for flame stabilization: (i) sudden expansion/dump plane/backward-facing step, (ii) bluff-body/V-gutter, (iii) swirl flow (Figure 2). While the first and third are combined in the case of a gas turbine

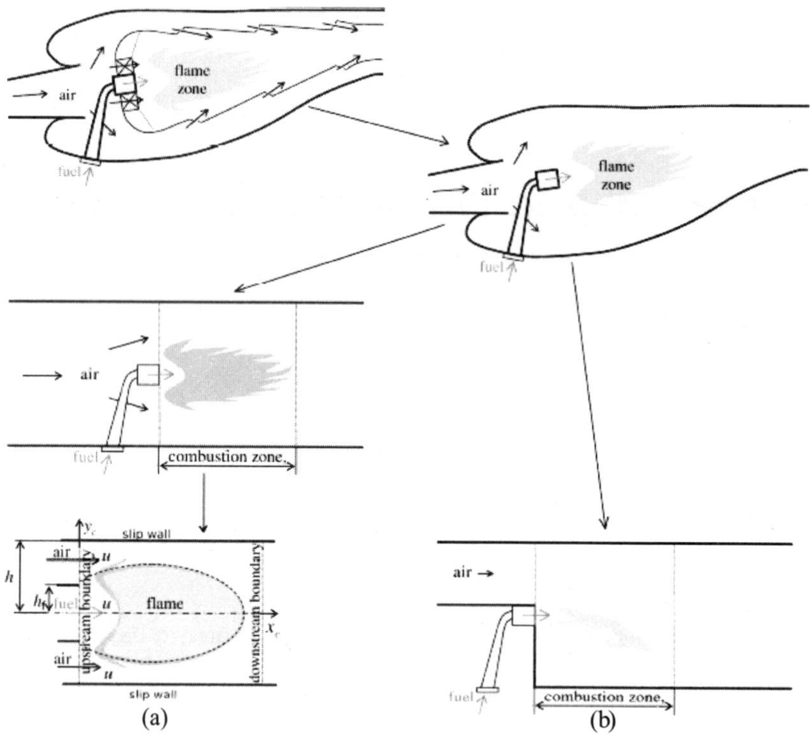


**Figure 1** Simplification of a gas turbine combustor geometry

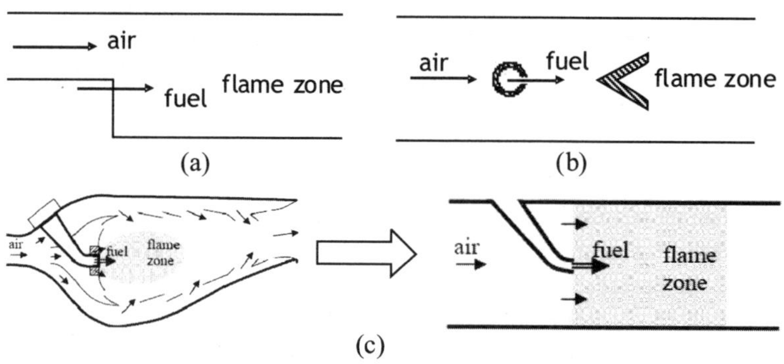


**Figure 2** Typical flame stabilization configurations in combustors

primary combustor, the first two are adopted in afterburner and ramjet combustors. Flow in these geometries involves shear layer fluctuations such as vortex roll-up similar to the Karman vortex roll-up in the wakes of bluff-bodies, and Kelvin-Helmholtz instability. Heat release rate fluctuations accompany such fluid dynamic fluctuations, and these heat release rate fluctuations excite acoustic oscillations in the combustion chamber. The acoustic oscillations, in turn, influence the unsteady fluid dynamics of the shear layers.

In the present paper, the backward-facing step and bluff-body geometries are experimentally investigated. Numerical investigation is confined to the backward-facing step geometry, and starts by considering the Burke-Schumann geometry, which is still simpler.

## EXPERIMENTAL DETAILS

The experimental set-up typically consists of a rectangular duct whose length can be varied by the use of extension ducts. One of these sections is a test section that contains either a backward-facing step or a V-gutter bluff-body. The fuel is injected at the backward-facing step, but through a manifold some distance upstream of the V-gutter, as in the case of practical afterburner combustors. The schematics of the two set-ups are shown in Figure 3.

The inlet air flow Reynolds number is varied in the 6000-60000 in the case of the backward-facing step combustor and 8000-80000 in the bluff-body combustor. The fuel-air equivalence ratio is varied in the range of 0.01-0.13 in both the cases. Methane is used as fuel in both. Besides the length of the duct, the axial location of the test section, the distance between the fuel injection manifold and the V-gutter, and the width of the bluff-body are varied.

Acoustic pressure measurements at various locations on the combustor are performed by piezo-electric transducers. High-speed chemiluminescence imaging is performed by a high-speed camera with 256 x 256 pixel resolution and a maximum framing rate of 2250 frames per second. An optical filter centred round 431 nm is used in this imaging. Particle image velocimetry (PIV) is performed

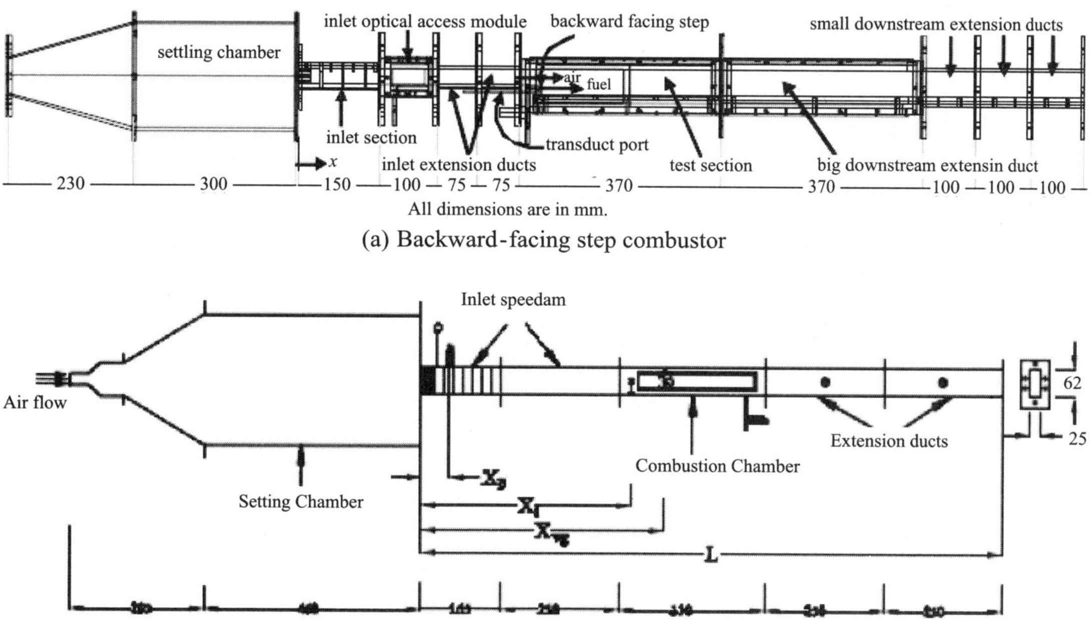


(a) Backward-facing step combustor

(b) Bluff-body combustor

**Figure 3** Schematics of experimental set-ups

with a double-pulsed Nd:YAG laser and an interline-transfer charge-coupled device (CCD) camera. MgO is used as seeding particles in the combustion experiments, whereas fine droplets of peanut oil are used in the cold flow experiments.

## EXPERIMENTAL RESULTS

### Backward-Facing Step Combustor

Figure 4 shows the dominant frequencies observed in the acoustic spectra measured at the base of the backward-facing step for different lengths of the combustor, at two different extreme fuel flow rates tested, in terms of the Helmholtz number *He* and the Strouhal number *St*. The Helmholtz number is defined as $He = fL/c$, where $f$ is the observed frequency, $L$ is the length of the combustor, and $c$ the speed of sound at a reference condition, the inlet in this case. The Strouhal number is defined as $St = fH/U$, where $h$ is the height of the backward-facing step, and $U$ is the average air-flow velocity at the inlet. The significance of the Helmholtz number is that it remains relatively constant with respect to the flow Reynolds number if the observed frequency is that of the natural acoustic mode of the duct, i.e., without the combustion heat release. The significance of the Strouhal number is that it remains relative constant relative to the flow Reynolds number if the

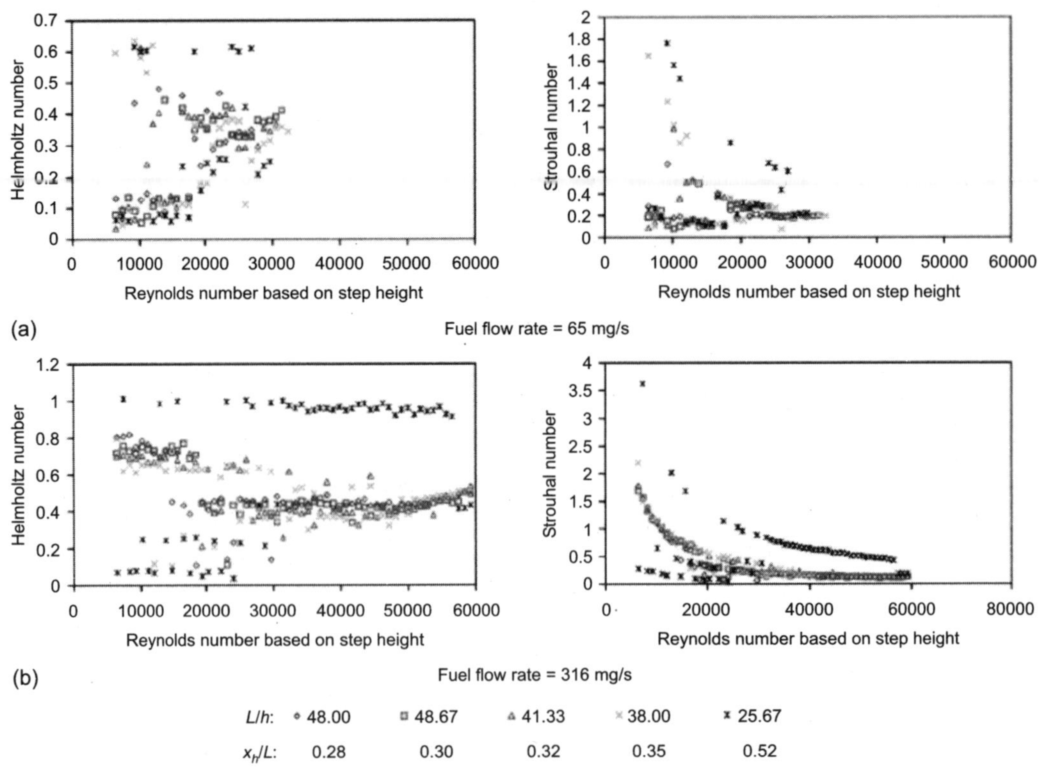


**Figure 4** Helmholtz and Strouhal numbers of dominant frequencies in the acoustic spectra in the backward-facing step combustor

observed frequency is that of the vortex shedding frequency in the recirculation zone downstream of the backward-facing step. On the other hand, the Helmholtz number increases linearly with the air-flow Reynolds number if the observed frequency is the vortex shedding frequency, and the Strouhal number decreases hyperbolically with the Reynolds number if the observed frequency is the natural acoustic mode of the duct. It can be seen in Figure 4 that at low fuel flow rate (Fig. 4a), the Strouhal number decreases hyperbolically, and the Helmholtz number remains constant, signifying that most of the dominant frequencies observed are that of the natural acoustic mode of the duct. On the other hand, at high fuel flow rates (Fig. 4b), the constant Helmholtz number trend transitions to that of a linearly increasing one at high air-flow Reynolds number, where the Strouhal number approaches a constant value after decreasing hyperbolically in the lower Reynolds number range. This signifies the onset of a flow-acoustic lock-on, wherein the dominant observed frequency is that of the vortex shedding in the shear layer downstream of the backward-facing step.

Figure 5 shows the acoustic amplitudes corresponding to the dominant frequencies observed in the previous figure, for two lengths and two fuel-flow rates, at all locations of the transducer in the combustor. It can be seen that in the case of the longer duct at large fuel-flow rate, the amplitude increases nonlinearly to high values at about the same Reynolds number as when the Helmholtz number transitions from a constant trend to a linearly increasing one noted above. This aspect is pursued later in greater detail in this paper.

Figure 6 shows the effect of fuel flow rate in greater detail. As the fuel flow rate is increased, the transition from a constant Helmholtz number to a linearly increasing one occurs at a higher Reynolds number. The corresponding increase in the amplitude also retains this trend.

On the contrary, the dominant frequencies observed under cold flow conditions (Figure 7) show constant trends, signifying the excitation of the natural acoustic mode without lock-on to the vortex shedding in the shear layer. Accordingly, the amplitudes recorded are so low that they have not been shown here. This shows that the heat release rate fluctuations during combustion conditions in the shear layer excite acoustic oscillations of high amplitude which in turn modulate the vortex shedding in the recirculation zone.

Pursuing the case of one dataset of Helmholtz number variation and its corresponding amplitude variation with Reynolds number, as indicated earlier, many mode transitions between constant and linearly increasing Helmholtz number variations can be observed in Figure 8, of which the one marked as transition 3 is accompanied by the nonlinear increase in the amplitude. The conditions corresponding to this transition is recorded in all the cases for which the experiments are performed, and marked in a domain of $St/He = (h/L)(1/M)$, where $M$ is the air-flow Mach number, versus the overall fuel-air equivalence ratio (Figure 9). It can be seen that the onset of this transition follows a hyperbolic trend. This offers a tool to characterize the combustion instability limit of a combustor.

Focus is now turned to the question of whether the lock-on in question is between the duct acoustics and the vortex shedding in the recirculation zone downstream of the backward-facing step or any other shear layer oscillations that exhibit a constant Strouhal number versus Reynolds number too. Figure 10 shows the time-resolved CH* chemiluminescence images over a cycle of dominant oscillations. It can be seen that prominent vortex roll-up can be observed in the recirculation zone downstream of the backward-facing step. Figure 11 compares the spectra of the

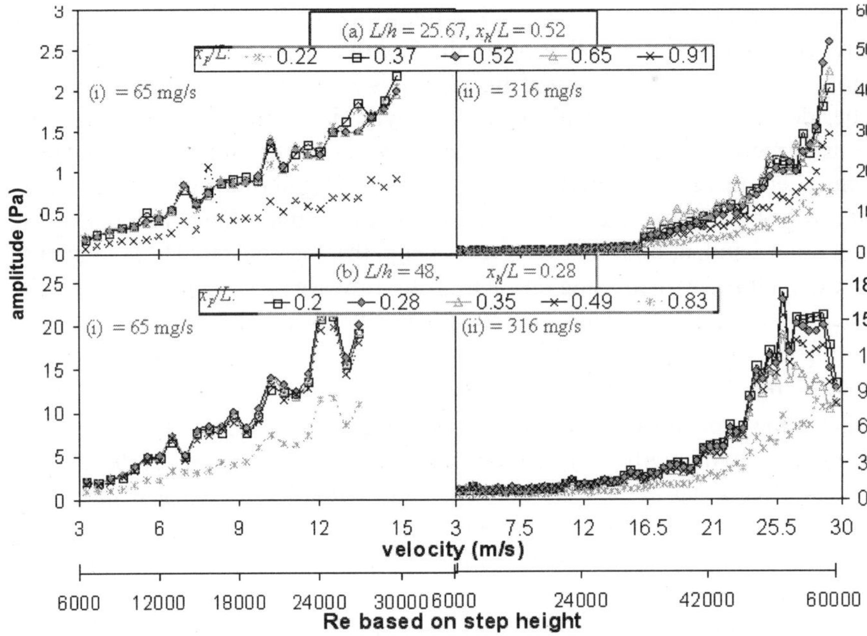


**Figure 5** Amplitudes of dominant acoustic frequencies in the backward-facing step combustor

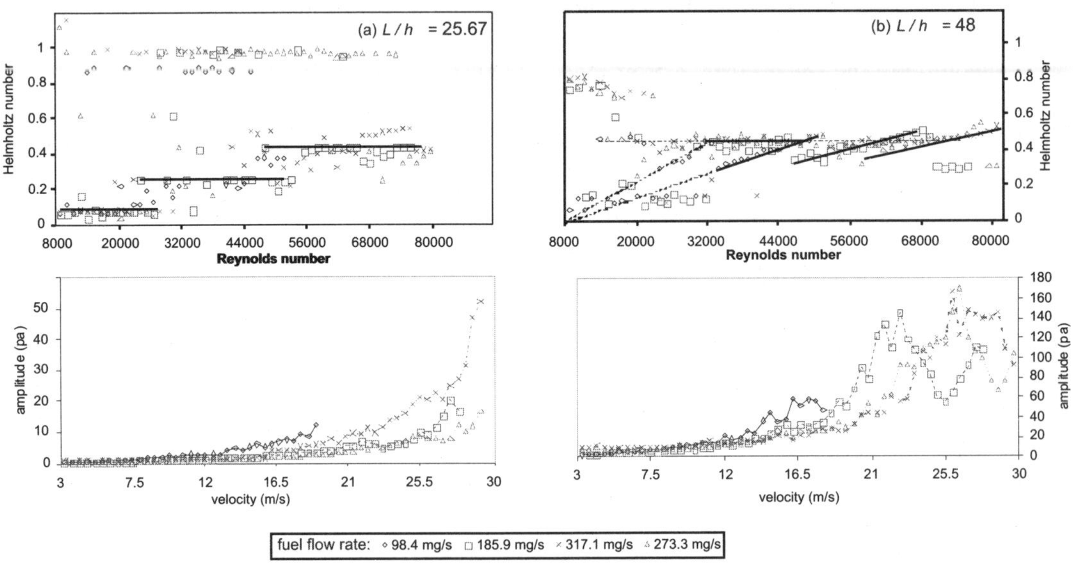


**Figure 6** Dominant frequencies and amplitudes at different fuel flow rates in the backward-facing step combustor

Coloured

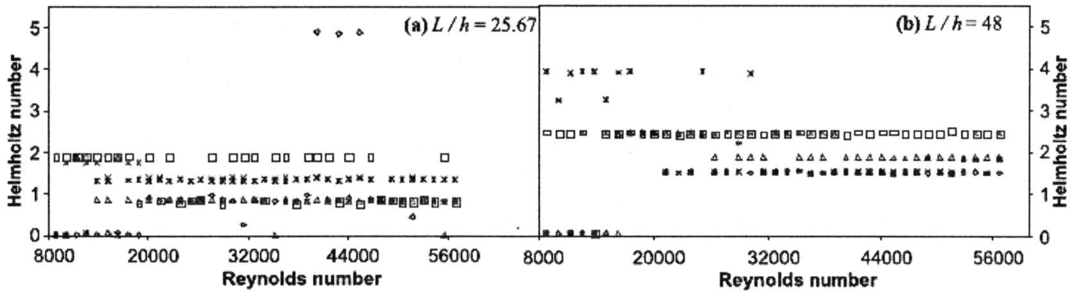


**Figure 7** Dominant acoustic frequencies under cold flow conditions in the backward-facing step combustor

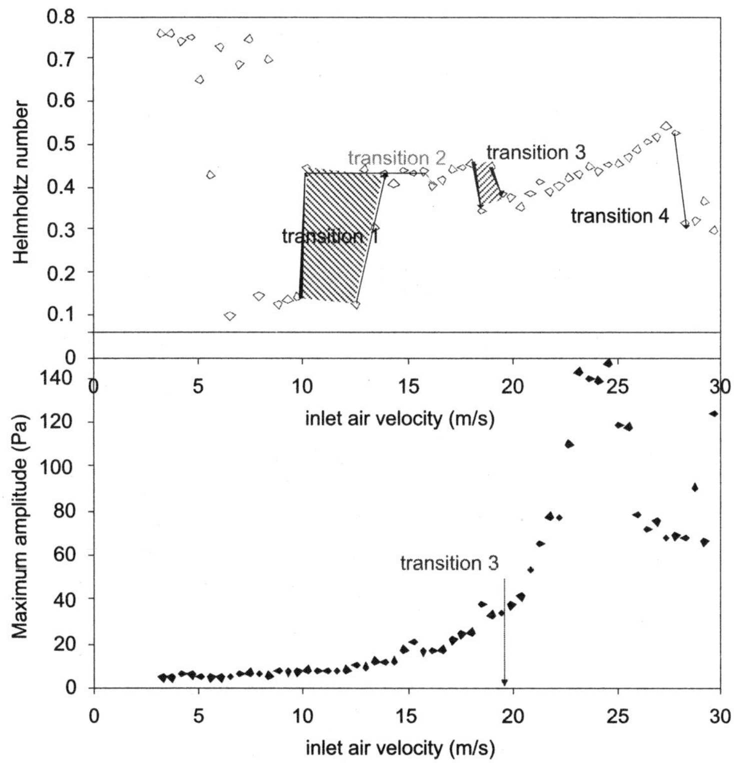


**Figure 8** Mode transitions

total chemiluminescent intensity in the combustion zone with the acoustic spectra observed simultaneously for two lengths of the combustor before and after the onset of flow-acoustic lock-on, all other conditions remaining the same. The dominant frequencies observed in the chemiluminescent intensity fluctuations and the acoustic fluctuations are the same; the chemiluminescent intensity peak at the dominant acoustic frequency particularly increases in amplitude after the onset of the flow-acoustic lock-on. The matching of the dominant frequencies

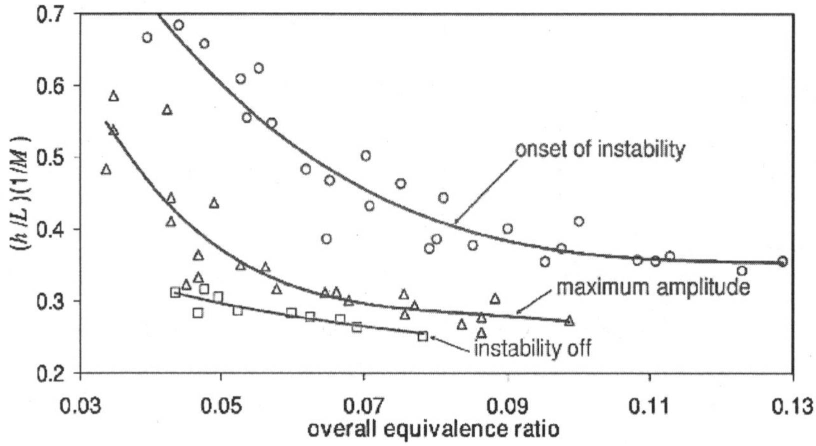


**Figure 9** Map of onset of instability for backward-facing step combustor

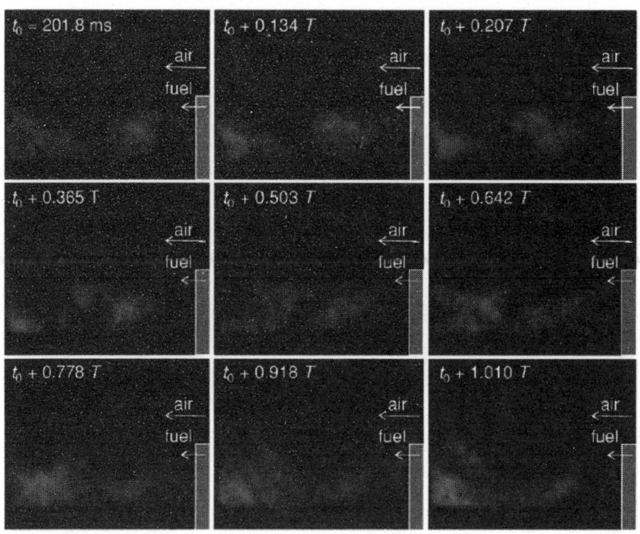


**Figure 10** Time resolved CH* chemiluminescence of flame in backward facing-step combustor

between the two kinds of spectra can be observed at a large number of flow and geometric conditions in Figure 12.

We now turn to the PIV measurements for information on the flow field, particularly vorticity contours, so as to examine the role of vortex shedding in this phenomenon. Before looking at the PIV under combustion conditions, the cold flow PIV data is presented in Figure 13 to assure the quality of the PIV measurements obtained here; for, these measurements are similar to those reported in the literature under similar conditions. In particular, the recirculation zone length remains nearly invariant with the flow Reynolds number in the range tested, as reported earlier [11].

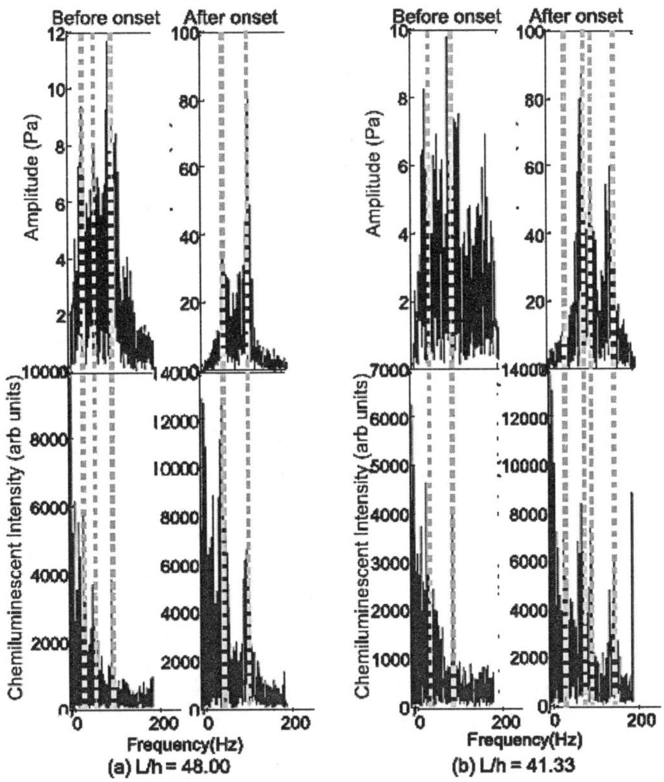


**Figure 11** Comparison of acoustic and total chemiluminescence spectra

Figure 14 shows the ensemble-averaged PIV images obtained under combustion conditions before the onset of flow-acoustic lock-on, just after the lock-on, and during a second lock-on at a lower vortex shedding mode. It can be seen that the recirculation zone becomes smaller with increase in air flow rate as the system moves from the non-lock-on condition to the lock-on condition (Figs. 14a and b), but a large-scale roll-up is again found at a still higher air-flow rate corresponding to the second lock-on condition, attesting to the vortex merger that leads to half the frequency of the first lock-on vortex shedding that signifies the lower vortex shedding mode.

At conditions prior to lock-on, the vortex shedding process occurs as seen in Fig. 14a, but it does not result in a dominant acoustic frequency, but remains sub-dominant. Counting the sub-dominant frequency as well and plotting alongside the dominant frequency in Figure 15, one can see that the sub-dominant frequency linearly increases with the Reynolds number and coincides with the dominant frequency at the onset of the flow-acoustic lock-on. This shows that the duct acoustics locks on to the vortex shedding process, and not the other way round. This is further confirmed by the fact that the phase difference between the dominant or the sub-dominant (vortex shedding) oscillations measured at adjacent locations in the test section of the combustor does not linearly vary as would the case with a convective disturbance as vortex shedding, but lies within a band of –20° to 10° (Figure 16). A small phase difference is indeed the case for the fundamental

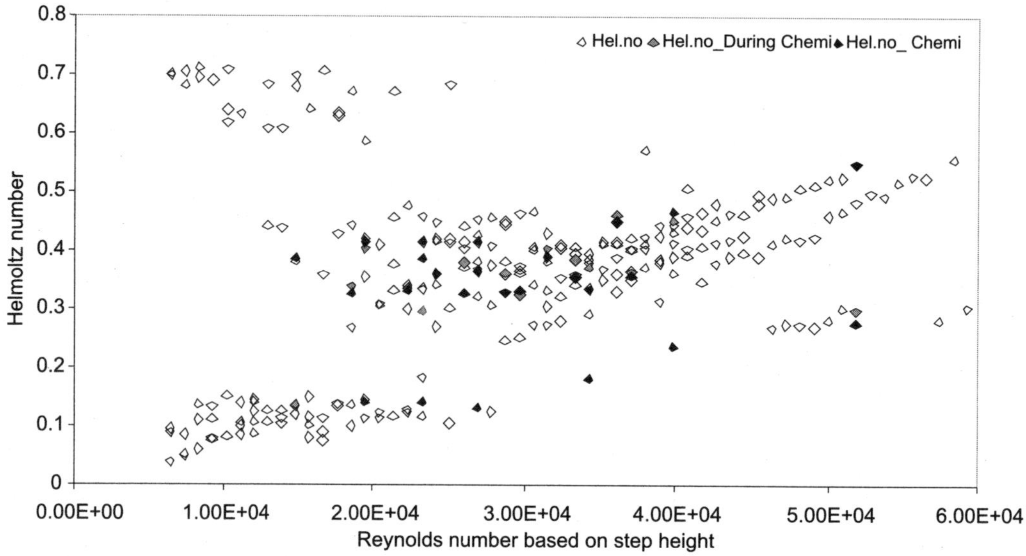


**Figure 12** Dominant frequencies in the chemiluminescence and acoustic spectra

acoustic pressure mode of a duct with both sides open, within the duct. This shows that the dominant oscillations recorded in the duct are acoustic oscillations, but they behave as the vortex shedding frequency as the flow Reynolds number is increased.

## Bluff-body Combustor

Turning to the results for the bluff-body combustor, Figure 17 shows the Helmholtz number and amplitude variations of the dominant acoustic frequencies with the flow Reynolds number. The transition from a constant to a linearly increasing variation of the Helmholtz number is observed in this case also. Correspondingly, the amplitude increases and then decreases in the range of the linear increase in the Helmholtz number. As with the backward-facing step combustor, the dominant frequency observed under cold flow conditions is a constant at all locations, except downstream of the bluff-body, where the fluid dynamic fluctuations due to the vortex shedding process registers a linearly increasing frequency (Figure 18).

The time-resolved chemiluminescence images show the formation of symmetric vortex shedding under the flow-acoustic lock-on conditions over one half of the dominant acoustic cycle, as can be observed in Figure 19. The spectra of the total chemiluminescent intensity fluctuations and the local chemiluminenscent intensity fluctuations at the point in the flow field marked by the white dot in Figure 19f are shown in comparison with the acoustic spectra obtained simultaneously downstream of the bluff body, in Figure 20. Here again, the frequencies of the peaks in the three spectra coincide, showing that the flame fluctuations are responsible for the acoustic oscillations.

The mode shape of the acoustic oscillations at different flow conditions is obtained by measuring at all locations along the combustor, as shown in Figure 21. This indicates the fundamental acoustic pressure mode of the combustor with both ends open, suggesting that the

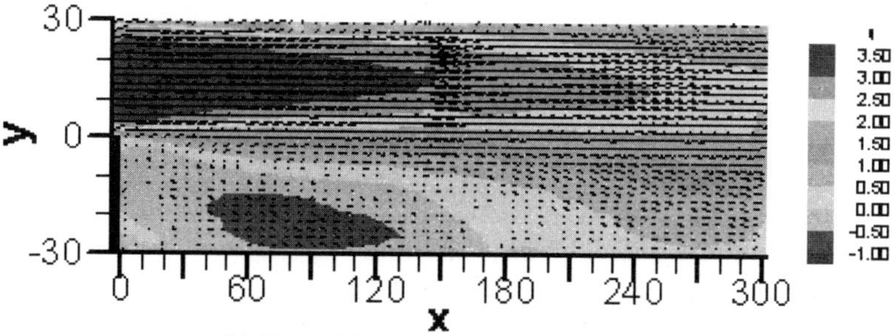


(a) Ensemble -averaged velocity field

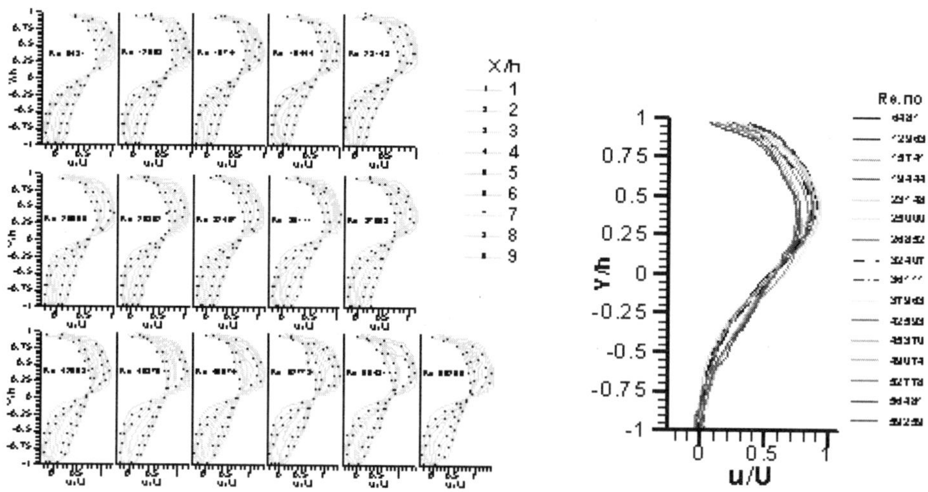


(b) Velocity profiles at various locations and different Reynolds numbers

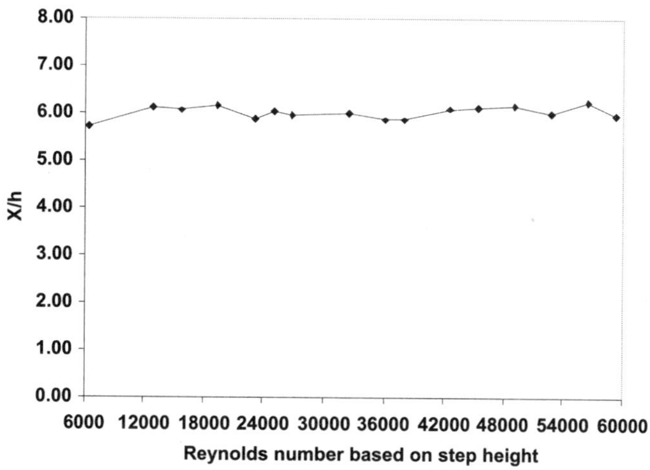


(c) Length of the recirculation zone

**Figure 13** Cold flow PIV measurements in backward-facing step combustor

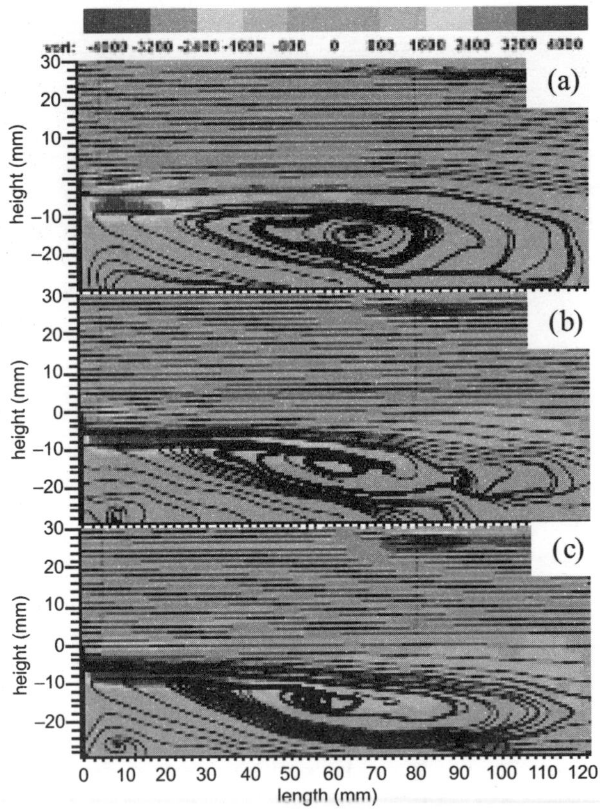


**Figure 14** Ensemble-averaged velocity field under three different conditions in the backward-facing step combustor

acoustic oscillations are prevalent under the flow-acoustic lock-on condition, but their frequency varies as the vortex shedding frequency, as the flow Reynolds number is increased, as before.

## NUMERICAL INVESTIGATIONS

The numerical investigation starts from a level that is more basic than the experimental conditions reported so far, in two ways: (a) it considers the classical Burke-Schumann diffusion flame, and builds up on complexity by relaxing the constant density assumption, and then moves on to investigate the backward-facing step flow, all in the laminar regime; (b) it then considers turbulent flow in a backward-facing step with fuel injection at the step, and mixing, adopting the mixed-is-burnt approach to obtain the heat release rate fluctuations.

### Laminar Simulations

First, we consider the Burke-Schumann geo-metry. Figure 22 shows the flame structure of the diffusion flame in this geometry, but with finite-rate single-step chemistry. The comparison here is

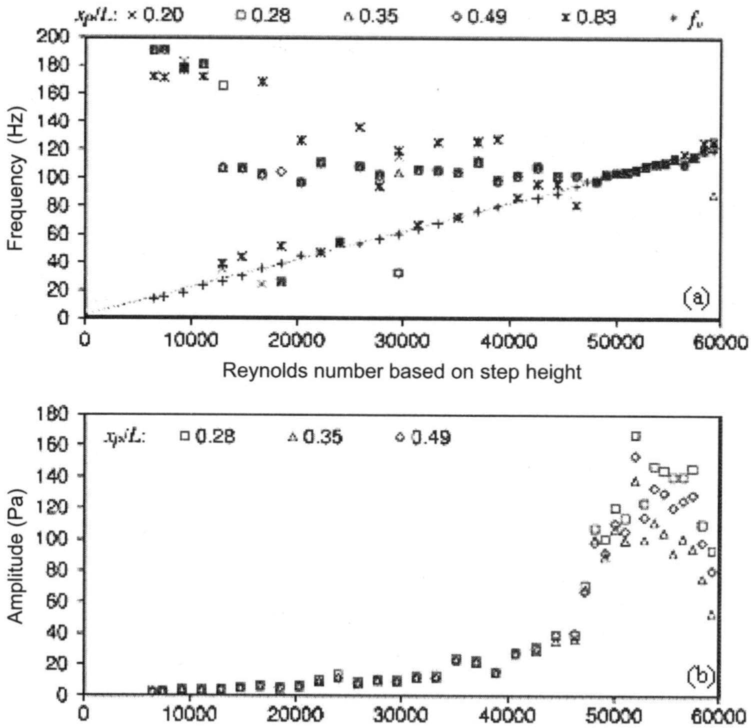


Figure 15 Sub-dominant frequencies and amplitudes

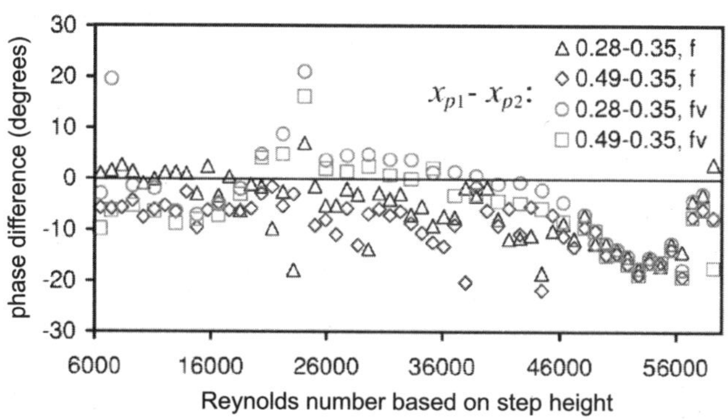


Figure 16 Phase difference in acoustic signals

between the constant density assumed in what is typically referred to as a thermo-diffusive model [12], and the density varying with temperature, but at constant thermodynamic pressure. Therefore, both are incompressible flow simulations. It can be seen that the streamlines diverge and converge in the latter case, whereas they remain uniform in the former, as expected. Correspondingly, the flow accelerates near the flame with variable density, and therefore the flame tends to extend out, and the leading edge of the flame locates itself farther away from the burner lip as well.

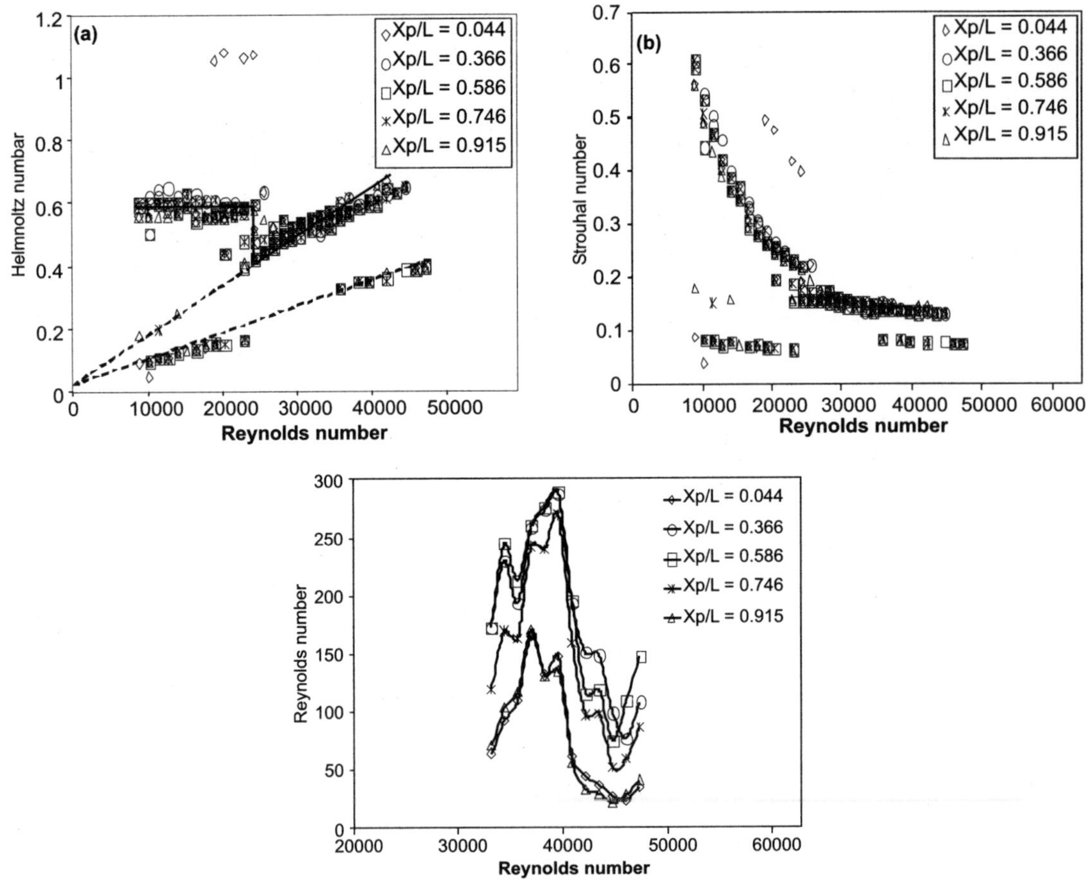


**Figure 17** Helmholtz and Strouhal numbers of dominant acoustic frequencies and their amplitudes in the bluff-body combustor

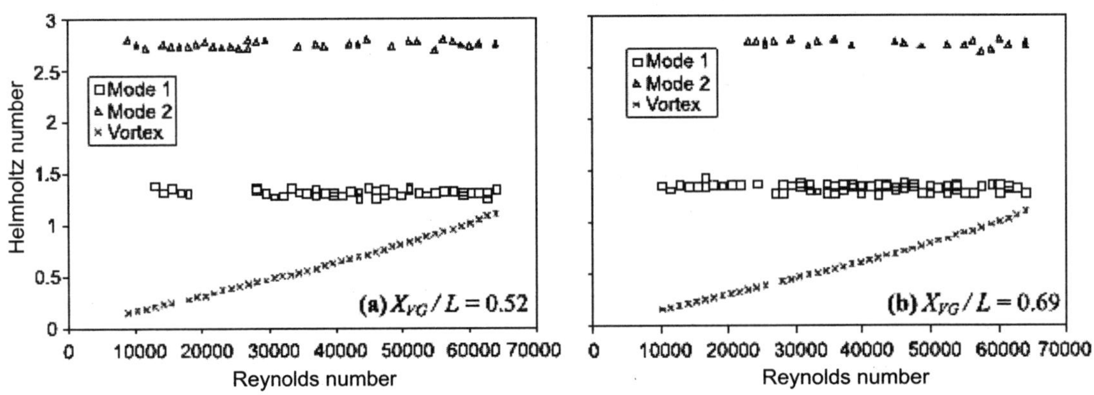


**Figure 18** Cold flow acoustic dominant frequencies in the bluff-body combustor

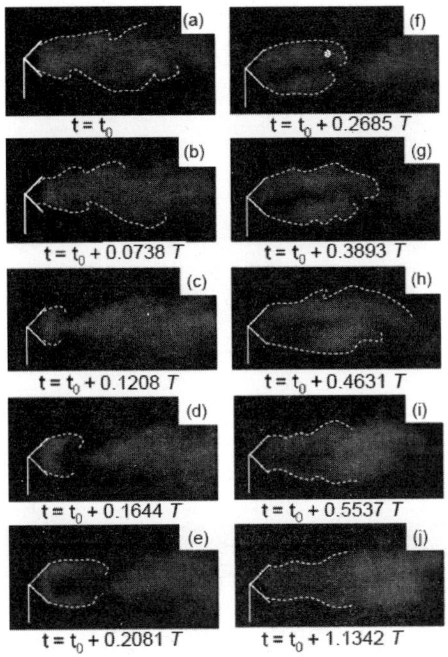


**Figure 19** High-speed chemilumine-scence images in the bluff-body combustor

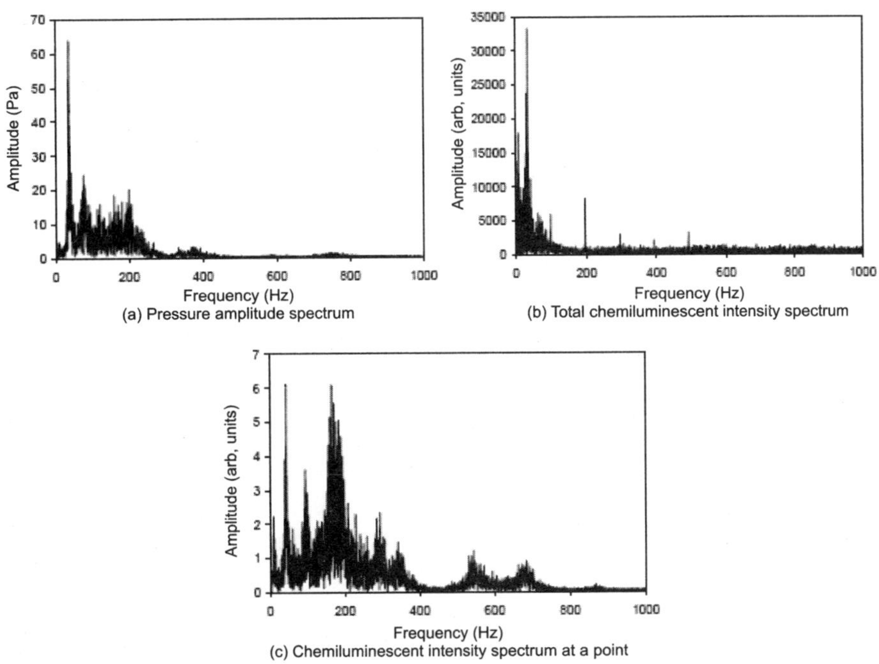


**Figure 20** Comparison of acoustic spectra with total and local chemilumine- scence spectra in bluff-body combustor

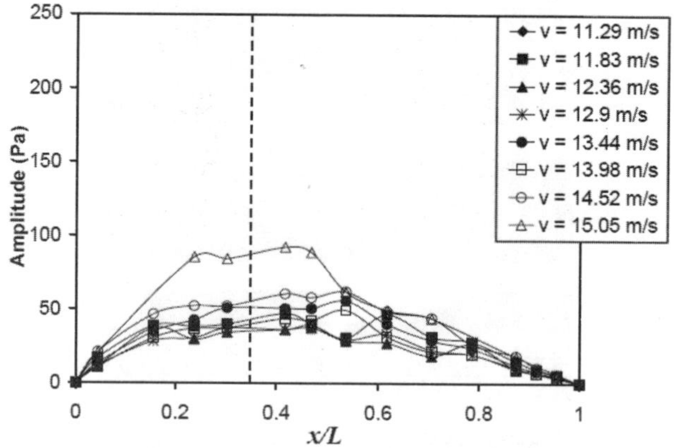


**Figure 21** Mode shapes in the bluff-body combustor under lock-on conditions

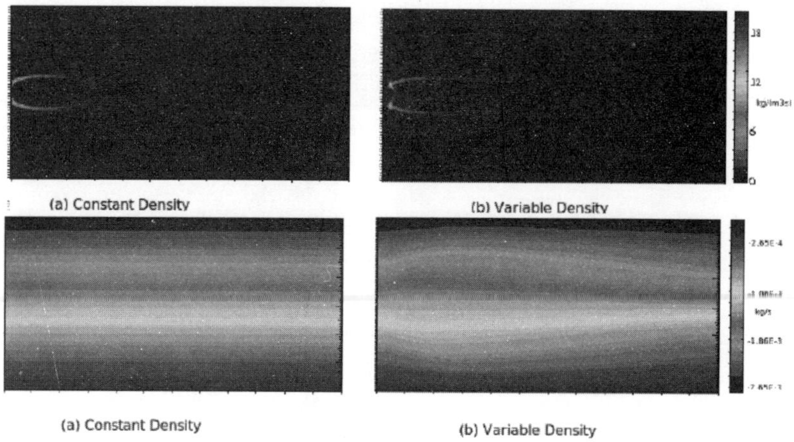


**Figure 22** Flame structure (reaction rate contours) and streamlines for Burke-Schumann geometry

When subjected to inlet flow oscillations, the response of the flame decreases with increase in the frequency of the oscillations, as reported for Burke-Schumann flame [13]. However, the effect of variable density is significant at higher frequencies, as can be seen in the response function plotted on a logarithmic scale in Figure 23a. Figure 23b shows the customary plot on a linear scale of response function, where this effect is seen diminished, but it also shows the response function for two different amplitudes of the inlet flow excitation, wherein no significant effect is observed. This shows that the flame's response is linear, despite the non-linearity in the chemical reaction and heat release rate terms.

Turning to the flame simulation in the backward-facing step geometry, the steady-state flame is anchored at some distance downstream of the step corner in different cases shown in Figure 24. The step height is at 0.03 units along the vertical in these figures. It can be seen that increasing the mean flow velocity or decreasing the equivalence ratio has the effect of locating the flame farther away

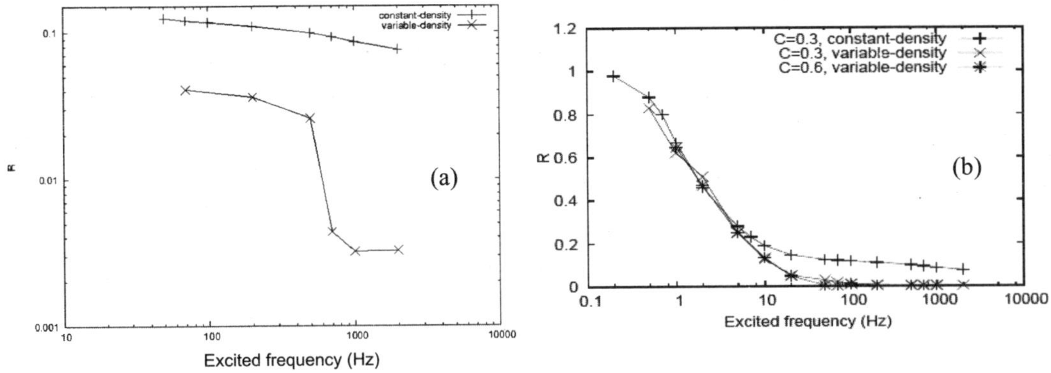


**Figure 23** Response function of non-premixed flame in Burke-Schumann geometry

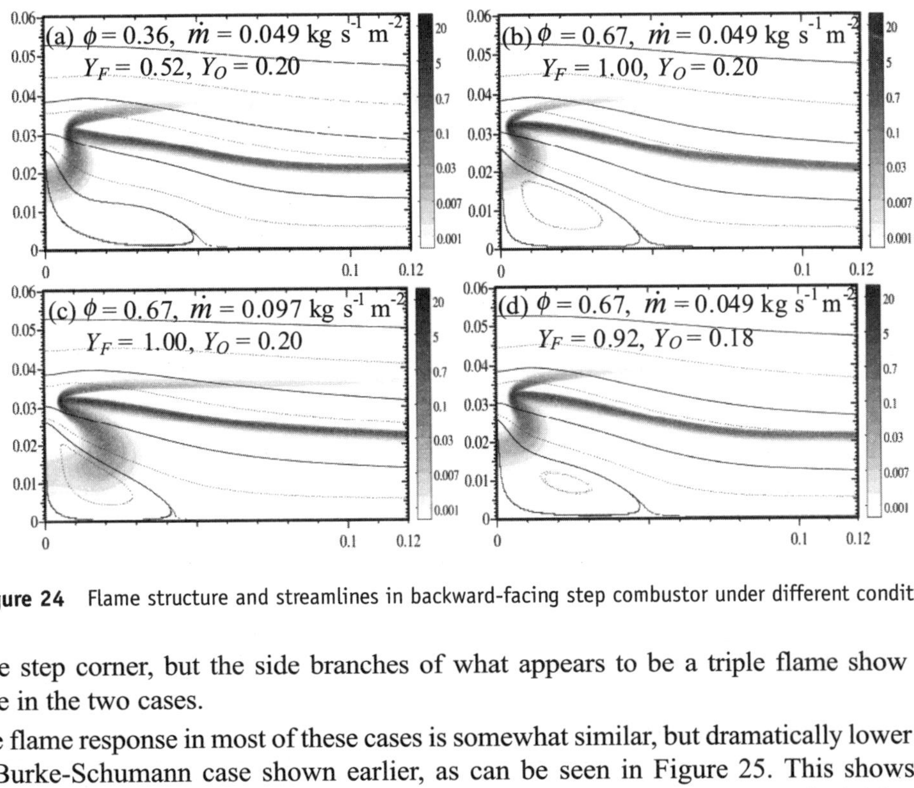


**Figure 24** Flame structure and streamlines in backward-facing step combustor under different conditions

from the step corner, but the side branches of what appears to be a triple flame show different structure in the two cases.

The flame response in most of these cases is somewhat similar, but dramatically lower than that of the Burke-Schumann case shown earlier, as can be seen in Figure 25. This shows that the presence of the shear layer in the backward-facing step geometry tends to diminish the flame response, a result that is not readily apparent, and not explicitly shown or discussed in earlier literature.

To understand the flame's response in the presence of the shear layer, it is pertinent to examine the shear layer's response to inlet flow fluctuations under cold flow conditions. Figure 26 shows streamline contours that exhibit a periodic lift-off of the recirculation zone at a preferred frequency

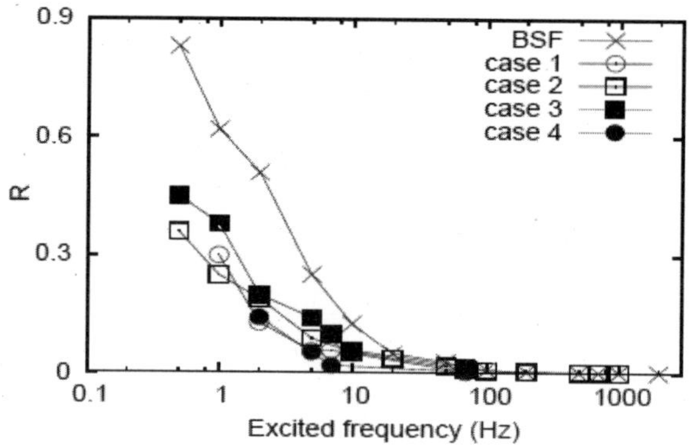


**Figure 25** Response function of backward-facing step combustor flame (as in Fig. 24), in comparison with Burke-Schumann flame (BSF)

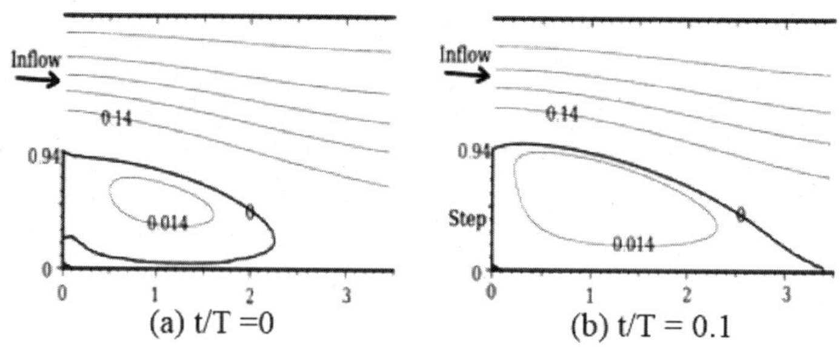


**Figure 26** Periodic lift-off of the recirculation zone in cold-flow past backward-facing step excited by noise in a limited frequency range

corresponding to $St$ = 0.2, when excited by a white noise in a frequency band outside of this frequency. Fourier decomposition of the flow field at this frequency shows the shedding of large-scale vortices (Figure 27). Single frequency excitation of the inlet flow over a range of frequencies shows preference of different frequencies by different parts of the shear layer, based on the response of the fluctuations induced in the transverse component of velocity due to the shear layer oscillations, as can be seen in Figure 28 for three Reynolds numbers. A similar figure for the fluctuations in the streamwise component of velocity, shown in Figure 29, exhibits alternating regions of constructive and destructive inter-ference between the vortical and potential components of the flow, when it is split into these two parts [14]. Because of this, there appears the possibility of hold the flame in a zone of silence downstream of the backward-facing step. This has specific implications to the gas turbine combustor geometry, as can be seen in Figure 30.

## Turbulent Simulations

Figure 31 shows the cold flow simulations of the flow past the backward-facing step by means of incompressible large-eddy simulations (LES). The comparison of the simulated results at two grid levels with the experimental data is shown in Figure 32. Figure 33 shows an instantaneous snap-

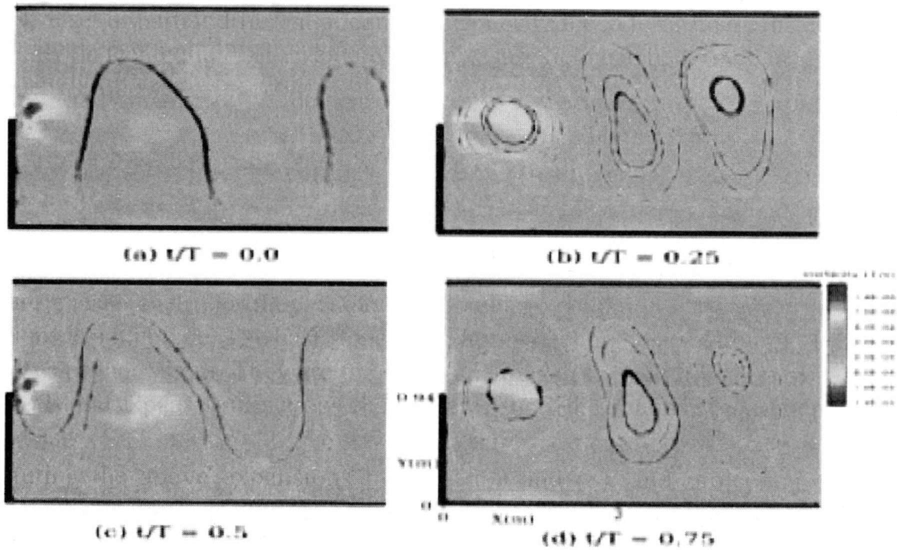

**Figure 27** Fourier decomposition of noise-excited backward-facing step cold flow at St = 0.2

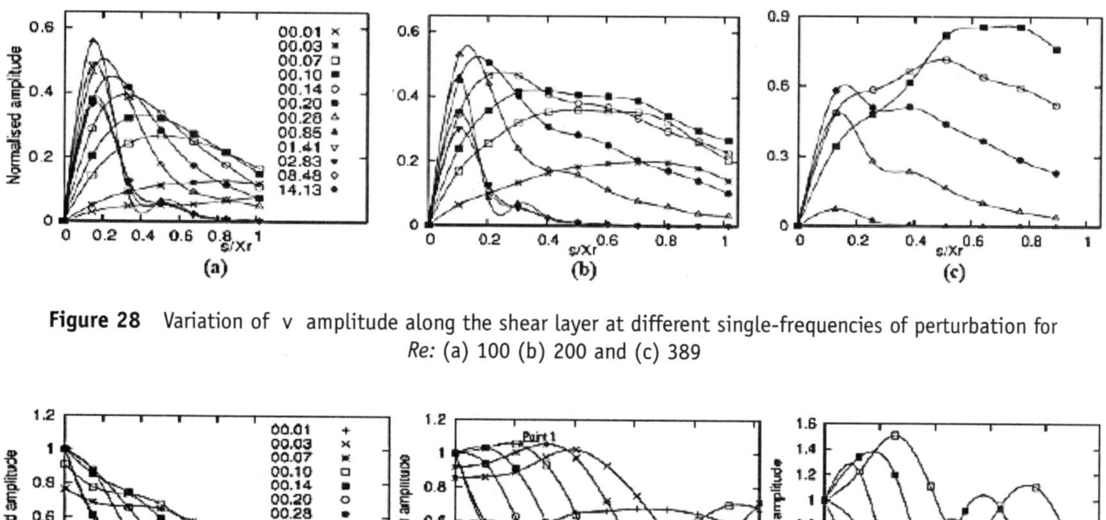


**Figure 28** Variation of v amplitude along the shear layer at different single-frequencies of perturbation for *Re:* (a) 100 (b) 200 and (c) 389

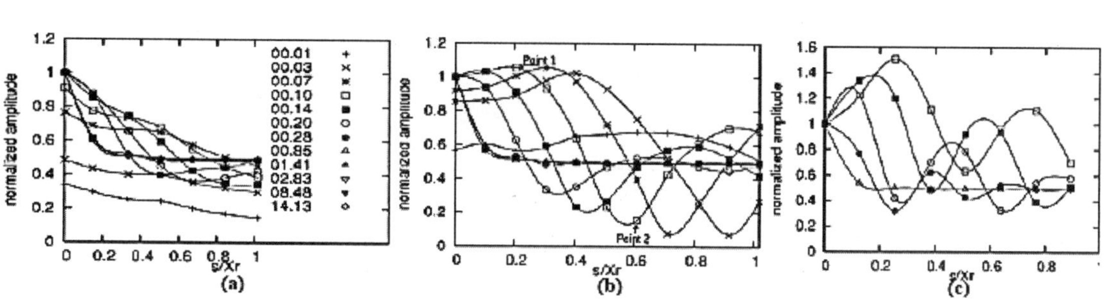


**Figure 29** Variation of u amplitude along shear layer at different single-frequencies of perturbation for *Re:* (a) 100 (b) 200 and (c) 389

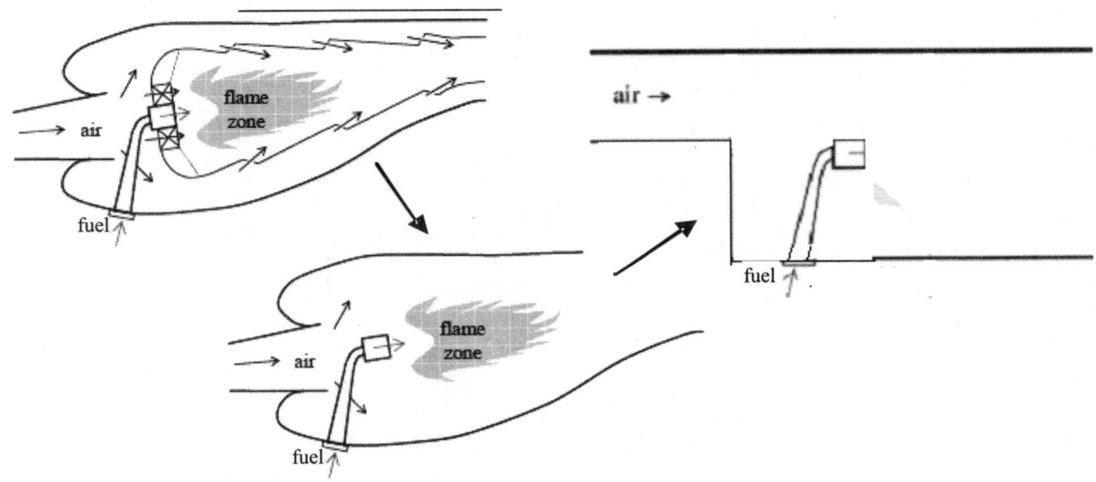


**Figure 30** Excitation at different frequencies along the shear layer downstream of the step: possibility of zone of silence at the flame location

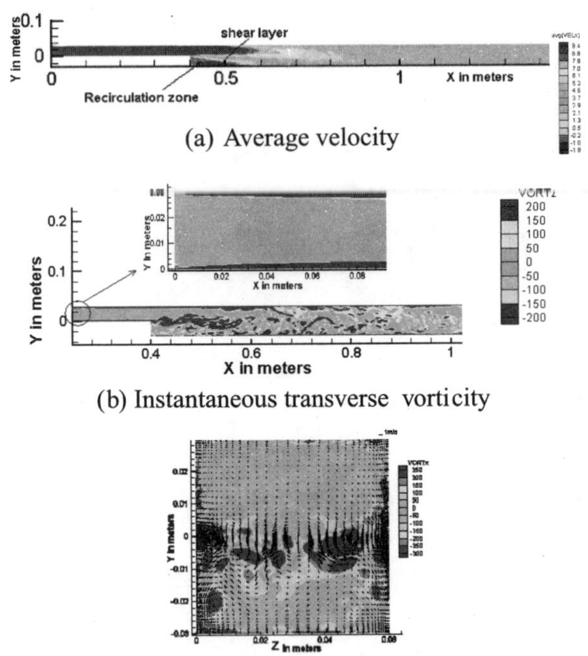


**Figure 31** Large eddy simulation of turbulent cold flow past a backward-facing step

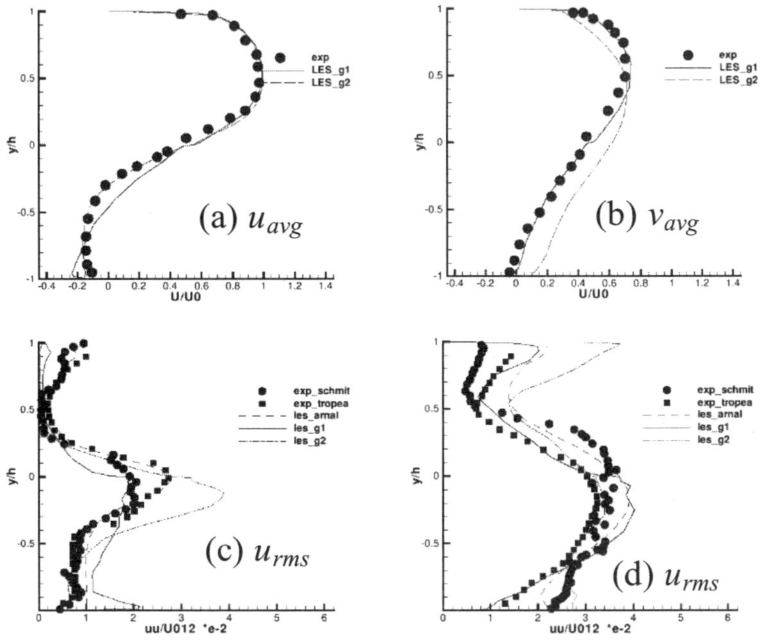


**Figure 32** Comparison of LES with experimental data

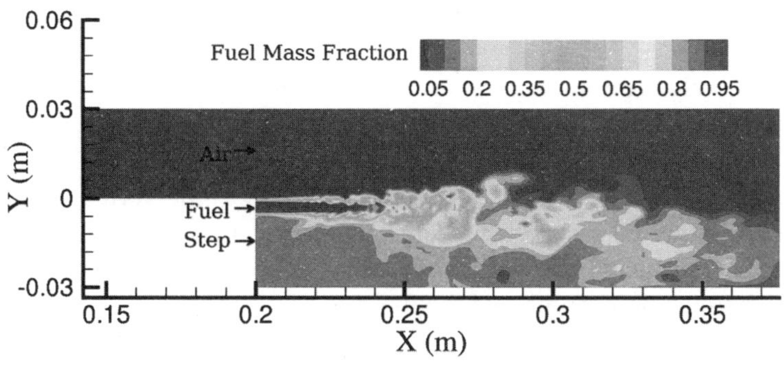


**Figure 33** Fuel mass-fraction contours from LES of fuel injection in backward-facing step flow

shot of fuel injection into the turbulent backward-facing step flow, simulated by LES. Adopting the mixed-is-burnt approach, it is possible to extract the heat release rate fluctuations from the mixture fraction distribution of the fuel-air mixture in the flow field at any instant. The fluctuation in the heat release rate in the combustion domain without any inlet flow fluctuations is shown in Figure 34. Note the intrinsic periodicity of the fluctuations, indicating a preferred frequency of the flame, because of the presence of the step. The integral effect of the heat release fluctuations over the full domain reduces the turbulent fluctuations in this quantity barring at the large-scale level over the dominant preferred frequency. These heat release fluctuations can be input into an acoustic solver, which could in turn be coupled to the incompressible LES computations, similar to [15].

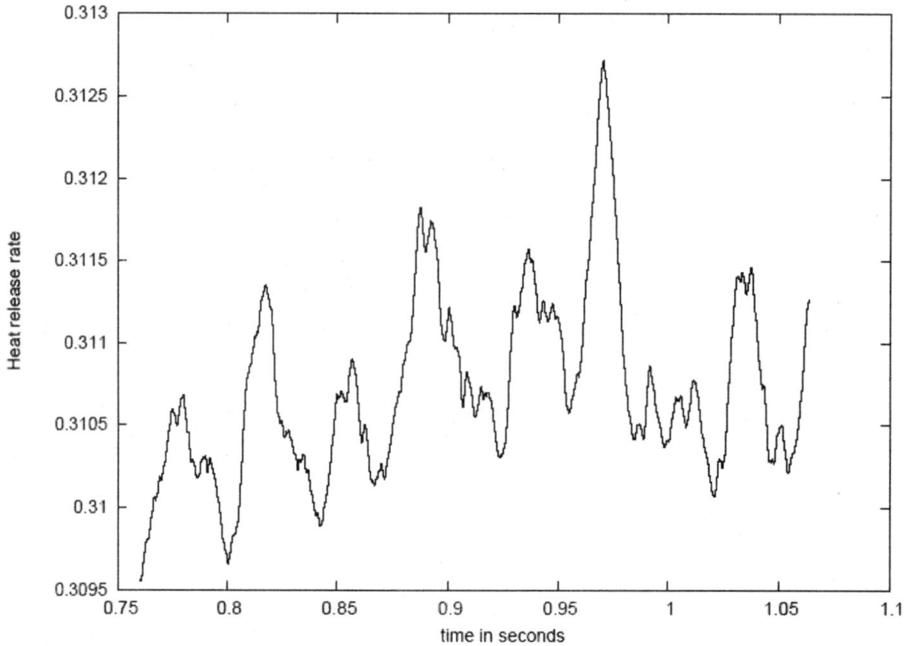


**Figure 34** Heat release rate fluctuations from fuel injection in turbulent backward-facing step flow

## Conclusions

This paper presents both experimental and numerical results pertaining to unsteady oscillatory combustion leading to or in the context of combustion instability in non-premixed combustors. The experimental evidence suggests that, for both backward-facing step and bluff-body combustors, the duct acoustic modes with the heat release feedback locks on to the vortex shedding modes in the shear layer in the recirculation zone where flame stabilization takes place. A map of instability is reported for the backward-facing step combustor which depends on the axial locations of the flame stabilization relative to the duct length and the fuel injection relative to the flame stabilization, and the shear layer instability modes.

The numerical work starts with more basic investigations than the complexity pertaining to the experimental conditions. It shows that the inclusion of variable density effects decreases the flame response in the high frequency range. Considera-tion of shear layer fluctuations further decreases the flame response in the low frequency range as well. Different frequencies are preferred by different parts of the shear layer, and could excite different modes for different fuel injection locations. The cold flow shear layer excitation results suggest the possibility of zones of silence where fuel injection would not excite acoustic oscillations. Large eddy simulations of turbulent flow past the backward-facing step also show preferred frequency of heat release fluctuation due to the large-scale shear layer dynamics in the recirculation zone.

In the near future, experiments with swirl flow and combinations of flame-holding geometries and fuel injection configurations would be interest. Further, instability mapping could be attempted in actively-controlled combustors also.

## References


1. Gutmark, E. J., Schadow, K. C., Sivasegaram, S., and Whitelaw, J. H. *Combustion Science and Technology* 79 (1991) 161-166.
2. Sterling, J.D., Zukoski, E.E. *Combust. Sci. Tech.* 77 (1991) 225-238.
3. Schadow, K. C. and Gutmark, E. J. *Progress in Energy and Combustion Science* 18 (1992) 117-132.
4. Fleifel, M., Annaswamy, A.M., Ghoneim, Z.A., Ghoniem, A.F., *Combust. Flame* 106 (1996) 487-510.
5. Bloxisidge, G.J., Dowling, A.P., Langhorne, P.J., *J. Fluid Mech.* 193 (1998) 445-473.
6. Lieuwen, T. C., Torres, H., Johnson, C., and Zinn, B. T., *Journal of Engineering for Gas Turbines and Power* 123(1) (2001) 182-190.
7. Ducruix, S., Schuller, T., Durox, D., Candel, S. *J. Propulsion* 19(5) (2003) 722-734.
8. Pankiewitz, B., Sattelmayer, T., *J. Eng. Gas Turbines* 125 (2003) 677-685.
9. Lawn, C.J., Evesque, S., Polifke, W. *Combust. Sci. Tech.* 176 (2004) 1331-1358.
10. R. Balachandran, B. Ayoola, C. Kaminski, A. Dowling, E. Mastorakos, *Combust. Flame* 143 (2005) 37-55.
11. Armaly, B. F., Durst, F., Pereira, J. C. F., Schonung, B. *J. Fluid Mech.* 127 (1983) 473-496.
12. Kurdyumov V. N., Matalon, M. *Combust. Flame* 139 (2004) 329-339.
13. Tyagi, M., Jamadar, N., Chakravarthy, S. R. *Combust. Flame* 149(3) (2007) 271-285.
14. Wilford, Z., Andreas, B., *Dynamics of Atmosphere: A Course in Theoretical Meteorology*, Cambridge University Press, 2003.
15. Tyagi, M., Chakravarthy, S. R., Sujith, R. I., *Combust. Theo. Modelling* 11(2) (2007) 205-226.

# Computational Combustion: Uncertainties in Our Knowledge: What We Don't Know

**Akshai Runchal**

Principal Engineer, Computational Com ACRi, 1931 Stradella Road, Los Angeles, CA 90077

E-mail: runchal@ACRiCFD.combustion

---

Computational combustion – a special field of Computational Fluid Dynamics (CFD) - has made great strides over the last 4 decades. Today computational combustion is routinely used in engineering practice. It is implicitly assumed that the computational models provide qualitative and quantitative information that can be used for design. While that is undoubtedly true in many cases, it is far from universally true. What is less widely talked about are the uncertainties inherent in our basic understanding of the physical processes that lead to combustion and the impact of these on the answers obtained from computational models. Those who deal with the design and performance issues related to combustion equipment need to account for these uncertainties much as the "safety factors" of handbooks of yesteryears.This talk will dwell on the inherent uncertainties that should be taken account of in evaluating the role of computational model in problems of practical design.

# Connect-pipe Testing and Evaluation of Scramjet Combustor

**V. Ramanujachari**
DRDL, Hyderabad

Hypersonic cruise vehicle designed to fly over the flight Mach number range of 6-7 is powered by a supersonic combustion ramjet where combustor entry Mach number ranges between 2 to 2.5. Two variants of scramjet combustors, viz., strut based combustor and ramp-cavity based combustor are designed, developed and tested in the connect pipe test facility. The high enthalpy air needed for the connect pipe test facility is generated in a air heater by the combustion of hydrogen and air with excess oxygen to keep the oxygen content of the vitiated air (products of combustion of the heater) at 23% by mass for subsequent combustion in the scramjet combustor with aviation turbine fuel. At present the air heater can heat up 4 to 5 kg/s of air to about 1800-2000K temperature. The operation of the connect-pipe test facility is fully automated as per the safety regulations.

The connect pipe facility consists of vitiated air heater, circular to rectangular transition duct and rectangular facility nozzle. The scramjet combustor is fitted to the facility nozzle. Both the test facility and the test article are extensively instrumented to acquire the pressure, temperature, flow rate data during the operation of the heater and the combustor for further evaluation. The static tests have been conducted for the combustors using strut based fuel injection and ramp-cavity based fuel injection. Both high-pressure fuel injection and effervescent atomization have been employed for the static tests.

As CFD is very expensive and time consuming, simple quasi-one-dimensional methodology has been developed for the performance evaluation and the relative performance of the strut based combustor and the ramp-cavity based combustor are discussed.

# Development of Water-mist Nozzles and Fire Suppression by Total Flooding

**Suresh Lal and J.C. Kapoor**

Centre for Fire Explosive & Environment Safety, Defence R&D Organization, Min of Defence, Delhi-110054

E-mail: slalcfees@gmail.com, jck_drdo@rediffmail.com

Following the ban under the Montreal Protocol, on the production and use of the ozone depleting Halons, there is world-wide R&D effort on-going to develop suitable and effective fire suppression media for replacing Halons in the fire suppression systems in civil & defence sector. Water mist technology has emerged as one of the most viable alternative because of the inherent advantages such as rapid heat absorption and fire suppression effectiveness, easy availability, non-toxicity and environmentally most compatible.

Although water is the oldest firefighting medium known to the mankind, its usage until now, has been in a gross and bulk mode in which only 5-20% of the mass is actually utilized in the processes i. e. heat absorption & cooling of the flame source, and denial of $O_2$, leading to fire suppression. Recently, it is being recognized that it is possible to maximize rate of heat & mass transfer, the two most important processes during interaction of water surface and fire front, by using water in a highly discretized form i.e. as micron sized water droplets. Possibility of application of this mode of water as replacement of Halons depended on developing technical capability of generating very fine water droplets at large enough release rates. Being a liquid, the challenge in most efficient use of water as fire suppressant, was to impart gas like properties to released water for rapid transport to all spaces in the enclosure. At the same time, it was also important to retain thermodynamic characteristics of the bulk water for utilizing its latent heat of evaporation in heat removal from the fire scene. It was further analyzed that cutting down the water droplets size to a few micron will also yield enhanced heat transfer rates, higher evaporation rates and hence faster heat dissipation. Another advantage that accrues from such situation is that the fast expanding water vapour cloud dilutes the concentration of $O_2$ and fuel vapour below critical ignition levels.

A project "Development of Water Mist-based Fire Protection Systems for Defence Applications" was initiated in our lab in 2004 and the work reported here has been carried-out

Invited Talk in 'International Workshop on Advances in Combustion' IIT, Kanpur 31 Dec 2007 to 2 January 2008.

under that programme. Since the important characteristics of water-mist e.g. transport behaviour, evaporation rate, and their life time in air are all functions of droplet size, it becomes the most critical parameters in determining the effectiveness of a water-based suppression system. Major objective of this project was to optimize water mist system parameters from a set of fire suppression experiments. Another important objective was to evolve a scaling law from fire suppression experiments carried-out at different scales of simulated fire conditions. This work will lead to the design and evaluation of real scale prototype water mist fire suppression systems for the intended applications.

## Nozzle Development & Characterization

Several water mist nozzles based on single fluid and twin fluid concepts and using different atomizing principles were developed and detailed characterization w.r.t. mist output and droplet size distribution were carried-out. During our nozzle development programme we started with establishing very fundamental criterion for twin-fluid nozzles i. e. separation distance between air and water inlet orifice as a function of the two diameters.

The twin fluid internally mixed atomizer is developed for controlled atomization. Figure 2 indicate the solid model of this atomizer and its droplet size characteristic at different operating conditions. This atomizer is capable of generating mist generation rate up to 1.9 lpm and droplet size as low as 20 μm as indicated in Fig 3.

**Figure 1** Atomizer characterization – Integrated facility for droplet size measurement and two axis atomizer traverse system

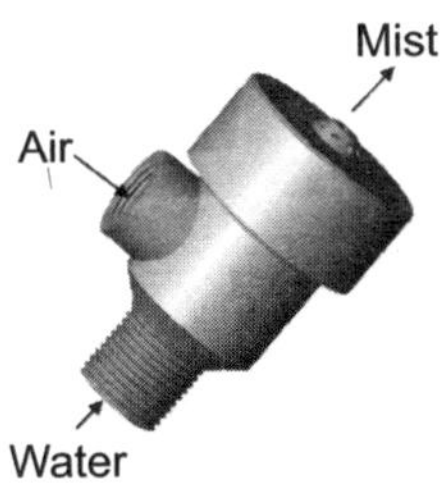


| Operating pressure (psi) | | Flow rate (lpm) | | $D_{10}$ (μm) | $D_{50}$ (μm) | SMD (μm) |
|---|---|---|---|---|---|---|
| Water | Air | Water | Air | | | |
| 20 | 25 | 1.9 | 115 | 57.48 | 148.6 | 90.3 |
| 20 | 40 | 1.7 | 145 | 50.67 | 127.4 | 75.3 |
| 20 | 50 | 1.6 | 165 | 43.9 | 116.6 | 64.9 |
| 20 | 60 | 1.4 | 185 | 37.7 | 103.5 | 54.0 |
| 20 | 70 | 1.25 | 195 | 28.48 | 90.7 | 41.3 |
| 20 | 80 | 0.9 | 215 | 16.74 | 65.0 | 27.1 |

**Figure 2** Droplet size distribution of Internally Mixed Atomizer

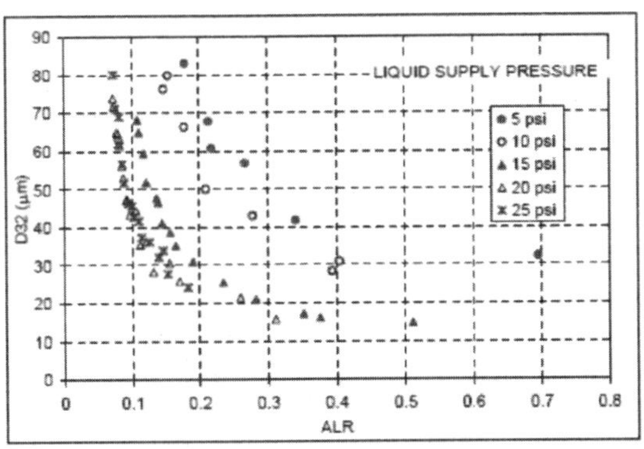


Variation of droplet SMD with ALR for different liquid supply pressures for IMK

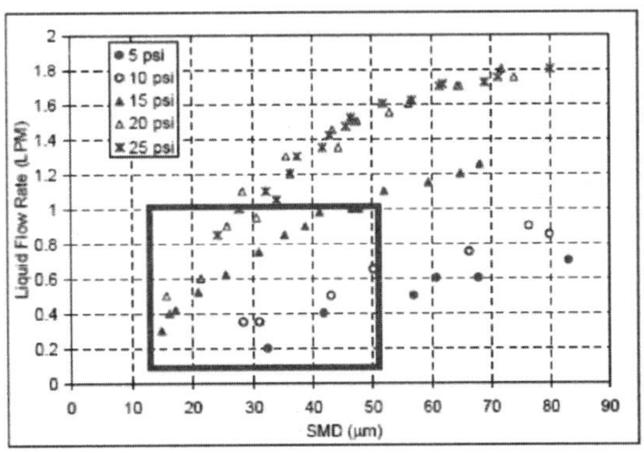


Performance map of the IMK

**Figure 3** Performance characteristics of internally mixed atomizer

Design evolution involved locating maximum orifices (3, 6, and 12 orifices) on the given nozzle head to obtain high mist generation rates of up to a few litres per minute in the desired size range. In one of the novel designs, we have been able to place up to 25 nozzle heads of 12 orifice type (300 orifices both of air and water) in a single square block of 78 × 78 mm, with common supply of compressed air and water, we call it multi nozzle array (5 × 5 array).

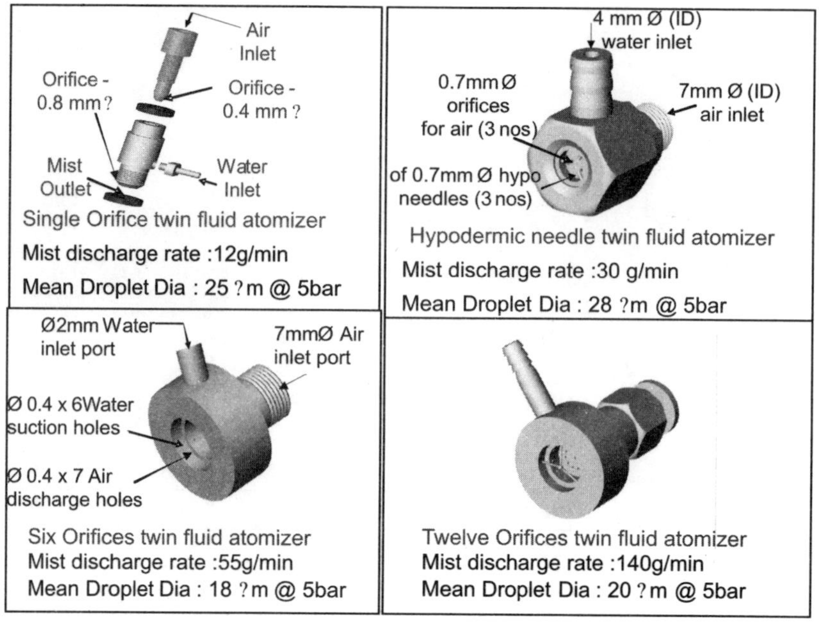


**Figure 4** Siphon fed twin fluid atomizers

The siphon–fed twin fluid atomizers are developed as indicated in Fig. 4 for the applications with requirement of low mist through-put with small droplet size distribution. These atomizers have used in cup burner test studies and fire suppression in electronic cabinet.

Another unique design that came out of this work is C-D nozzle as indicated in Fig. 5 with air and water feed through two co-centric tubes placed at the centre of the circular discs forming circular converging-diverging sections, and water mist issuing from the periphery.

Work presented here also covers the effervescent type, and externally mixed nozzles designed and fabricated by IIT Kanpur under a research project from DRDO and characterized at our labs for mist droplet size distribution. With such a wide range of nozzles, it is possible to generate water mist with average mist diameter from 20 to 80 micron with different generation rates. These nozzles have been used for carrying out experimental studies on determination of suppression time as function of droplet sizes, fire sizes and test enclosures sizes

## Cup Burner Studies

Laboratory studies of fire suppression using water-mist were carried out at three levels. In first stage, utilizing standard cup burner apparatus modified for controlled supply of mist to the burn-tube, critical mass concentration 160 gm/m$^3$ was determined for extinction of diffusion flame (0.50 kw).

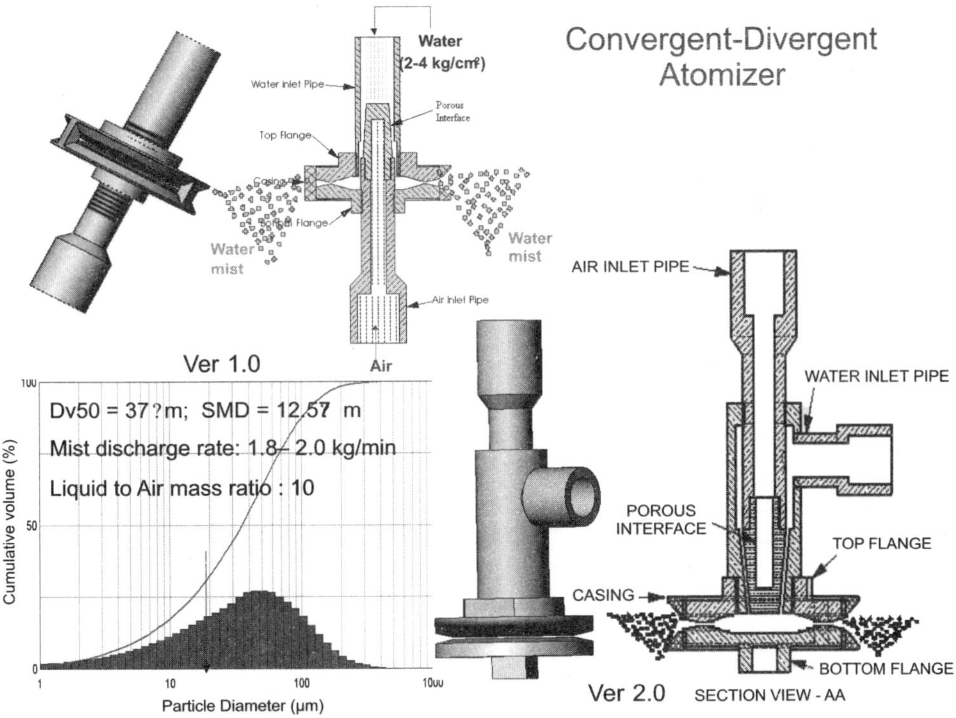


**Figure 5** Convergent - divergent atomizer

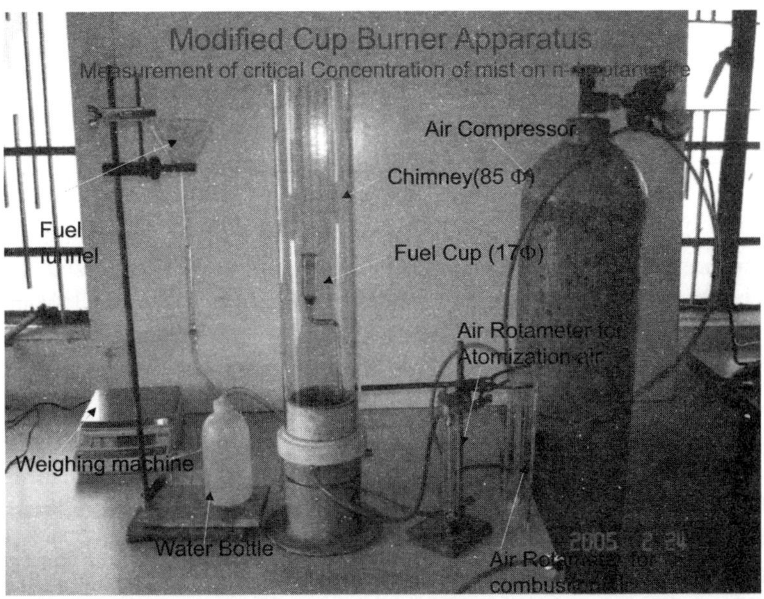


**Figure 6** Cup burner apparatus for measuring critical concentration

## 1 m³ Enclosed Fire test Studies

In second stage, n-heptane fires of 5, 17, 22, 34 kw, in an instrumented 1 $m^3$ test enclosure were subjected to water-mist with similar characteristics. Extinction times of <40 second with continuous mist release and <25 second with pulsed release were obtained using approx 150 $gm/m^3$ and 80 $gm/m^3$ mist concentration. The results are quite encouraging as they demonstrated the capability of small scale experiments to predict the critical water mist concentration required for fire extinguishment.

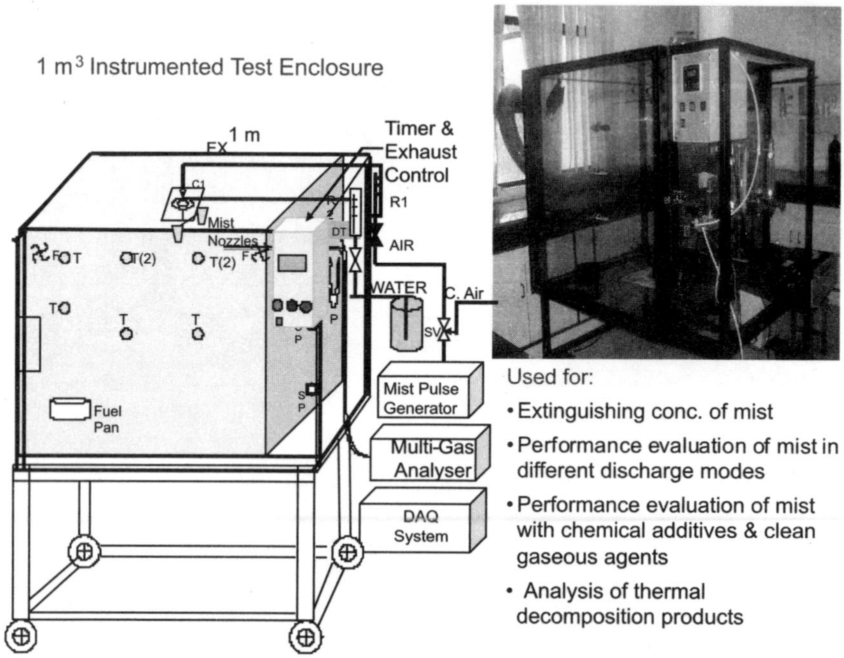


**Figure 7** 1 $m^3$ small scale fire test set-up

In 1 $m^3$ test enclosure, effect of three chemical additives i.e. sodium bi-carbonate mono-ammonium phosphate and ammonium acetate (1%, 2% and 5%) in water used for mist generation was studied. Experiments were conducted on all fire sizes of 5 kW to 34 kW in continuous and pulsed modes of mist discharge. Marginal improvement in extinguishment time was achieved for 2% MAP in both discharge modes, whereas no improvement was observed with other chemical additives.

An important area where water mist based system has great potential and work was carried out is the data-processing and telecommunication centres. A real-size electronic cabinet (0.45 m3) was obtained from Deptt. of Tele-communication and experiments were conducted with fine water mist (mean dia <20 mm) for extinguishment of simulated fire. The experiments demonstrated the extinguishment of 2kW simulated fire within 12 sec using only 4.7 gm of water in pulsed discharge mode of mist without damaging the live PCBs.

**Table 1** Fire suppression performance characteristics during continuous and pulsed discharge mode in 1 $m^3$ chamber

| *Fire Size (HRR)* | *Pre-burn duration (s)* | *Fire Suppression Time (s)* | | | *Water Mist Discharged @ 4.6 gm/s (gm)* | | |
|---|---|---|---|---|---|---|---|
| *kW* | | *Cont.* | *Pulsed* | *MAP 5%-Pulsed* | *Cont.* | *Pulsed* | *MAP 5% Pulsed* |
| 5 | 10 | 55 | 42 | 40 | 207.0 | 84.0 | 82.5 |
| | 20 | 53 | 40 | 37 | 198.0 | 80.0 | 76 |
| | 30 | 49 | 37 | 34 | 179.4 | 74.0 | 70 |
| 17 | 10 | 42 | 38 | 37 | 193.2 | 84.2 | 76 |
| | 20 | 36 | 33 | 31 | 165.6 | 73.0 | 63 |
| | 30 | 31 | 29 | 23 | 142.6 | 64.2 | 47 |
| 22 | 10 | 25 | 23 | 20 | 115 | 53.0 | 46 |
| | 20 | 18 | 17 | 15 | 82.8 | 39.0 | 34 |
| | 30 | 16 | 12 | 11 | 73.6 | 27.6 | 25 |
| 34 | 10 | 17 | 18 | 16 | 78.2 | 41.4 | 36 |
| | 20 | 15 | 14 | 13 | 69.0 | 32.2 | 30 |
| | 30 | 10 | 11 | 10 | 46.0 | 25.3 | 23 |

## Large Scale Fire Test Studies

Large scale experiments with varying location and directionality of mist release on fire of size up to 800kW were carried out in demonstration scale fire test enclosure of $40m^3$ (3.50m × 3.50m × 3.25m). The schematic of the large fire test facility in show in figure 7 he fire extinction time for different pool fire with all three different types of the system is indicated in the Table 2. For fire under the atomizer location, fire was suppressed in 6- 12 secs for 40 μm droplets.

Figure 9 represents the temperature profile at three different locations,. The maximum rise in temperature came down very sharply after injection of fine mist. The analysis of test results obtained from these three scales of experimentation indicate that total mass of water-mist required for fire suppression follows a linear scale-up law w.r.t. volume of the enclosure. This is a highly significant finding of this study and will be very for useful contribution in design of water-mist fire protection system for practical applications.

The concentration of oxygen came down to 18% with air as atomizing gas and 15% with nitrogen as atomizing a gas as indicated in figure 10.

Another application for which use of water-mist was explored is fuel air explosion suppression. A lab scale 8 litre explosion suppression vessel with automated control and recording of data was assembled and commissioned recently. Results from preliminary experimental runs indicate that reduction of up to 30-40% of the dry explosion the flame velocity & intensity of explosion is achieved using ~20 micron water mist.

## Conclusions

1. The interaction of water mists with a diffusion flame in an enclosed space was studied. If water mist with enough volume flux was applied to the diffusion flame, smaller droplets played the dominating role in fire extinction. The effective water flux is important

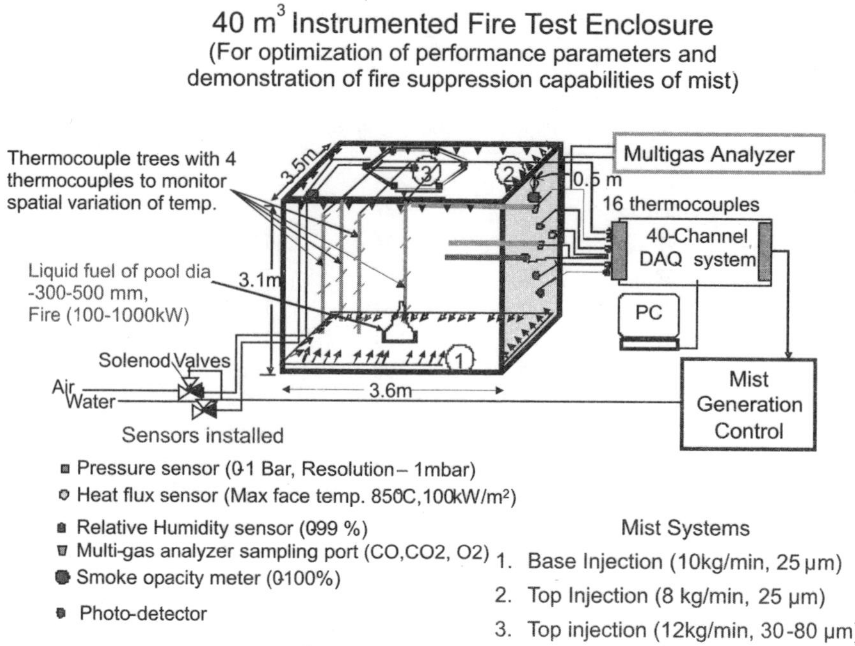


**Figure 8** Large scale fire test facility for fire test studies

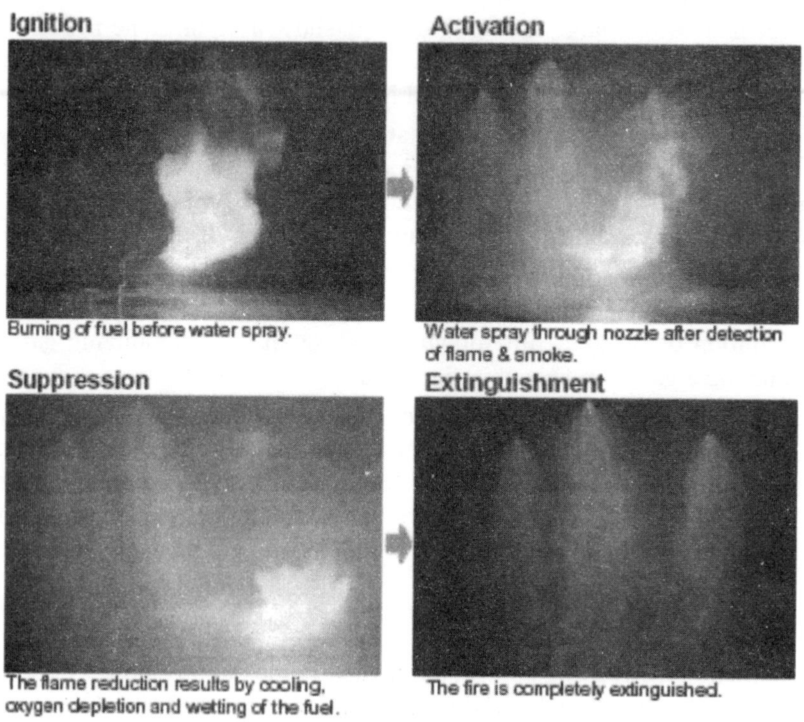


**Figure 9** Suppression of pool fire with counter flow fine mist

**Table 2** Fire extinction time for pool fires

| *Evaluation Parameter* | *Mist System-1* | | *Mist System-2* | | *Mist System-3* | |
|---|---|---|---|---|---|---|
| Atomization Technique used | Twin Fluid Internally mixed Atomizers | | Twin Fluid Air Blast Atomizers | | Twin Fluid Air Blast Atomizers | |
| Mist Injection Mode (Continuous) | Top injection | | Top injection | | Base injection | |
| Operating characteristics: | | | | | | |
| Water Supply Pressure (Pw) | 4.5 Bar | | 2.0 Bar | | 2.0 Bar | |
| Air Supply Pressure (Pa) | 15 Bar | | 15 Bar | | 15 Bar | |
| Air Flow rate (Qa) | 1400 lpm | | 1500 lpm | | 1500 lpm | |
| Water Flow Rate (Qw) | 8 lpm | 10 lpm | 8 lpm | 10 lpm | 8 lpm | 10 lpm |
| Droplet Size Distribution (SMD) | 31 μm | 40 μm | 25 μm | 30 μm | 25 μm | 30 μm |
| Test Fire Scenario: | | | Fire Extinction Time | | | |
| i) 300 kW in Centre | | 12 | | 12 | | 39 |
| ii) 300 kW in Corner | 39 | 35 | 43 | 49 | 48 | 55 |
| iii) 500 kW in Centre | 7 | 7 | 8 | 8 | 33 | 33 |
| iv) 500 kW in Corner | 35 | 35 | 15 | 12 | 37 | 37 |
| v) 800 kW in Centre | 6 | 6 | 14 | 12 | 34 | 24 |
| vi) 800 kW in Corner | 24 | 22 | 25 | 20 | | 30 |
| vii) Two pans of 300 kW & 500 | | 12 | | 12 | | - |
| kW placed closely: in Centre | | 18 | | 10 | | - |

**Table 3** Fire suppression performance using different atomizing

| *Features* | | *Water Feed rate Qw (lpm)* | *Droplet Size (μm)* | | *Fire Extinction time (secs)* | *Actual Qty of water used (lpm)* | *Mist Conc. (gm/m3)* |
|---|---|---|---|---|---|---|---|
| Gas Pressure | 15 bar | | SMD | VMD | | | |
| Atomizing Gas Used | $N_2$ | 8 | 26 | 60 | 24 | 3.2 | 80 |
| | | 10 | 31 | 40 | 36 | 6 | 150 |
| | Air | 8 | 26 | 60 | 50 | 6.6 | 165 |
| | | 10 | 31 | 70 | 64 | 10.6 | 265 |
| Obstructed | $N_2$ | 8 | 26 | 60 | 42 | 5.6 | 140 |
| | | 10 | 31 | 70 | 51 | 8.5 | 212.5 |

parameter for real applications, because it has a physical meaning in such situation, as the cooling of the fire plume rather than of fuel surface is believed to be the dominant extinguishment mechanism. The entire region within the fire extinction limit can be divided into two parts, sudden cooling & gradual cooling and water mist plays dominating role in both the regimes.

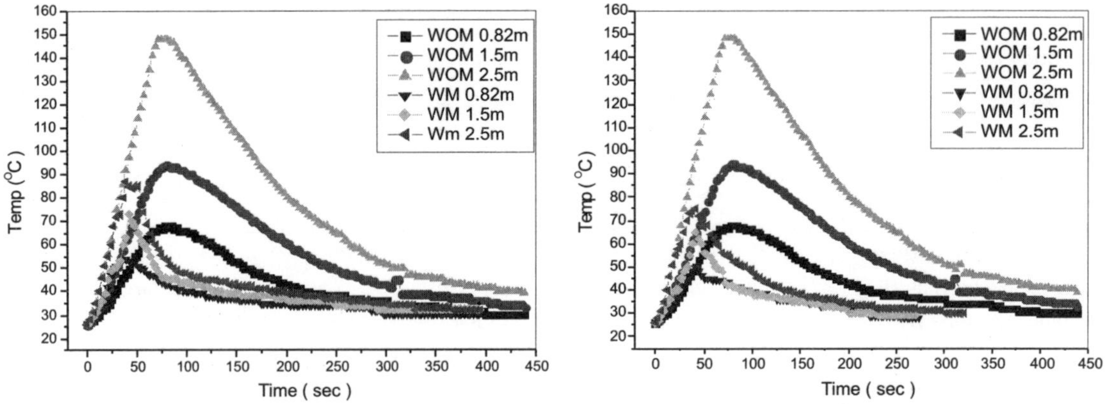


Temperature profile for 800 kW HRR with air as atomizing gas

Temperature profile for 800 with N2 as atomizing gas kwHRR

**Figure 10** Temperature profile at different locations in the enclosure

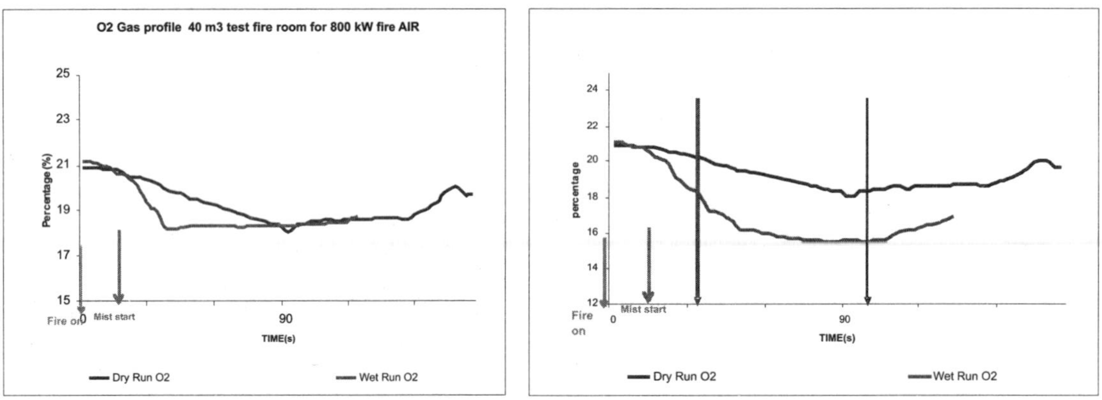


Oxygen concentration profile with air as atomizing gas for 800 kW HRR

Oxygen concentration profile with N2 as atomizing gas for 800 kW HRR

**Figure 11** $O_2$ concentration in the 40 $m^3$ with different atomizing gas

2. The critical concentration of mist for suppression of unobstructed fire was 80-130 gm/$m^3$ and 125-270 gm/$m^3$ with nitrogen and air respectively. The critical concentration of mist for suppression of obstructed of pool fires was measured as atomizing gas and 150-183 gm/$m^3$ and 250-370 gm/$m^3$ with nitrogen and air respectively.
3. The experimental results indicate that 40 μm mist is more effective as compared to larger droplets in suppressing a pool fires. As HRR of pool fire increase, it becomes easier to extinguish and requires lesser mist concentration. Obstructed fires require more concentration as mist enter only through open spaces to reach the combustion zone, which leads to higher extinction time as compared to unobstructed pool fires.

4. Larger the fire size, less is the time required to extinguish the fire. This can be attributed to the larger heat available for evaporation of the water mist droplets leading to attaining the inerting environment faster by formation of water vapor and causing dilution of oxygen concentration below that required for sustained combustion reaction. The oxygen concentrations measured typically in the ranged from 17-18 percent by volume, which was further lowered up to 16 percent with use of nitrogen as atomizing gas.
5. Faster reduction in compartment temperature with water mist indicate that total flooding is very effective in cooling hot gases, which accumulate near the ceiling. This can be attributed to the better mixing of water mist with hot gases during top injection and hence larger heat extraction from plume and faster evaporation of water mist droplets. This leads to faster extinguishment of fire due to (a) cutting down the thermal feedback to the fuel zone (b) diluting entraining air into the flame zone and plume with water vapour and inert combustion products, and (c) achieving unsustainable combustion conditions faster by diluting the fuel-oxygen/air concentration in the fire zone with water vapor.

## References


[1] Braidech MM, Neale JA, Matson AF, Dufour RE. The mechanism of extinguishment of fire by finely divided water. Underwriters Laboratories Inc. For the National Board of Fire Underwriters, 1955. p. 73.

[2] Rabash DJ, Rogowski ZW, Stark GWV. Combust Flame 1960; 4:223–4.

[3] Mawhinney JR, Dlugogorski BJ, Kim AK. A closer look at the fire extinguishing properties of water mist. Proceedings of the Fourth International Symposium on Fire Safety Science. International Association for Fire Safety Science, 1994.

[4] Sheinson RS, Ayers S, Williams FW, Adiga KC, Hatcher Jr RF. Feasibility evaluation study of very fine water mist as a total flooding fire suppression agent for flammable liquid fires. In: Proceedings halon options technical working conference, Albuquerque, New Mexico, 2004.

[5] K.C. Adiga, Robert F. Hatcher Jr, Ronald S. Sheinson, Frederick W. Williamsb, Scott Ayersc, Fire Safety Journal 42 (2007) 150–160 A computational and experimental study of ultra fine water mist as a total flooding agent.

[6] Muller-DethlefsK, Schalder AF. Combust Flame 1976; 27:205–15.

[7] Zhang C, Atreya A, Lee K. Proceedings of the 24th Symposium (International) on Combustion. 5–10 July 1992. University of Sydney, Australia, 1992. pp. 1049–57.

[8] Rasbash DJ, Rogowski ZW, Stark GWV. Mechanisms of extinction of liquid fires with water sprays. Combustion and Flame 1960;22:223–34. 316 G. Heskestad / Fire Safety Journal 38 (2003) 301–317.

[9] Braidech M, Neale JA, Matson AF, Dufour RE. The mechanism of extinguishment of fire by finely divided water. Report No. 10, The National Board of Underwriters, 1955.

[10] Nakakuki A, Takahashi M. Liquid fuel fires—extinction with water sprays. Bull. Fire Prevention Soc Japan 1971; 20:37–53.

[11] Wolfe J, DeSipio P. Results of fire extinguishment testing with fine water mist. Report No. NAWCADWAR-94123-60, Naval Air Warfare Center, Aircraft Division Warminster, prepared for Naval Air Systems Command, Washington, DC, July 30, 1944.

[12] Kim Myung-Bae, Jang Yong-Jae, Yoon Myoung-O. Extinction limit of a pool fire with water mist. Fire Safety J 1997; 28:295–306.

[13] NFPA 750. Standard on water mist fire protection systems, 2000.

[14] Suresh Lal, A. Kushari, J.C Kapoor and S. Maji., "Characterization of an Externally Mixed Air – Assisted Atomizer", 10th International Congress on Liquid Atomization and Spray Systems, Aug 27 – Sep 01,2006, Kyoto, Japan.

[15] A. Kushari, Y. Neumeier, O. Israeli, E. Lubarsky and B. T. Zinn, "Internally Mixed Liquid Injector for Active Control of Atomization Process", *Journal of Propulsion and Power*, Vol. 17, No. 4, 2001, pp. 878-882.

[16] Dougal Drysdale, An Introduction to Fire Dynamics, A Wiley-Interscience publication.

[17] Particle Size Analysis, Guide to Laser Diffraction Methods, ISO / DIS 13320

[18] Yao B, Fan W, Liao G. Experimental study on the interaction of water mists with fires in the confined space. Proceedings of the Third Asia-Oceania Symposium on Fire Science and Technology, 10-12, June, 1998, Singapore.

[19] Downie B, Polymeropoulos C, Gogos G. Interaction of a water mist with a buoyant methane diffusion flame. Fire Safety J 1995;24:359}81.

[20] Fristrom RM, Westenberg AA. Flame structure. New York: McGraw-Hill, 1965.

[21] Baulch DL, Drysdale DD. Combust Flame 1974;23:215–25.

[22] Puri R, Santoro RJ. Fire Safety Science. Proceedingsof the Third International Symposium, 8–12 July 1991, University of Edinburgh, Scotland, 1991. pp. 595–604.

# Air Quality and Climate Impacts of Aviation Emissions: Challenges and Solution Strategies

**Mohan Gupta, Curtis Holsclaw and Lourdes Maurice**
Office of Environment and Energy, Federal Aviation Administration, USA

## INTRODUCTION

Aviation has helped the world become a global society over the last 50 years by providing mobility and means of transport in a safe and efficient, fast and affordable way – the essential ingredients of economic stability, growth and security. About 2.2 billion passengers fly on scheduled passenger aircraft, and air traffic is expected to increase to two to three times the present level over the next two decades, albeit this is likely to be tempered by the significant recent spike in oil prices[1]. Trends in aviation growth have generally followed and outpaced the trends in economic growth. During the timeframe from 2005 to 2025, aviation traffic is projected to increase at an annual rate of 4.6% as compared to 3.5% for the annual world GDP[2]. Most of this growth in aviation is expected to originate from travel within and to/from the developing world, although again the impact of higher fuel costs is yet to be assessed.

Traditionally, growth in aviation is governed by various factors including economic growth, fuel prices, airline productivity gains and airport/airspace capacity. However, concerns about aviation's impacts on the environment, which have accompanied its growth, impact human health and welfare and could potentially restrict the ability of the aviation system to grow. The extent of emissions growth and resulting environmental impacts due to the projected growth in aviation will be influenced by several factors including the mix of future aircraft fleet as well as implementation of new aircraft technologies and alternative fuels, efficient operational procedures and air traffic control, market based options and regulatory interventions. Nonetheless, there is a need to maintain an appropriate balance between environmental protection, sustained growth of aviation and economic benefits.

Noise and air quality environmental impacts of aviation need to be better characterized and uncertainties associated with these characterizations need to be reduced to levels that enable defining attainable goals and developing effective solutions in a cost-beneficial manner. Historically, technological advances have made the major contribution toward reductions in aircraft noise and emissions. For example, currently operational aircraft are about 70% more fuel efficient and generate equally less noise than the ones that were in use during 1980s[2]. However, further improvements in new technologies, operational procedures and air space management guided by fully evaluated interdependencies and tradeoffs between noise and emissions and their environmental impacts are needed to sustain aviation growth.

The focus of this paper is on aviation emissions and their impacts on air quality and climate. We also provide insights on how integrated solution strategies are being implemented and will be expanded in the future to mitigate these impacts.

## AIRCRAFT EMISSIONS

Aircraft engines produce emissions that are similar to other emissions produced during the combustion of fossil fuels. However, there is a key difference between emissions from aircraft and those from other sources in that aircraft release emissions both near the surface and aloft along the flight trajectory. The bulk of aircraft emissions (~90%) occur at higher cruise altitudes above the airport terminal area which is typically considered to be 3,000 feet above ground level. Like waterborne transport, international aviation also joins national boundaries through direct release of emissions along the flight path and also at the points of origin and destination. Aircraft fuel combustion and emissions are functions of various parameters including airframe and engine type, the number of flights, engine operating conditions and payload. Common byproducts of aircraft engine combustion are: carbon dioxide ($CO_2$), water vapor ($H_2O$), nitrogen oxides (NOx or $NO+NO_2$), carbon monoxide (CO), sulfur oxides (SOx), unburned hydrocarbons (UHC, including hazardous air pollutants or HAPs) and particulate matter (PM) as shown in Figure 1. The exhaust gas is roughly 70% $CO_2$, a little less than 30% $H_2O$ and less than 1% comprising of emissions of other air pollutants.

The International Civil Aviation Organization (ICAO) has established standards for HC, CO, $NO_X$ emissions and smoke number at four different modes of engine operation during the landing and takeoff (LTO) cycle: takeoff (100%, 0.7 min), climb (85%, 2.2 min), approach (30%, 4 min) and taxi (7%, 26 min). The numbers in bracket represent the engine thrust setting and time duration corresponding to each mode of operation. These certification based time-in mode and respective engine thrust levels may differ from those that are experienced in real operational activities. There are also known start up emissions but their characterization is not well established.

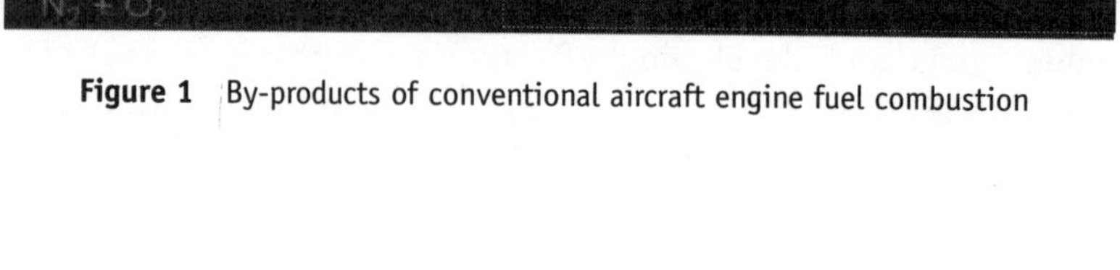


**Figure 1** By-products of conventional aircraft engine fuel combustion

In general, specific rates of direct release of different air pollutant emissions from aircraft are dependent upon operating mode. For example, NOx emissions are higher at higher engine thrust. NOx emissions play an important role in air quality (through pathways that lead to the formation of PM and ozone) and climate impacts (through changes in greenhouse gases such as ozone and methane). Production of NOx could be reduced by controlling combustor temperatures and decreasing the residence time during high temperature operations. However, high temperature combustion is desirable for the maximum thermodynamic efficiency and consequently for lowest fuel consumption. Clearly, there is always a need to achieve an optimum balance between the two – more importantly given high fuel prices and environmental demand to reduce $CO_2$ emissions due to climate impact concerns, as noted later in this paper. There are also similar interdependencies and potential tradeoffs among emissions of HC, NOx and PM. For example, the production of volatile PM emissions is greatest at low thrust setting while non-volatile black carbon emissions are predominate at higher engine power settings. It is therefore important to characterize the relative environmental impact of each air pollutant as emissions control and impact mitigation strategies are devised.

## ENVIRONMENTAL IMPACTS

There are environmental impacts associated with a given configuration of aviation system. There are often also interdependencies and tradeoffs among noise, emissions and fuel burn whenever any changes are introduced in the system. Therefore, it is important to fully characterize the underlying impacts including uncertainties in order to help define the goals of environmental protection and inform future actions. Normally, there is not a single option that would uniquely address all concerns for environment and energy. The use of this information becomes more important when multiple options are considered to develop optimum solutions in a cost-beneficial manner. There are ranges of costs and benefits associated with each option. For example, emissions of all air pollutants are not equally important from an environmental impact point of view and the costs associated with measures to reduce emissions of one air pollutant versus another could also differ significantly. Likewise, given associated non-linearities, the costs and benefits would potentially differ when emissions of an air pollutant are reduced from two different starting points.

In 2004, ICAO adopted three environmental goals: (1) limit or reduce the number of people affected by significant aircraft noise; (2) limit or reduce the impact of aviation emissions on local air quality; and (3) limit or reduce the impact of aviation greenhouse gas emissions on the global climate[2]. Under the ICAO, the Committee on Aviation Environmental Protection (CAEP) is charged with addressing environmental concerns associated with aviation. The current ICAO/ CAEP practice relies upon technological feasibility, economic reasonableness and the environmental benefits to assess the various options considered to mitigate environmental impacts. Even though the use of the term ‘impacts’ is very much at the foundation of the ICAO environmental goals, environmental impact analyses and cost-benefits aspects of options considered are not generally included in the analyses used to inform the decision-making. Realizing the importance of considering impacts analysis in its decision making to reduce noise and emissions in a cost-beneficial manner, the CAEP recently convened a workshop on “Assessing current scientific knowledge, uncertainties and gaps in quantifying climate change, noise and air quality aviation impacts". This workshop recommended CAEP to move from emissions inventory *only* to environmental impacts as the basis of its decision-making. This report will soon be released by the ICAO/CAEP.

To meet the aviation demand projected over the next two decades, the United States is transforming its aviation system in the form of the dynamic, flexible and scalable Next Generation Air Transportation System (NextGen) while being responsive to continuously evolving social, economic, political and technological changes[3]. Environmental is at the heart of NextGen with an objective of environmental protection that would allow sustained aviation growth. The stated environmental goals of NextGen include: (1) reduce significant community noise and air quality emissions in absolute terms and the rate of aircraft fuel burn, (2) limit or reduce the impact of aviation greenhouse gas emissions on global climate; and (3) proactively address emerging environmental issues (e.g. water quality). The United States has developed a National Plan for Aeronautics Research and Development and Related Infrastructure[4] that will help to address the NextGen goals. One of the goals of this plan is to develop technologies and operational procedures to decrease the environmental impacts of aviation system. The objectives of this goal are being supported through the development of improved tools and metrics to quantify and characterize aviation's environmental impact, uncertainties and tradeoffs as well as interdependencies among various impacts.

In support of attaining U.S. National Aeronautics and NextGen environmental goals, the Office of Environment and Energy of the U.S. Federal Aviation Administration (FAA/AEE) is actively involved in development of an integrated suite of tools that will help to analyze aviation system, environmental impacts and related tradeoffs and interdependencies based on the cost-benefit approach (see Figure 2). The Environmental Design Space model (EDS) estimates aircraft performance tradeoffs for different technology assumptions and policy scenarios. The Aviation Environmental Design Tool (AEDT) produces estimates of noise and emissions inventories at global, regional and local levels using detail input on fleet descriptions and flight schedules. AEDT development leverages FAA's current environmental tools[5]; namely INM (Integrated Noise Model), NIRS (Noise Integrated Routing System), EDMS (Emissions and Dispersion Modeling System), MAGENTA (Model for Assessing Global Exposure to Noise from Transport Aircraft) and SAGE (System for assessing Aviation's Global Emissions). The Aviation Portfolio environmental Management Tool (APMT) includes both economic and environmental impact analyses capabilities. For the environmental impact analyses, APMT takes the outputs from AEDT and performs comprehensive analyses for climate, ambient air quality and noise impacts.

The FAA and other organizations are pursuing various research activities under PARTNER (Partnership for AiR Transportation Noise and Emissions Reduction)[6]. PARTNER, a leading world-class aviation cooperative research organization, is sponsored as a Center of Excellence by FAA, NASA and Transport Canada. PARTNER research portfolio on multiple fronts includes characterization of aircraft emissions and their related air quality, health and climate impacts. PARTNER is also actively engaged in assessing and advancing alternative aviation fuels for commercial aircraft. Below we provide a brief account of the research approach that FAA is undertaking to characterize environmental impacts of aviation emission and reduce associated uncertainties. This research also provides the scientific and technical input to support the aviation emissions impacts analysis component of APMT.

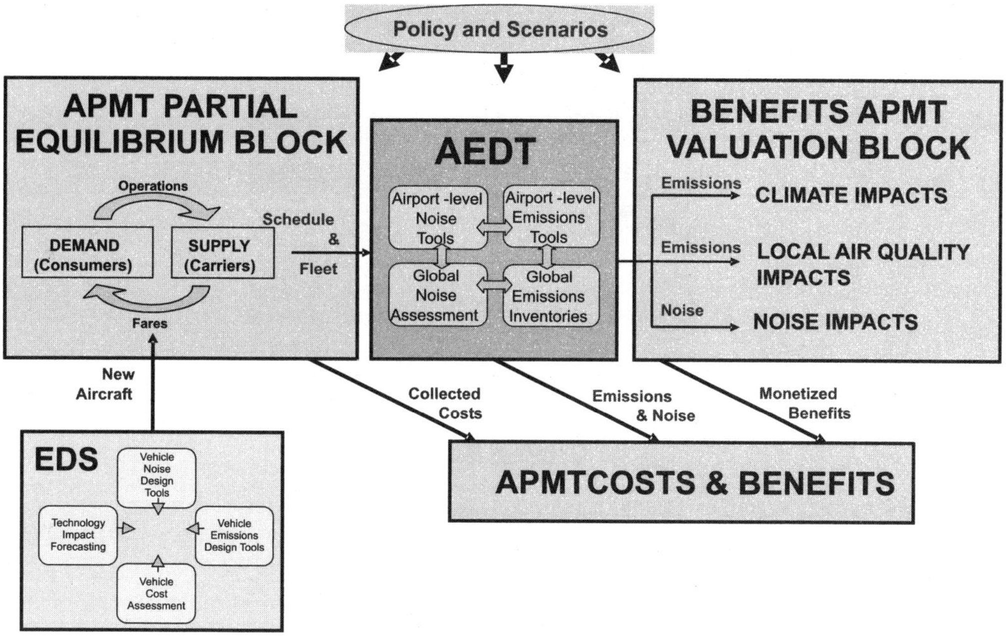


**Figure 2** Schematic of the FAA-NASA-transport Canada integrated suite of tools for better understanding of cost-benefit aspects of aviation related environmental issues

## AIR QUALITY IMPACTS

Although aircraft noise issues remain the primary environmental challenge for airports in the U.S., aviation related air quality issues are getting increased attention. At present emissions from aviation operations comprise a relatively small portion of the total gaseous and particulate emissions from all sources in the U.S. With projected growth in aviation (albeit this may be tempered by increasing oil prices), emissions of some pollutants from aviation sources will likely increase against the generally decreasing emissions from background sources. Therefore, the air quality impacts of aircraft engine exhaust emissions may become a fundamental constraint on the growth of air transportation.

Figure 3 displays the partial distribution of top 50 airports that are located within non-attainment air quality regions either for ozone or $PM_{2.5}$. The number of these airports in non-attainment areas is anticipated to increase particularly as more stringent air quality standards are implemented. For example, the EPA has recently revised the 8-hour ozone standard from 80 ppb to 75 ppb[7]. This change has increased the number of counties in ozone non-attainment from 85 to more than 300.

At the airport, aircraft are not the only source of emissions. There are numerous other sources (including stationary sources, ground support equipment, ground access vehicles, and auxiliary power units) that also contribute to overall emissions originating from airport activities. In general, aircraft contribute the most to airport emissions. In a "business as usual" scenario (that is no new technologies or more efficient operations are introduced), increases in aircraft operations also leads

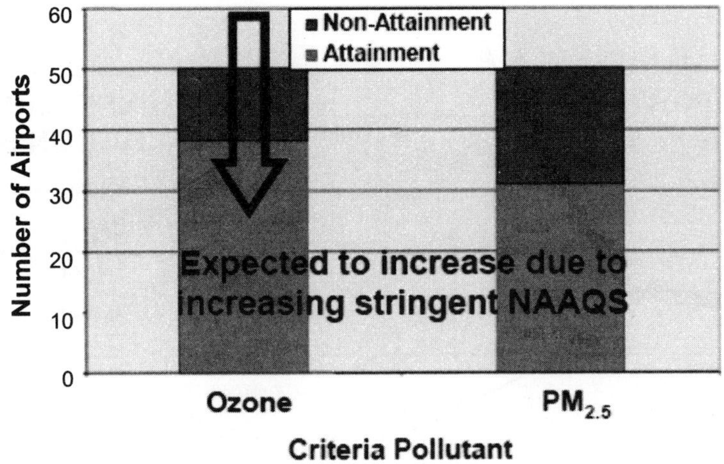


**Figure 3** Number of U.S. airports located in counties that are presently in non-attainment for ambient air quality standards for ozone and/or $PM_{2.5}$

to increases in non-aircraft emissions from the airport. Air quality environmental concerns tend to pertain to effects due to aircraft emissions during the LTO cycle. These emissions typically occur up to 3,000 feet (or 0.984 km) above ground level. Potential adverse effects of air pollutants on air quality primarily pertain to human health and welfare. There are no known unique emissions and health impacts associated with aviation emissions. Once released in the atmosphere, aviation emissions evolve in similar ways as those due to non-aviation sources.

Air quality and health impacts of airport/aircraft emissions are both local and regional (that could extend as far as ~300 km away from the airport) in nature primarily due to direct emissions of air pollutants (such as PM, HAPs) and secondary formation of air pollutants (such as ozone, PM and HAPs) respectively. The secondary air pollutants are formed due to long-range transport and chemical as well as microphysical transformation of directly emitted air pollutants while interacting with background air. The extent of their dispersion is governed by the background meteorology and associated atmospheric lifetimes. It is important to quantify the present and future trends in 'incremental' air quality impacts due to aviation emissions within the context of those due to background air. These trends in impacts will evolve in the future due to changing aviation and background emissions, potential variability in meteorological conditions driven by climate change as well as changing population distribution within the regions of interest. There are indications that higher altitude aircraft emissions may potentially contribute to surface air quality not only in the regions of flight corridors but also on a hemispheric basis[8]. However, more studies are needed to develop reliable understanding and estimates of the impacts of cruise aircraft emissions on surface air quality.

The FAA is undertaking a comprehensive approach following the flow chart shown in Figure 4 to best characterize aircraft emissions and to quantify air quality and health impacts through a comprehensive suite of measurement and modeling projects. Measurements of emissions at the engine exit plane and evolution of the plume as well as air quality monitoring at the community

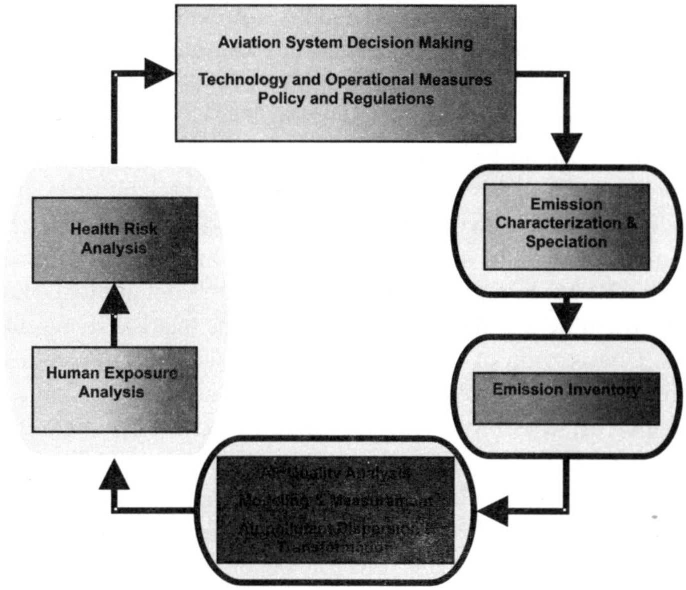


**Figure 4** A schematic to relate aircraft emissions to health impacts analysis

scale around the airport are being pursued. The purpose is to correlate the observed variability in air quality with airport level activity, hence emissions, as well as other factors including variability in meteorology and background air. Given the sub- and large-scale changes in air quality due to aviation emissions, air quality simulations using both plume models (e.g. AERMOD[9]) as well as variable grid size urban/regional scale model (e.g. CMAQ - Community Multiscale Air Quality Modeling System[10]) are being used. The FAA also supports the Airport Cooperative Research Program (ACRP) managed by the Transportation Research Board[11]. For the air quality specific projects, the ACRP approach is similar to that shown in Figure 4. Both FAA and ACRP air quality research activities are coordinated with a common objective to characterize the aircraft emissions and the related air quality impacts.

AERMOD is a steady-state plume model that simulates air dispersion based on planetary boundary layer turbulence structure and scaling concepts, including treatment of both surface and elevated sources, and both simple and complex terrain. Concentration distribution of air pollutants (considered as inert species) simulated by AERMOD are generally applicable within a 50 km radius region around the emission source. AERMOD is an EPA (U.S. Environmental Protection Agency) recommended model for plume dispersion and is integral to FAA models such as EDMS and AEDT. CMAQ is a state-of-the-art comprehensive one-atmosphere air quality modeling system that combines current knowledge in atmospheric science and air quality modeling with multi-processor computing techniques in an open-source framework to deliver fast, technically sound estimates of ozone, particulates, toxics, and acid deposition.

The FAA has funded a number of activities to better characterize aircraft particulate emissions and speciation of aircraft hydrocarbon emissions, to improve AERMOD dispersion capability using LIDAR measurements of aircraft specific plume characteristics and to better represent 4-

dimensional nature of aircraft emissions in air quality simulations. LIDAR (Light Detection and Ranging) is an active optical remote sensing technology that measures properties of scattered light due to particles within the path of propagating laser pulse. LIDAR instruments are capable of measuring 3-dimensional geometry of plume evolution.

Figure 5 shows the variable grid size options for CMAQ model that are being presently exercised. The use of hybrid models that combine subscale variability in air quality changes as simulated by the AERMOD model with the larger urban and regional scale changes simulated by CMAQ is also being explored. In addition, PARTNER has recently developed a parametric response surface model (RSM) to examine perturbation to ambient particulate matter concentration distribution using an ensemble of CMAQ model simulations that are representative of wide range of aircraft emissions. Comparison of the performance of this RSM model has shown remarkable results against the results obtained from the CMAQ simulation for the same emission scenarios. The purpose for pursuing these multiple options is to configure the optimum framework that can be efficiently and routinely used for analyses of aviation emissions air quality impacts.

Progressive results from multiple studies related to aircraft emissions performed to date conclusively indicate that the role of secondary air pollutants should not be ignored during air quality impact analyses[12]. In the case of PM, this role becomes even more important as the distance from the source region increases as shown in Figure 6. The secondary formation of PM is due to chemical and microphysical transformation of directly emitted gaseous emissions and their interactions with the background air during their transport from source region to receptor region. Model simulations also indicate that ozone, a secondary air pollutant, decreases near the region of NOx emissions due to ozone-NOx titration; however, it increases as the distance from the source region increases, due to complex interplay of NOx and hydrocarbons chemistry and transport. The

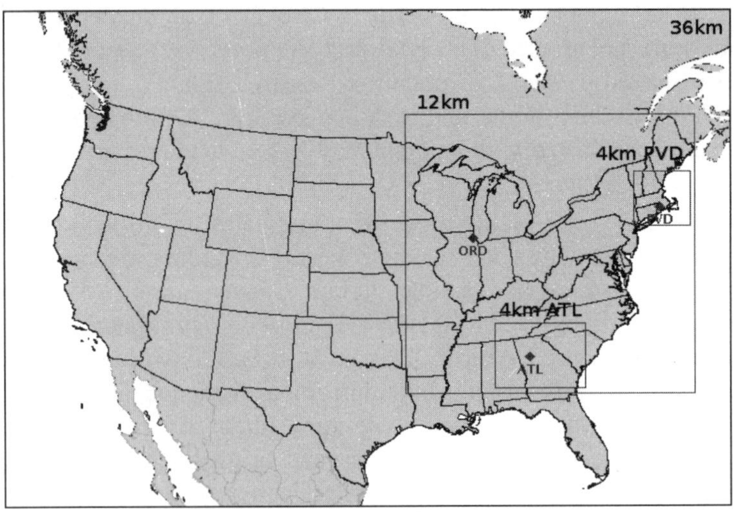


**Figure 5** An example of nested multiple grid resolution options that are currently being exercised to evaluate the air quality impacts of aviation emissions

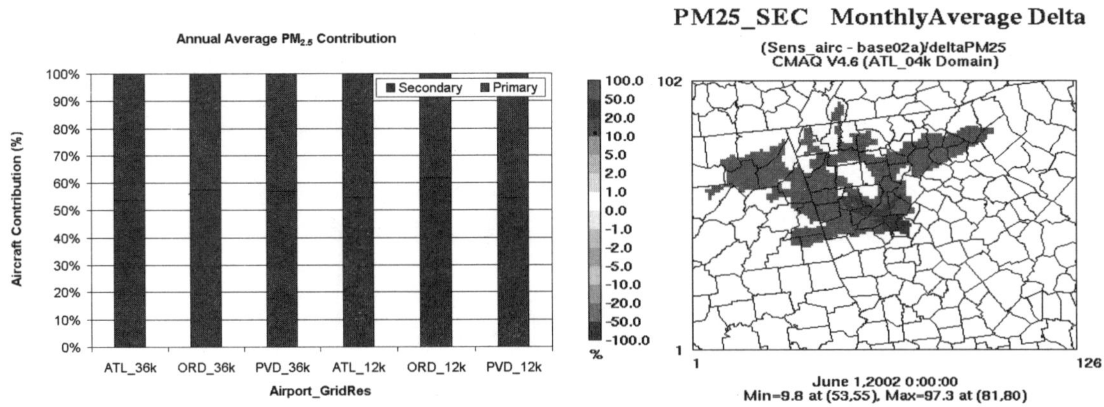


**Figure 6** Relative contributions from direct emissions and secondary formation to the overall change in PM concentration due to aircraft emissions (a) within the grid that contains the airport and (b) away from the source region

qualitative nature of this finding is generally applicable to emissions from all other sources and should not be considered uniquely applicable to aviation alone.

Simulated and/or measured changes in air quality (e.g. PM, ozone, HAPs) due to combined primary emissions and secondary formation along with emissions, population distribution and potency values are collectively used to develop health exposure risk estimates for aviation related air pollutants of interest. As an example, Figure 7 displays the total population risk distribution for changes in PM concentrations due to aircraft emissions from three airports in the eastern U.S. In this case, the annual averaged change in surface PM concentration was calculated using the CMAQ model with $12 \times 12 km^2$ grid resolution in horizontal direction. There are several hot points mainly around the airport and at the places corresponding to local maxima of the product of change in concentration and localized population.

This health risk analysis framework is also being used to prioritize the air pollutants that need to be addressed based on their health risk exposure and also to inversely determine the minimum amount of emissions needed to exceed the health risk concentration threshold. For aircraft emissions alone, the analyses performed to date indicate that the health impacts are dominated by PM (as the sum of direct emissions as well as secondary formation) followed by ozone and HAPs. This path of relating emissions to health risk exposure is quite routinely used in the public health community. Our primary objective is to quantify the 'incremental' changes due to aviation emissions and within the context of background emissions for both present and future conditions.

The schematic shown in Figure 5 is also being used to investigate how projected change in future aircraft activities under one of the NextGen scenarios for years 2015 and 2025 will change aircraft emissions relative to emissions for year 2004 and how these aircraft emissions would impact background air quality as well as health risk exposure. This study will also implicitly consider future changes in non-aircraft emissions as well as potential changes in meteorology, ambient temperature and water vapor due to changing climatic conditions.

In summary, significant progress is being made on multiple research fronts to best characterize the air quality impacts of aviation emissions and related uncertainties that would ultimately help to

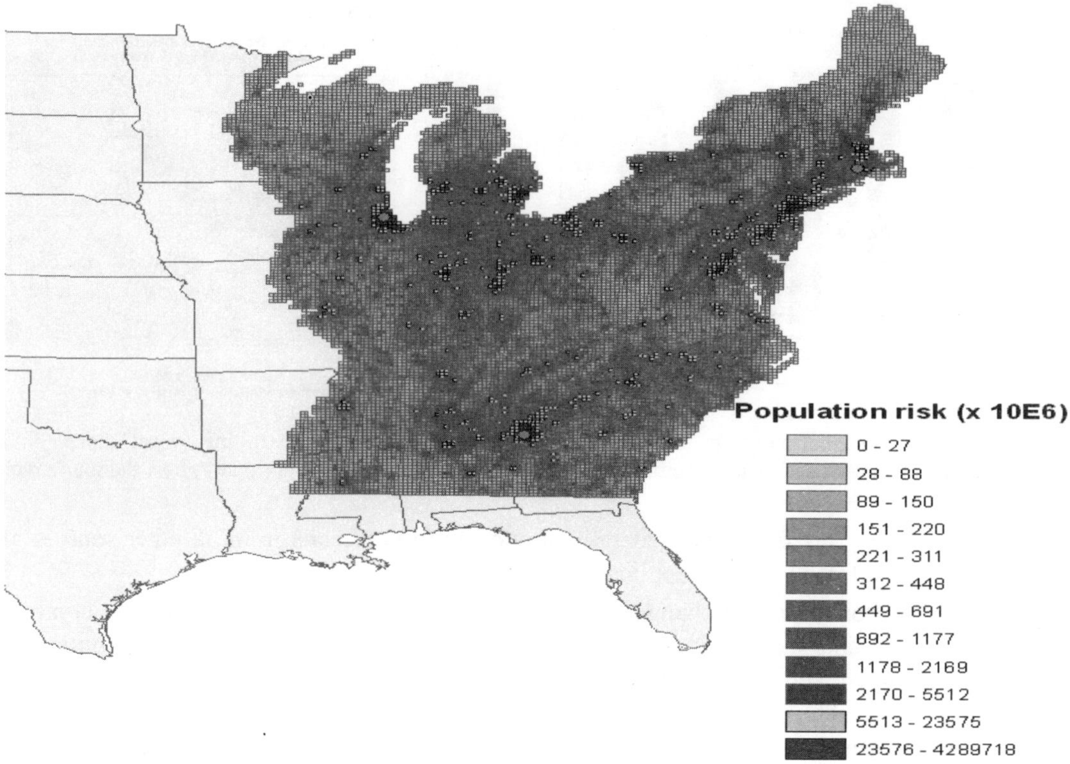


**Figure 7** Maps of total population risk by grid cell using 12 km resolution CMAQ runs. The values in each grid cell represent number of mortalities per year m ultiplied by $10^6$

inform the preferred future directions for mitigation of impacts. Note that some of the results presented here are preliminary and will likely be revised as further improvements are made in the input values and analyses approach.

## CLIMATE CHANGE IMPACTS

Aircraft typically operate at cruising altitudes of 8-13 km where they release emissions which may alter the composition of atmosphere and potentially contribute to climate change. Aircraft engines produce emissions that are similar to other emissions from fossil fuel combustion. These emissions undergo complex chemical as well as microphysical transformation processes and subject to atmospheric transport processes ranging from plume to global scales. Better characterization of these processes through measurements and numerical representation in atmospheric models is an active area of scientific research.

At present, aviation contributes 2-3% of anthropogenic GHGs emissions. Aviation accounts for 13% of the total GHG emissions from the transport sector. However, given the projected increase in aviation, the aircraft emissions are likely to grow in absolute terms and as a percentage of the total.

The extent of changes in emissions and resulting climate impacts due to aviation will be influenced by several factors including the mix of future aircraft fleet as well as implementation of engine and fuel technological advances and insertion into the fleet, more efficient operational procedures, market based options and regulatory interventions.

It is important to note that, like other source emissions, there are 'direct' and 'indirect' chemical and climate impacts associated with aircraft emissions. There is, however, an exception when it comes to the formation of condensation trails (or contrails) and induced cirrus clouds. The formation of contrails is the most physically visible perturbation to the atmosphere. When aircraft encounter sufficiently cold, ice-supersaturated air masses, the water vapor released by engines triggers the formation of contrails which may also increase cirrus cloudiness. Both contrails and induced cirrus cloud could potentially exert indirect impacts on the climate. The release of nitrogen oxides emissions contributes to enhanced photochemical production of tropospheric ozone ($O_3$) and loss of methane ($CH_4$). Both $O_3$ and $CH_4$ are also known as greenhouse gases. Therefore, aircraft NOx emissions indirectly affect the climate. Aircraft emissions of $CO_2$, $H_2O$ and carbon soot directly affect the climate. Gaseous emissions of SOx undergo gas-to-particle conversion and the resultant sulfate aerosols directly affect the climate. Figure 8 displays a sequence that relates aircraft direct emissions to change in chemical composition, climate response and impacts of that climate response on social welfare.

Following a request from ICAO, the IPCC developed the only report to date assessing sector specific (aviation) impacts of emissions on atmospheric composition and climate change[13]. Recently, the IPCC Fourth Assessment Report (AR4)[14] characterized net aviation contribution at about +3% (nearly half due to $CO_2$ alone) of the overall climate change for 2005 as shown in

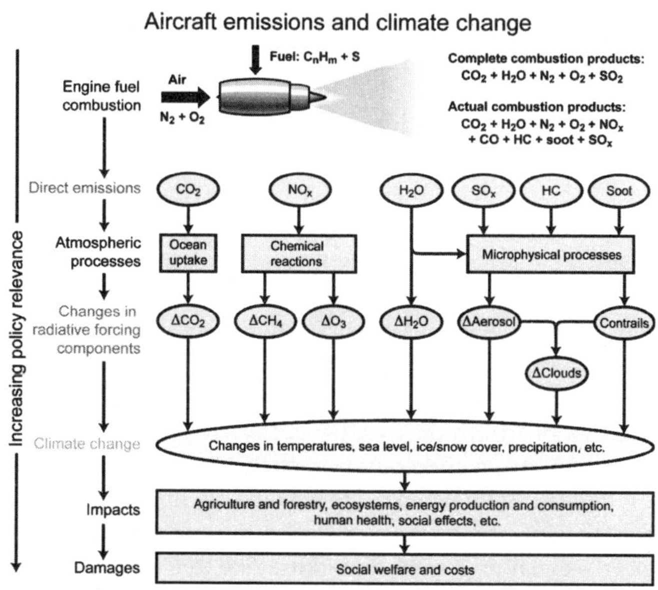


**Figure 8** A schematic to relate aircraft emissions to climate change and impacts on social welfare

Figure 9[15]. Note that this estimate does not include effects due to contrail-induced cirrus clouds due to the very low level of scientific confidence associated with these effects. Radiative forcing (RF) is used as a metric in Figure 9 to represent instantaneous measure of climate response due to direct and indirect atmospheric perturbations brought by release of emissions. Positive RF represents heating of the Earth's atmosphere and a negative forcing represents a cooling. Note that there are negative RFs associated with changes in $CH_4$ and sulfate aerosols caused by aircraft gaseous emissions of NOx and SOx respectively.

Given the release of emissions at higher altitude, in general, the non-$CO_2$ aircraft emissions are more efficient in perturbing the atmosphere on per unit emissions basis as compared to emissions at the surface. However, this is not the case for $CO_2$ emissions. Due to its long atmospheric lifetime (>100 years), no matter what the source of $CO_2$ emissions is or where in the atmosphere it is emitted, $CO_2$ has the same climate effect.

As shown in Figure 9, there are wide ranges of spatial scales associated with these non-$CO_2$ perturbations due to aviation emissions which are primarily governed by the atmospheric lifetime associated with changes in gaseous, contrails/cirrus clouds and aerosol species. As noted in Wuebbles *et al.* (2008)[17], the RF has been widely used as a climate change metric for long-lived and well mixed greenhouse gases. However, there are continued questions about its viability and usefulness for other short-lived (e.g. hours to days for contrails and aerosols) and spatially inhomogeneously distributed changes for ozone, aerosols, contrails and clouds[18]. Therefore, within the context of aviation, there is a clear need for identifying a suitable metric that can inter-relate aviation induced climate responses caused by long- and short-lived changes and compare them against the climate impacts due to non-aviation emission sources.

Note that there is medium to low level of scientific understanding in quantifying the climate response due to non-$CO_2$ perturbation caused by aviation emissions[15]. The science behind climate change effects from aviation is complex and evolving. In fact, a report to the U.S. Congress on Aviation and the Environment stated that the climate change impact of aircraft emissions is the topic of greatest uncertainty and contention[19]. This complexity can give rise to inaccurate and misleading interpretation of the dimension of current and future contributions from climate change. The IPCC (1999) report[13] clearly stated that further work is required to reduce scientific and other uncertainties to better design options for reducing emissions, to better inform decision-makers and to improve the understanding of the social and economic impacts associated with the demand for air transport.

According to IPCC, climate change is already having significant impacts in certain regions of the world and on most ecosystems. Economic assessment of climate change indicates that the cost of inaction will most probably exceed the cost of taking the early action. The IPCC AR4 noted that the mitigation of $CO_2$ emissions from the aviation sector can come from improved fuel efficiency based on advanced aircraft/engine technology, operational procedures and air traffic management. However, the overall mitigation will also need to account for non-$CO_2$ climate impacts of aviation as these effects are better understood. It is important to note that in case of aviation, non-$CO_2$ climate impacts are nearly equal or greater than those due to $CO_2$ alone. To support the NextGen environmental goals discussed above, under the auspices of Joint Planning and Development Office the FAA and NASA jointly convened an international workshop on the Impacts of Aviation

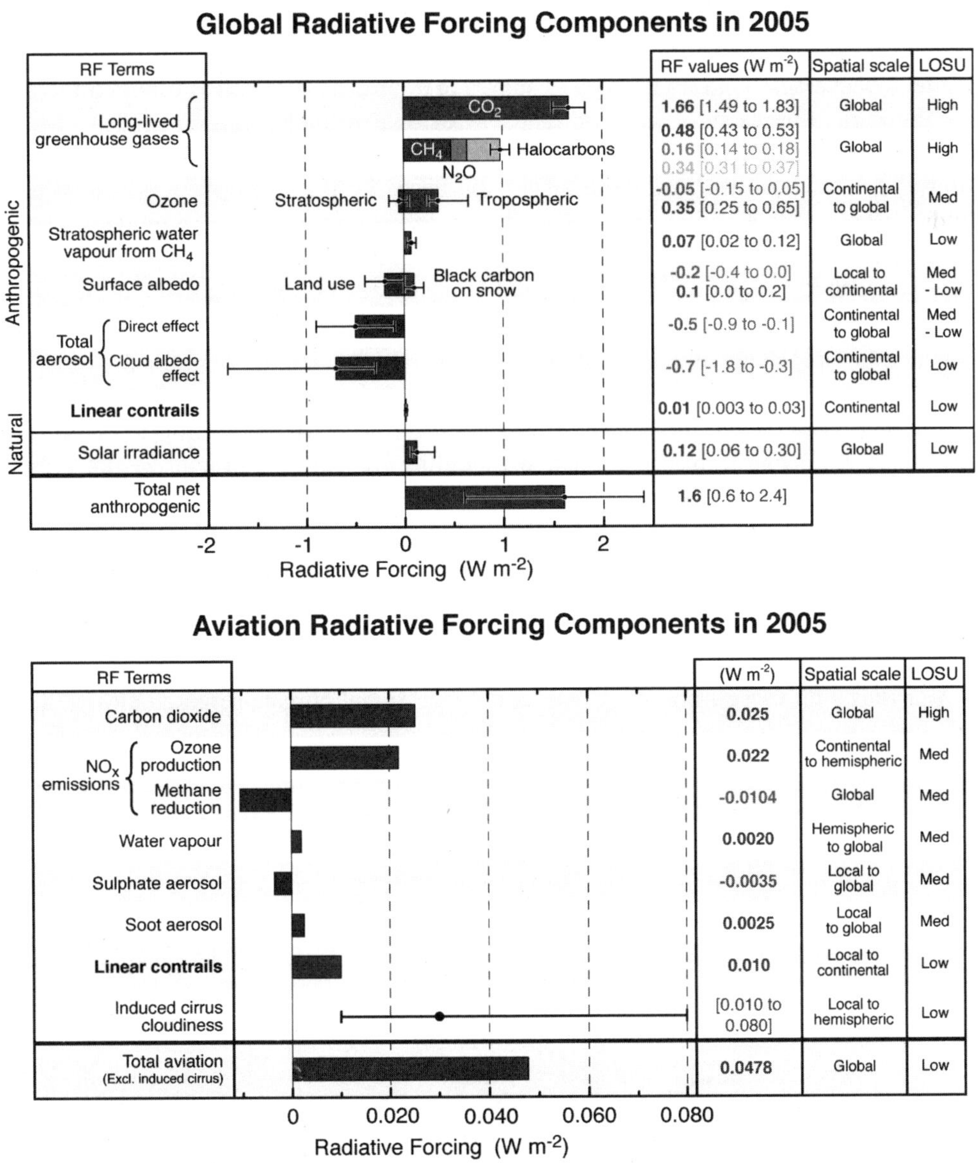


**Figure 9** Global radiative forcing from (a) all aggregated forcing terms based on IPCC AR4 and (b) from aviation alone based on Sausen et al. (2005)[16]. Columns represent spatial scale and level of scientific understanding. This figure is adopted from Forster and Rogers (15)

on Climate Change. The workshop report[20] clearly concluded that even though significant scientific advances have been made since the 1999 IPCC report, there remain large uncertainties and gaps in scientific understanding in quantifying the climate impacts of non-$CO_2$ aviation emissions and strongly recommended a need for focused research efforts in the U.S.

The FAA and Transport Canada, through PARTNER, are jointly pursuing practical and need-driven multiple research approaches to best address climate concerns of aircraft emissions based on the climate impact assessment framework as shown in Figure 10. PARTNER is pursuing a project to quantify the climate impacts of aviation using a regional to global nested model[21] for chorded emission data along the flight track over the entire year by integrating the early wake plume evolution of contrail with the grid scale dispersion based on the atmospheric physical processes. This study will simulate all components of climate impacts within the single modeling framework in the interactive manner while allowing dynamical, radiative and chemical feedbacks. The model simulates atmospheric response for background atmosphere and the aircraft-induced perturbation simulated by the model will be compared against the available observation data.

As stated before, characterization of climate impacts of aviation is one of the most complex issues and results from one particular study may not be representative of the entire picture. Therefore, there is a need for wider involvement from various disciplines (e.g. emission measurements, atmospheric observation and modeling) to better characterize the impacts and associated uncertainties. From this perspective, the FAA has developed a program – the Aviation Climate Change Research Initiative (ACCRI) – with participation from other federal research agencies such as NASA, NOAA and EPA. The objective of ACCRI is to identify and address key scientific gaps on a priority basis while providing timely scientific input to inform optimum actions and policies. Pursuant to the availability of research funds, the ACCRI program is designed to provide support for policy relevant, solution-focused research activities and to coordinate as well as link its activities with those from other U.S. and international climate research efforts. The short-

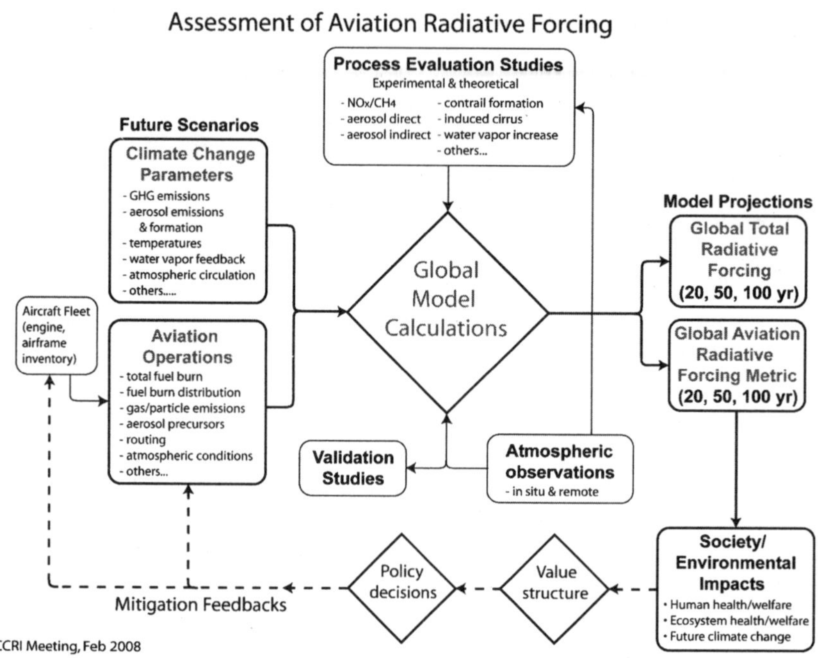


**Figure 10** An integrated framework for evaluating climate impacts of aviation

and long-term priorities of ACCRI research elements will be driven by the realistic achievability of the objectives and their usefulness which will make difference in improving the impact estimates using the modeling, measurement and data analyses capability. Under ACCRI, several subject-specific whitepapers on various aspects of aviation and climate change have been prepared and a science meeting has been convened to develop the way forward. These whitepapers and the meeting report will be made available via the internet at the following website (http://www.faa.gov/about/office_org/headquarters_offices/aep/aviation_climate/).

## Mitigation of Environmental Impacts Through a Balanced Approach

Protecting the environment is at the heart of the NextGen Plan. A balanced and integrated approach is needed to ensure energy availability and efficient, sustained aviation growth while protecting the environment. To better inform the decision-making process, the FAA, in coordination with other federal agencies and stakeholders, has developed a comprehensive 5-pillar strategy to achieve the short- and long-term NextGen environmental goals which includes:

(1) Characterizing the environmental impacts of aviation and reducing associated uncertainties through practical, needs-driven research programs such as ACCRI, development of analysis tools and research activities within PARTNER.

(2) Development and implementation of air traffic management technologies that will reduce congestion and fuel use (hence emissions). Some of the approaches that are being developed and implemented include 'Continuous Descent Approach (CDA),' Reduced Vertical Separation Minimums (RVSM) and satellite controlled air traffic management for direct routing.

There are other measures such as fleet optimization, direct routing, optimum flight path adaption guided by real time meteorological information, speed reduction and flying at optimum cruise altitude, reduction in ground delays and airport congestion, payload management planning and voluntary emission reduction programs for use of low emission technologies for airport operations etc. that can also help to reduce the fuel combustion and emissions.

(3) Development and integration of progressive advances in engine and airframe technologies into fleet. The FAA, in collaboration with NASA, has developed a program – Continuous Low Energy, Emissions and Noise CLEEN) – to expedite the migration of proof of concept knowledge (from lower Technology Readiness Levels (TRLs)) to higher TRLs (6-7) in order to incentivize insertion into the fleet of technologies that will enhance energy efficiency and reduce emission and noise.

(4) Assess the potential of use of alternate jet fuel for reduction of emissions and enable these fuels for commercial use. The U.S. commercial aviation supply chain (airlines, manufacturers, airports and the FAA) has established the Commercial Alternative Aviation Fuel Initiative (CAAFI) to coordinate the development and commercialization of "drop-in" alternative aviation fuel including renewable options.

(5) Develop market based measures and policy initiatives. In general, market forces help to drive fuel efficiency given the rising fuel cost making it in the interest of airlines to be fuel

efficient. There are many options (e.g. tax incentives, emission trading, carbon offset, market driven use of congested airspace etc.) that could potentially help to reduce the fuel use. However, each of these options and related pros and cons need to be fully understood before designing and implementing any one or mix of these options. For some of these options, it is important to develop international cooperation for common grounds particularly given the global nature of aviation and the fact that environmental protection entails a highly complex system of interdependent technologies, operations, policies and market conditions.

## SUMMARY

Albeit presently tempered by high oil prices, expansion of aviation is likely to continue and as in the past may continue to outpace the economic growth. Growth in aviation activities will likely increase the emissions and environmental impacts related to air quality and climate (as well as noise) unless a balanced and integrated approach is undertaken to mitigate these impacts. There is a need to characterize aviation emissions impacts including associated uncertainties so that the realistic goals and actions to mitigate impacts can be defined. There are multiple options available to consider and not a single option will uniquely help to achieve the targeted goals. Solutions with multiple options generally come with tradeoffs and interdependencies which must be fully understood and minimized before optimum cost-beneficial solutions are designed and implemented in a timely manner. The FAA fully understands these complexities and is proactively working with several national and international partners in multiple directions towards meeting the NextGen environmental goals so that the improved aviation infrastructure is not constrained by environmental concerns and can be used to its full potential.

## References

1. Hileman, J, Ortiz, D., Brown, N., Maurice, L., and Rumizen, R., "The Feasibility and Potential Environmental Benefits of Alternative Fuels for Commercial Aviation," International Congress of Aeronautical Societies, Anchorage, Alaska, September 2008.
2. ICAO Environmental Report, 2007 (http://www.icao.int/icao/en/env/pubs/Env_Report_07.pdf) International Civil Aviation Organization, ICAO, 2007
3. Integrated work plan for the Next Generation Air Transportation System (http://www.jpdo.gov/iwp/IWP_Version_02_Master_w-o_appendix.pdf) Joint Planning and Development Office, version 0.2, Feb. 2008.
4. National Plan for Aeronautics Research and Development and Related Infrastructure (http://www.ostp.gov/galleries/default-file/Final%20National%20Aero%20RD%20Plan%20HIGH%20RES.pdf), December, 2007.
5. FAA aviation environmental models (http://www.faa.gov/about/office_org/headquarters_offices/aep/models)
6. PARTNER, the Partnership for AiR Transportation Noise and Emissions Reduction (www.partner.aero)
7. National Ambient Air Quality Standards-NAAQS (http://www.epa.gov/air/criteria.html)
8. Tarrason, L., J.E. Jonson, T.K. Berntsen and K. Rypdal, Study on air quality impacts of non-LTO emissions from aviation, Norwegian Meteorological Institute Report No. 3, 2004.

9. AERMOD Model (http://www.epa.gov/scram001/dispersion_prefrec.htm#aermod)
10. CMAQ Model (http://www.cmaq-model.org/)
11. Transportation Research Board - Airport Cooperative Research Program (http://www.trb.org/CRP/ACRP/ACRP.asp)
12. Arunachalam, S., B.H. Baek, A. Holland, Z. Adelman, F.S. Binkowski, A. Hanna, T. Thrasher and P. Soucacos (2008). An Improved Method to Represent Aviation Emissions in Air Quality Modeling Systems and their Impacts on Air Quality, In Proceedings of the 13th Conference on Aviation, Range and Aerospace Meteorology, New Orleans, LA, January 2008.
13. Aviation and the Global Atmosphere (http://www.grida.no/climate/ipcc/aviation/index.htm) Intergovernmental Panel on Climate Change, 1999.
14. Climate Change 2007 - The Physical Science Basis, Contribution of Working Group I to the Fourth Assessment Report of the IPCC (http://www.ipcc.ch/ipccreports/ar4-wg1.htm), 2007.
15. Forster, P. and Helen Rogers, Metrics for comparison of climate impacts from well mixed greenhouse gases and inhomogeneous forcing such as those from UT/LS ozone, contrails and contrail-cirrus, subject-specific white paper submitted in support of ACCRI (http://www.faa.gov/about/office_org/headquarters_offices/aep/aviation_climate/media/ACCRI_SSWP_VII_Forster.pdf)
16. Sausen, R. *et al.*, Aviation radiative forcing in 2000: An update on IPCC (1999), Meteorologische Zeitschrift, 14, 555-561, 2005.
17. Wuebbles, D., M. Gupta and M. Ko, Evaluating the impacts of aviation on climate change, EOS Trans., American Geophysical Union, Vol. 88, 14, 157-160, 2007.
18. National Academy of Sciences, Radiative forcing of climate change: Expanding the concept and addressing uncertainties, Washington, D.C., 2005.
19. Aviation and the Environment: A national vision statement, framework for goals and recommended actions, Report to the United States Congress (http://web.mit.edu/aeroastro/partner/reports/congrept_aviation_envirn.pdf), 2004.
20. A report of findings and recommendations, Workshop on the Impacts of aviation on climate change (http://www.faa.gov/airports_airtraffic/environmental_issues/media/Aviationclimworkshop-final.pdf), 2006.
21. Jacobson, M. Z., GATOR-GCMM: A global through urban scale air pollution and weather forecast model. 1. Model design and treatment of subgrid soil, vegetation, roads, rooftops, water, sea ice, and snow, *J. Geophys. Res.*, *106*, 5385-5402, 2001.
22. Aviation Climate Change Research Initiative – ACCRI (http://www.faa.gov/about/office_org/headquarters_offices/aep/aviation_climate/)

# Mercury Emission and Control for Coal-fired Power Plants Using Coal and Cattle Biomass

**Kalyan Annamalai*, Udayasarathy Arcot V.[1], Hyukjin Oh[1] and John M. Sweeten[2]**

[1]Department of Mechanical Engineering, [2]Texas A&M AgriLife Research at Amarillo & Vernon

Texas A&M University, College Station, TX 77842

*E-mail: kannamalai@tamu.edu

## 1. INTRODUCTION

The drive for clean air has caused an increasing concern for control of toxic emissions from coal combustion systems including $NO_x$, $SO_2$, particulate matter, and mercury (Hg). In particular, the Hg emission has been targeted for control because of its unique characteristics such as high volatility, bio-accumulation and other toxic properties which could result in adverse health effects in human ecology, which include lung injury, nervous system failure, memory loss, skin rashes, and muscle weakness [1]. Organic mercury is typically more toxic than inorganic mercury. Organic mercury compounds are those containing covalent bonds between carbon and mercury. Some of the effects of organic mercury include changes in vision, sensory disturbances in the arms and legs, cognitive disturbances, dermatitis, and muscle wasting. Exposure of childbearing-aged women is of particular concern because of the potential adverse neurological effects of Hg in fetuses [2]. During outbreaks of methylmercury poisonings some mothers with no symptoms of nervous system damage gave birth to infants with severe disabilities, it became clear that the developing nervous system of the fetus may be more vulnerable to methylmercury than is the adult nervous system [3]. Table 1 shows the percentage of women with blood mercury concentration greater than 5.8 μg/L (this is an estimated level assumed to be with no appreciable harm).

**Table 1** Percentage of women aged 16 – 49 years with blood Hg levels e" 5.8μg/L [3]

| *Race/Ethnicity* | *No* | *% with Hg levels e" 5.8μg/L* | *(95% CI*)* |
|---|---|---|---|
| Mexican American | 1,106 | 1.70 | (1.04 – 2.79) |
| White, non-Hispanic | 1377 | 5.77 | (3.71 – 8.97) |
| Black, non-Hispanic | 794 | 4.82 | (2.55 – 9.11) |
| Total | 3,637 | 5.66 | (4.04 – 7.95) |

*Confidence Interval

The chemical form of Hg in flue gas and its transformations during combustion dictates the mechanism of Hg capture. This is technically termed as Hg speciation: (1) Elemental mercury ($Hg^0$) – volatile metal which exist in gas phase at flue gas temperatures and is insoluble in water. It can not be captured by traditional pollution control devices such as particulate control units or wet flue gas desulphurization (WFGD) units, (2) Oxidized mercury ($Hg^{2+}$) – normally exist in gas phase, and can be captured by WFGD units, since they are highly water soluble, and (3) Mercury in particulate form ($Hg^P$) – exist in solid phase and can be easily captured at traditional particulate control devices such as electrostatic precipitator (ESP) or fabric filters (FF). Alkali and metal processing, incineration of coal, medical and other waste, mining of gold and mercury are major contributors of Hg emission. Many power plants have existing Hg capture as co-benefit of air pollution control technologies for $NO_x$, $SO_2$ and particulate matter. This includes capture of oxidized Hg in WFGD units. The use of selective catalytic reduction (SCR) units used for $NO_x$ control enhances oxidation of $Hg^0$ to its soluble ionic form $Hg^{2+}$ resulting in removal at WFGD systems. Alternative technologies which emerged recently include the use of activated carbon injection and advanced sorbents to capture Hg from flue gases at the FF used to collect ash. The U.S. EPA reported 48 tons out of 158 tons of mercury were released from coal-fired combustion sources, 28 tons were released from industrial and commercial boilers, 30 tons were released from municipal waste combustors, and the rest were from medical waste, manufacturing, and natural sources in 1997 [4, 5]. The reported Hg concentrations are about 0.04 μg/g in Colombian coal, 0.03 – 0.25 μg/g in Australian coal, 0.09 – 0.51 μg/g in eastern US coal, and 0.14 – 1.78 μg/g in Polish coal [6]. The average Hg concentrations of 0.070 μg/g in bituminous coal, 0.027 μg/g in sub-bituminous coal and 0.118 μg/g in lignite coal [6]. It was estimated that typically 0.24 μg/g of mercury occurs in Appalachian coals, 0.14 μg/g in Interior Eastern coals and 0.21 μg/g in Illinois Basin coals [7]. This may be misleading because, in order to obtain similar energy outputs, more low-rank coal has to be burned than a higher-ranked coal. This can result in a net mobilization of more total Hg to the environment. A better way to compare mercury data for coal is on an equal energy basis.

Mercury can exist in flue gas in any of these various forms with wide variations, and the speciation depends broadly on fuel type composition, combustor type and operating conditions, and pollution control devices present. In general, Hg emissions from coal combustion sources are approximately 20 – 50% $Hg^0$ and 50 – 80% $Hg^{2+}$, which may be predominantly mercuric chloride ($HgCl_2$), while $Hg^P$ constitutes less than 5% [8]. Experiments [9] conducted to study the fate and behavior of mercury in power plants showed that 43% (± 30%) of Hg present in the coal is found in the flue gases in vapor phase at the exit of pollution control devices, while the remaining 54% (± 24%) was captured at WFGD and ESP units. With the presence of HCl, $Hg^0$ (partly) is converted into $HgCl_2$ at temperatures less than 500 – 800°C. According to the one of the test conducted it was found that 53% of the Hg presented in a water soluble form, mostly in the form of $HgCl_2$. However, it is still in the vapor phase due to the high temperature of flue gases (140 – 150°C). Distribution of Hg species in coal combustion flue gases has been calculated using equilibrium calculations by Mojtahedi et al. [10] and Senior et al [11], which shows that all of the Hg exists in the form $HgCl_2$ below 450°C. And above 700°C, 99% of the Hg exists as gaseous Hg, and the rest is in the form of HgO [11]. Equilibrium is not attained in flue gas since the environment is highly transient and also due to fact that flue gas cools rapidly as heat is transferred from water to steam. Experiments conducted by Lindqvist et al. validate this proposition and confirmed that Hg exists in elemental form only at temperatures above 700°C [12]. Hall [13] said that the re-oxidation reaction is stated to occur rapidly at about 500°C and is described as Eq. (1). It occurs at temperatures between 400 to 700°C. Below 400°C, atomic chlorine (Cl) is responsible for further Hg oxidation as described in Eq. (2) and (3).

$$2Hg(g) + 4HCl + O_2 \Leftrightarrow 2HgCl_2(g,s) + 2H_2O(g) \quad (1)$$

$$Hg + Cl_2 \rightarrow HgCl + Cl \quad (2)$$

$$HgCl + Cl_2 \rightarrow HgCl_2 + Cl \quad (3)$$

$$Hg + Cl \rightarrow HgCl \quad (4)$$

$$HgCl + HCl \rightarrow HgCl_2 + H \quad (5)$$

$$HgCl + Cl_2 \rightarrow HgCl_2 + Cl \quad (6)$$

$$HgCl + Cl \rightarrow HgCl_2 \quad (7)$$

However, in case of flow reactor, where the temperature is very high at the upstream, stable HCl decomposes and supplies Cl which aids in formation of intermediate HgCl. As these species move downstream where temperature is lowered below 700°C, HgCl oxidizes further to form stable $HgCl_2$ which is favored at lower temperatures as presented in Eq. (4) to (7) [14]. It can be said that for oxidation of Hg in presence of HCl, high temperatures is required to decompose HCl to produce Cl, and occurrence of intermediate HgCl. While lower temperatures are required to further oxidize HgCl to $HgCl_2$. Thus the extent of oxidation depends on the concentration of chlorine in flue gases. The fraction of elemental Hg emission of coal-fired boilers decreases with increase in Cl content of coal [15]. The Cl content in Bituminous coals range from 200 to 2000 ppm (dry basis) while for low rank coals (sub-bituminous and lignite) it ranges from 20 to 200 ppm an order of magnitude lower. Thus the low rank Sub-bituminous and lignite coals reveal lower Hg capture (3 – 72%) in co-benefit systems than higher rank bituminous coal (9 – 98 %) [16]. Hg removal plotted against coal chlorine content reveals increasing Hg capture with Cl due to HgC12 formation. As Cl is low in sub-bituminous and lignite coals, the Hg exists primarily as elemental Hg, which is difficult to capture.

Since there is considerable concern regarding the potential global environmental impact of fossil fuels used for power generation, co-firing biomass with coal in traditional coal-fired boilers or using biomass as a reburn fuel in advanced coal-fired boiler configurations represent two options for combined renewable and fossil energy utilization. To add to the above, it can also be considered the best solution to combat the challenging waste disposal problem, with 110 million tons of dry animal manure produced annually in the US. It is known from previous works of Annamalai et al. on co-firing cattle biomass (CB) with coal that, it has great potential in reducing fossil fuel based $CO_2$, reduction in $NO_x$, reduction in fuel costs since biomass is cheaper than coal, and minimization of soil, water, and air pollution. Apart from the above, the CB has very high amount of chlorine content compared to coal with relatively good heating value. This gives a potential use for blending coal with biomass and co-firing it in existing boilers to increase the chlorine content in the fuel, and hence achieve higher Hg oxidation and hence its capture to reduce $Hg^0$ emissions. Recent research activities of reburning CB with coal [6] have shown remarkable results of reducing $NO_x$ emissions by almost 90%.

## 2. OBJECTIVES

The current paper focuses on a novel method of co-firing coal with CB which helps in oxidization of $Hg^0$ into its soluble ionic compound, mainly $HgCl_2$ which can be captured using traditional pollution control devices. Specifically dairy biomass (DB) will be used as co-fired fuel with coal in this study to investigate the reduction of $Hg^0$ and to determine operating conditions for optimum Hg reduction. Survey data from Information Collection Request (ICR) shows that high rank coals such as bituminous coal have large chlorine content, while low rank coals such as sub-bituminous or lignite coal have lower percentages of chlorine. This goes to highlight that there may be less amount of chlorine species during combustion of low rank coals, which results in lower Hg oxidation, and hence cause pollution threat. This research draws interest from two aspects: (1) It makes use of energy potential from CB which would be blended with coal to add to the heating value of the blended fuel and act as an oxidizing agent to aid Hg capture and (2) It addresses the waste disposal problems in cattle farms. The tasks that follow the current objectives are: (1) Obtain thermo-chemical properties of coal and DB used as fuels, including Hg and Cl contents, (2) Setup a wet chemistry system to measure both elemental and total Hg content in the flue gas, and (3) Conduct parametric studies on equivalence ratio (ER), different types of fuels and different blend ratio of fuels, and (4) Determine the optimum operating condition for maximum Hg reduction.

## 3. EXPERIMENTAL PROCEDURE

Extensive studies on Hg oxidation were performed using two different experiment facilities. A fundamental study of Hg oxidation was performed using a plug flow reactor (PFR) facility and then a small-scale (30 kW or 100,000 BTU/h) coal-fired boiler burner facility was used for the laboratory tests on Hg emission by the effects of co-firing DB with coal.

### 3.1 Plug Flow Reactor (PFR) Facility

First, the fundamental study of Hg oxidation was also performed with the PFR as shown in figure 1. It is composed of four distinct systems: (1) sources of simulated flue gas, (2) a Hg generation system, (3) an electrical heating furnace, and (4) a Hg measuring system. Different types of gas streams were produced, and they were preheated to about 60ºC and well-mixed. The Hg generation system contained a certified $Hg^0$ permeation tube. Flexible electric heating tapes were wrapped around a quartz tube, and the quartz tube was covered by insulation and sealed into an aluminium box. A temperature controller was used to keep the temperature of the Hg carrier gas constant near the permeation tube. The Hg vapor was mixed with the main simulated gas stream before the entrance to an electrical furnace (or PFR). The total flow rate of the simulated gas stream was 1100 SCCM (standard cubic centimeter per minute at 25°C and 1 atm). The PFR has three heating regions. The reactor temperature of each heating area could be increased to a maximum of 1200ºC. The temperature at the furnace exit was typically below 500ºC. After the gases exit the furnace, they are naturally cooled to near the ambient temperature, and the temperature at the entrance to the Hg vapor monitor (Mercury Instrument VM 3000) is about 25ºC. In the PFR experiments, concentrations of $Hg^0$ were only measured by the Hg vapor monitor. A direct indication of Hg concentrations in the flue gas stream was displayed instantaneously at the screen and was recorded at the workstation computer.

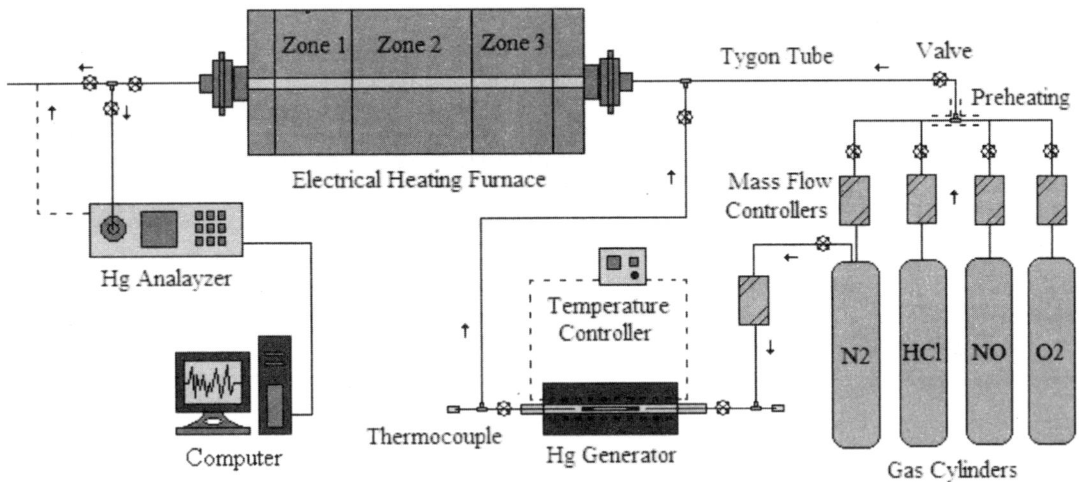


**Figure 1** Schematic of the experimental setup for the fundamental study of Hg oxidation

## 3.2 Co-Firing Facility for the Small-Scale Tests

Second, in order to validate reduction in Hg emission, co-firing experiments were conducted on the 100,000 BTU/hr (29.3 $kW_t$) small-scale coal-fired boiler burner facility as shown in figure 2. The furnace consisting of a 15.24 cm (6 in) diameter, 182.88 cm (6 ft) long downward fired combustor, is capable of burning blends of coal and biomass in co-firing scenario. Air required for burning is supplied by primary and secondary air. Primary air is the carrier air that carries fuel blends from fuel feeder to the main burner. The ER during combustion is varied by varying secondary air supply. Along the axial length of the reactor number of thermocouples are attached that monitor temperatures at each level of the furnace instantaneously. At the bottom end of the reactor flue gas is sampled through a sampling port that is located 78" below the main burner. A portable commercial flue gas analyzer based on electrochemical and nondispersive infrared (NDIR) sensors is used to measure different gas species such as CO, $CO_2$, $O_2$, $NO_x$, $SO_2$ and $C_xH_y$. Mercury Instrument VM 3000 along with custom-built wet chemistry system is used to measure Hg species in the flue gas using cold vapor atomic absorption (CVAA) principle as shown in figure 3.

Conducting experiments during coal and fuel blend combustion followed three distinct steps: preparation phase, firing phase and measurement phase. In preparation phase, the furnace is preheated to a temperature of about 2000 F by burning natural gas only at near stoichiometric condition with air being supplied from secondary air supply channel. This process of preheating takes about 3 to 4 hours until steady state is attained. In the meantime, the fuel feeder is loaded with required solid fuel, and feed rate is calibrated manually by measuring mass of fuel flow in specific amount of time (normally in 1 minute). Once steady state is attained, the second phase of firing solid fuel can begin. The natural gas supply is turned off and feeder motor is started to feed solid fuel that is carried by carrier air (also called primary air) to the furnace. Air flow rates are adjusted by varying the secondary air flow rate, the means by which desired ERs are achieved (from lean to

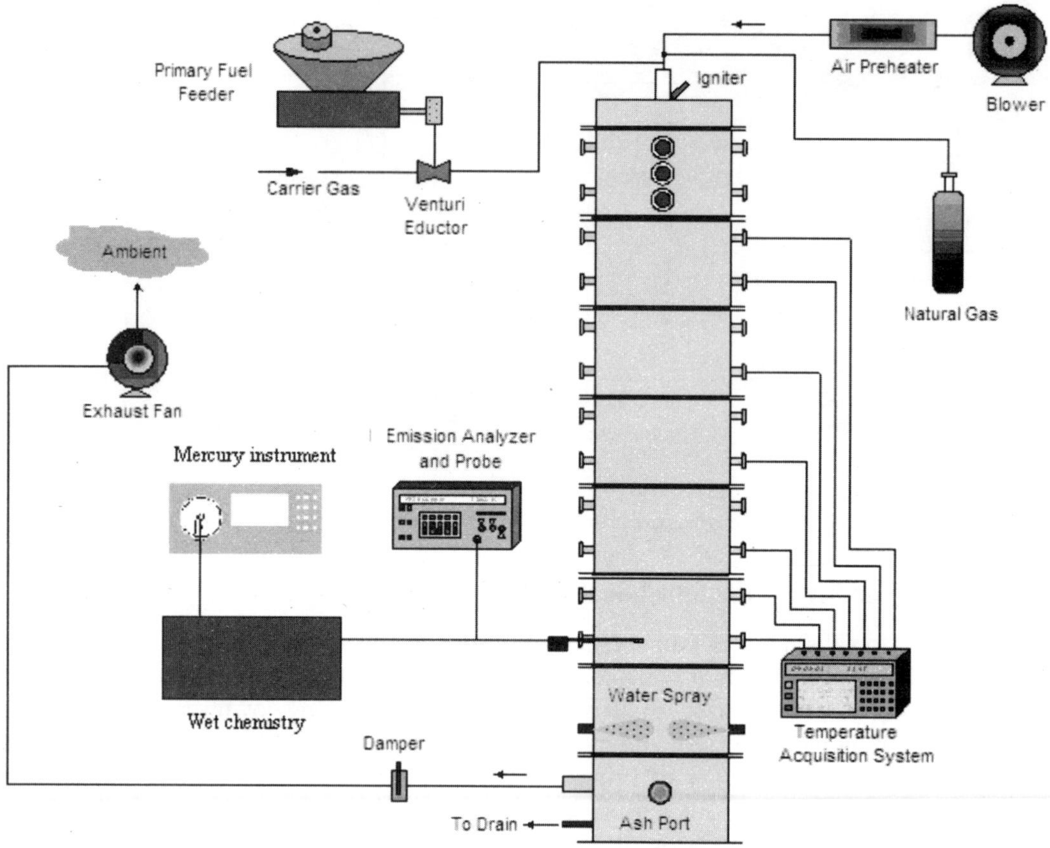


**Figure 2** Schematic layout of the furnace and accessories

rich combustion). Once steady state is attained (which takes roughly 10 minutes), measurement phase begins, when sampling probe is plugged into the sampling port to make measurements of flue gas species using GreenLine analyzer. Once these readings are taken, measurements of Hg are made using VM3000 analyzer and wet chemistry system. This measurement step is followed for every ER.

## 4. RESULTS AND DISCUSSIONS

### 4.1 HCl Effects on Hg Oxidation using the PFR

Effects on Hg oxidation in the presence of HCl in the gas stream (pure $N_2$) were investigated. Hg oxidation was defined as the ratio of the removed to supplied $Hg^0$ concentrations. The removed concentration was obtained by the difference of the measured input and output concentrations of $Hg^0$. Figure 4 shows effects of HCl concentrations (0 – 90 ppm) and reactor temperatures (700 – 1200°C) on Hg oxidation. The baseline input concentration of $Hg^0$ was measured to be

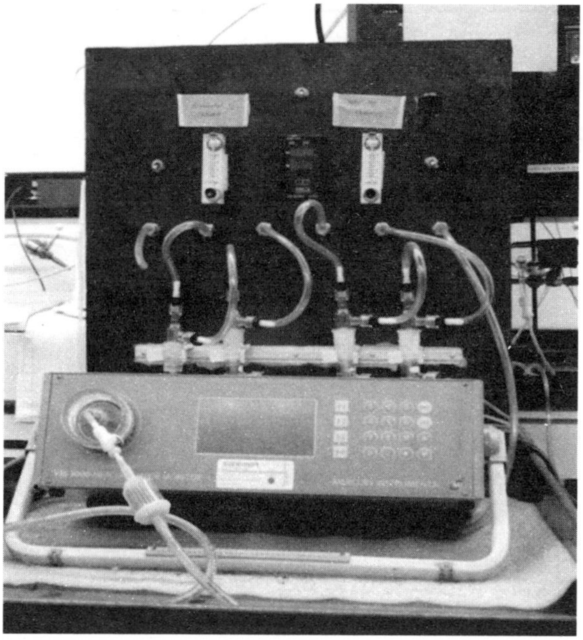

Figure 3 Mercury instrument VM 3000 along with custom-built wet chemistry system

approximately 514 μg/m$^3$ (62.7 ppb) at 700°C. For the effects of HCl concentrations in the first part of figure 4, there was no Hg oxidation measured up to 70 ppm HCl; however, 17% Hg oxidation was found with the addition of 90 ppm HCl at 700°C. The effect of temperatures was significant on Hg oxidation. The results are presented in the second part of figure 4 and listed in Table 2. Below 700°C, no effect was observed on Hg oxidation while above 700°C more Hg oxidation took place with an increase in the reaction temperature. About 88% Hg oxidation was achieved at 1200°C. It was also noted that the addition of small amounts of $O_2$ (1 – 5%) promoted Hg oxidation in the presence of HCl.

## 4.2 Co-Firing Tests on Hg Emission

### *4.2.1 Tested Fuels*

Fuels used during experiments were Texas Lignite Coal (TXL), Wyoming Subbituminous Coal (WYC), Separated Solids Partially Composted Dairy Biomass (Sep. Sol. PC-DB), High Ash Partially Composted Dairy Biomass (HA PC-DB), and their blends. The fuels used for experimentation were tested for their combustible properties and elemental constituents. Three individual samples of each fuel were prepared and sent to a commercial laboratory for analysis. The average of three samples for each fuel used for this study is reported in Table 3. The derived properties extracted from the analysis results are presented in Table 4 to compare the ash, chlorine and Hg content or loading with different fuels on as received and a dry ash free (DAF) basis.

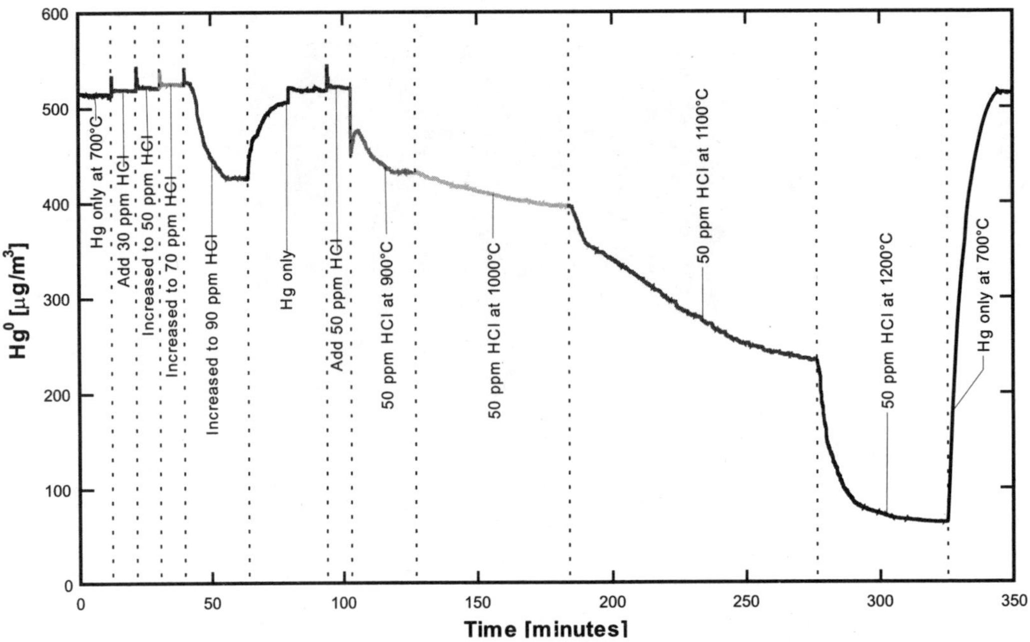


**Figure 4** Hg oxidation in the presence of HCl concentrations by the effects of various HCl concentrations (0 – 90 ppm) and reactor temperatures (700 – 1200°C)

**Table 2** Hg oxidation in the presence of HCl

| *Reactor Temperature (°C)* | *Exit $Hg^0$ [μg/m$^3$]* | *Hg Oxidation [%]* |
|---|---|---|
| 700 | 520 | 0 |
| 900 | 432 | 17 |
| 1000 | 396 | 24 |
| 1100 | 234 | 55 |
| 1200 | 65 | 88 |

It can be seen that WYC has twice as much Cl content as that of TXL on heat basis, while its Hg content is three times lower than that of TXL, which means WYC produces lesser Hg emissions compared to TXL. It can also be seen that DB has much higher chlorine content (1610 ppm to 3980 ppm), which is 14 to 39% higher compared to coal (70 ppm to 190 ppm), while its heating value ranges from 23 to 70% to that of the coal. When DB is blended with coal in different proportions, it tends to increase the chlorine and ash content in the coal based fuel, but decreases the heating value.

#### *4.2.2 Base Case with TXL*

Measurements of Hg made when 100% coal-fired is reported in this section and is termed base case with which other blend ratios will be compared to judge the reduction in Hg emissions. In all cases, total and elemental Hg were measured but Hg in particulate form was not measured. Figure 5 shows

**Table 3** Proximate and ultimate analysis of fuels used

| *Fuel Source* | *Texas Lignite and Wyoming Powder River Basin Coal provided by TXU Energy, Dallas* | | *Dairy Farm in Comanche County, Texas* | |
|---|---|---|---|---|
| *Reference* | *[8]* | | *[12]* | |
| *Ananlysis Lab* | *Hazen research inc., Golden, CO* | *Hazen research inc., Golden, CO* | *Hazen research inc., Golden, CO* | *Hazen research inc., Golden, CO* |
| *Sample ID* | *TXL 113-115* | *PRB 116-118* | *128-130* | *131-133* |
| *Date of sampling* | *10/10/2005* | *10/10/2005* | *5/15/2006* | *5/15/2006* |
| *Date of analysis* | *11/29/2005* | *11/29/2005* | *10/23/2006* | *10/23/2006* |
| Ash | 11.46 | 5.64 | 14.93 | 59.91 |
| Dry Loss (% Moisture) | 38.34 | 32.88 | 25.26 | 12.21 |
| FC | 25.41 | 32.99 | 12.95 | 3.85 |
| VM | 24.79 | 28.49 | 46.86 | 24.04 |
| Carbon, C | 37.18 | 46.52 | 35.21 | 18.04 |
| Hydrogen, H | 2.12 | 2.73 | 3.71 | 1.45 |
| Nitrogen, N | 0.68 | 0.66 | 1.93 | 1.15 |
| Oxygen, O (diff) | 9.61 | 11.29 | 18.60 | 7.07 |
| Sulfur, S | 0.61 | 0.27 | 0.43 | 0.19 |
| HHV (kJ/kg) | 14,287 | 18,193 | 12,844 | 4,312 |
| Chlorine, Cl %, (ppm) | 0.007 (70 ppm) | 0.019 (190 ppm) | 0.161 (1610 ppm) | 0.398 (3980 ppm) |
| Mercury, Hg g/kg (ppb) | 0.00013 (130 ppb) | 0.00007 (70ppb) | 0.00004 (40 ppb) | 0.00003 (30 ppb) |

**Table 4** Few derived properties of fuels

| | *Txlig-3 samples* | *Wy Coal-3 samples* | *DB-Sep solids-PC-2006* | *DB-HA-PC* |
|---|---|---|---|---|
| HHV (kJ/kg) | 14286.82 | 18193.02 | 12844.17 | 4312.40 |
| HHV, Boie Equation (kJ/kg) | 14582.32 | 18347.96 | 14799.12 | 7336.17 |
| HHV, DAF (kJ/kg) | 28459.80 | 29593.38 | 21473.37 | 15466.57 |
| Chlorine, Cl % (ppm) | 0.007 (70 ppm) | 0.019 (190 ppm) | 0.161 (1610 ppm) | 0.398 (3980 ppm) |
| Cl DAF % (ppm) | 0.0139 (139 ppm) | 0.0309 (309 ppm) | 0.2691 (2691 ppm) | 1.4274 (14274 ppm) |
| Cl, g/GJ | 4.899620184 | 10.44356385 | 125.3486795 | 922.9190957 |
| Mercury, Hg g/kg (ppb) | 0.00013 (130 ppb) | 0.00007 (70 ppb) | 0.00004 (40 ppb) | 0.00003 (30 ppb) |
| Hg DAF g/kg (ppb) | 0.000258 (258 ppb) | 0.0001138 (113.8 ppb) | 0.0000668 (66.8 ppb) | 0.0001075 (107.5 ppb) |
| Hg, g/GJ | 0.009099295 | 0.003847629 | 0.003114253 | 0.006956677 |
| Ash Loading (kg/GJ) | 8.02371134 | 3.10008948 | 11.62050617 | 138.9199528 |

the variation of elemental and total Hg at various ER for TXL and WYC. The $Hg^0$ for TXL and WYC is 0.8 and 0.7 μg/m$^3$ at stoichiometry, and it fluctuates to a maximum of 1.2 μg/m$^3$ for TXL. The total Hg for TXL is higher than that for WYC, which means the oxidized fraction of mercury is greater for TXL than that for WYC. This may be predominantly due to presence of higher ash content in the fuel which provides greater site and chance for Hg oxidation. It can be observed that

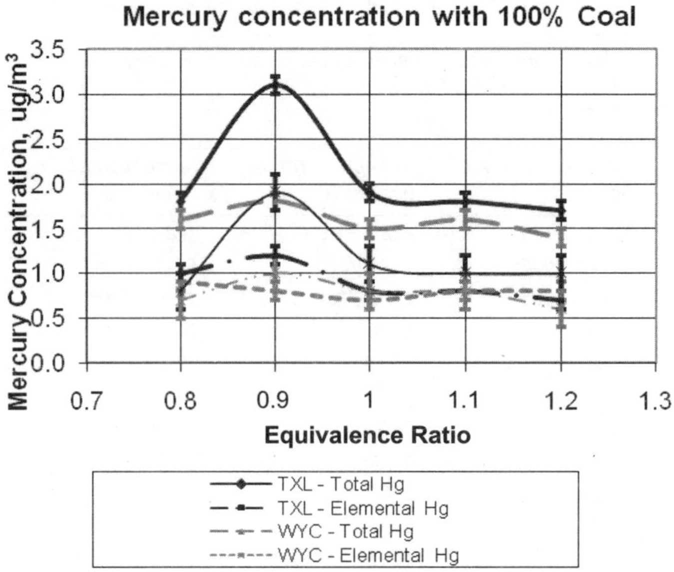


**Figure 5** Base case results on elemental and oxidized Hg for coal

during fuel-lean combustion (ER < 1.0) there is mercury variation seen more obviously with total Hg concentration, while minor variations were found at stoichiometry (ER = 1.0) and fuel-rich conditions (ER > 1.0). This may be attributed from the fact that at richer conditions there is presence of unburnt carbon which adsorbs and re-emits Hg at a fairly stable rate while during fuel-lean combustion the excess air burns unburnt carbon present on the refractory walls and causes release of more Hg.

#### *4.2.3 Blending TXL with DB*

In order to validate the hypothesis that blending of coal with DB can reduce Hg emission, experiments were performed by blending TXL with DB. The blending causes increased chlorine content in the fuels which aids Hg oxidation and hence reduce emission of $Hg^0$. Figure 6 shows the variation of $Hg^0$ for TXL with different blends of DB at various ERs. The chlorine content in the blend increases with increase in DB proportion, so that there is increased chance of more available chlorine species at combustion temperatures. This increase in chlorine species react with Hg as temperature drops down along the furnace and forms stable and soluble $HgCl_2$, and fewer elemental Hg is emitted.

#### *4.2.4 Validation of simulation*

Simulation studies on isolated particle combustion in flue gases conducted previously by Puchakayala et al. [17], predicted effective Hg oxidation when coal is fired with feedlot biomass (FB). He showed that presence of high chlorine concentration in flue gases substantially reduces emissions of $Hg^0$. The results of blending FB with coal in proportions of 10:90 and 20:80 revealed about 65 – 80% of mercury was converted to $HgCl_2$, while for pure coal only 9% mercury was oxidized. Experimental results shown in figure 7 shows that 88% of mercury is oxidized when 80:20 blend is used and 50% of mercury exists in oxidized form when pure coal is fired.

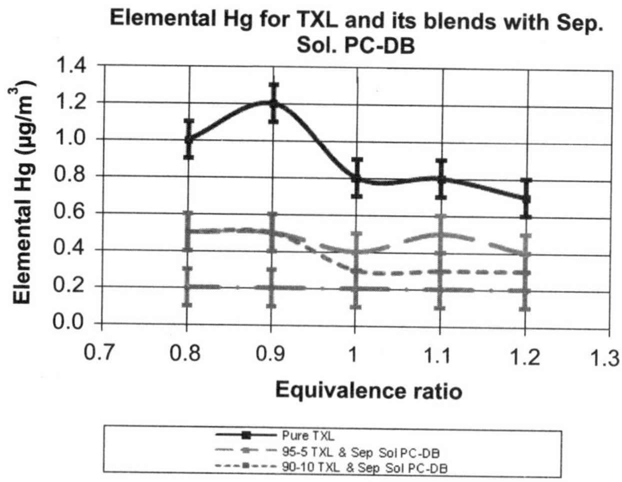


**Figure 6** Elemental Hg for TXL and its blends with Sep. Sol. PC-DB

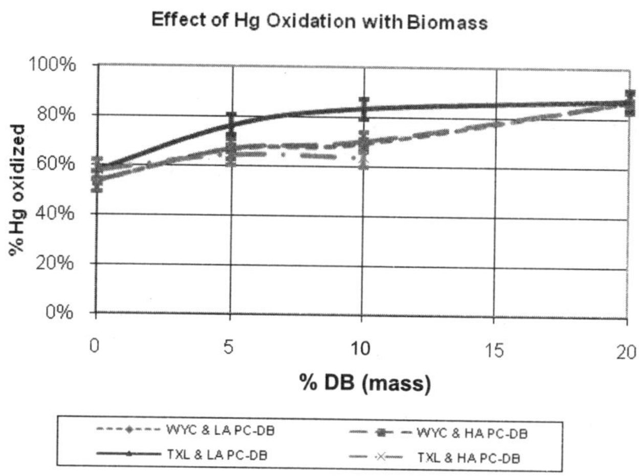


**Figure 7** Experimental results of Hg oxidation with DB

## SUMMARY AND CONCLUSIONS

A novel method of co-firing coal with CB which helps in oxidization of $Hg^0$ into its soluble ionic compound, mainly $HgCl_2$, was conducted. Prior to the co-firing tests, a fundamental study of Hg oxidation was examined using plug flow reactor.

1. The addition of HCl into the PFR showed a significant effect on Hg oxidation, and the Hg oxidation increased with an increase in the reactor temperature.
2. It seems that the unburnt carbon may adsorb and re-emit Hg at a stable rate in fuel-rich conditions while the excess air burns unburnt carbon and it causes high Hg release in the fuel-lean combustion.
3. Higher chlorine content in the fuels by blending DB with TXL aided Hg oxidation and hence reduced Hg emission.

## Acknowledgements

This material was prepared with the support of DOE- Golden, Colorado and partly from the Texas Commission on Environmental Quality (TCEQ). However, any findings, conclusions, or recommendations expressed herein are those of the author and do not necessarily reflect the view of DOE or TCEQ.

## References

[1] Mercury Study Report to Congress, Volume V: Health Effects of Mercury and Mercury compounds, US EPA, December 1997.

[2] http://www.epa.gov/hg/effects.htm

[3] K.R. Mahaffey, R.P. Clickner and C.C. Bodurow, Blood Organic Mercury and Dietary Mercury Intake: National Health and Nutrition Examination Survey, 1999 and 2000.

[4] J.D. Kilgroe, C.B. Sedman, R.K. Srivastava, J.V. Ryan, C.W. Lee, and S.A. Thorneloe, Control of Mercury Emissions from Coal-Fired Electric Utility Boilers. US EPA, US Government Printing Office, Washington, DC, December 2001.

[5] H. Agarwal, H.G. Stenger, S. Wu, and Z. Fan, Effects of $H_2O$, $SO_2$ and NO on Homogeneous Hg Oxidation by $Cl_2$. *Energy and Fuels* 2006, 20, 1068-1075.

[6] P. Chu and D.B. Porcella, *Water, Air, Soil Pollut.* 1995, 80, 135-144.

[7] P. Chu, B. Nott and W. Chow, Results and Issues from the Pisces Field Tests, in: 2$^{nd}$ International Conference on Managing Hazardous Air Pollutants, Washington, DC, 1993.

[8] A. Carpi, Mercury from Combustion Sources: A Review of the Chemical Species Emitted and Their Transport in the Atmosphere, June 1996.

[9] R. Meji, H.J.V. Leo and W. Henk, *J. Air Waste Manage. Assoc.* 2002, 52, 912-917.

[10] W. Mojtahedi, R. Backman and K. Larjava, Technical Research Centre of Finland, Publications 42, 1987.

[11] C.L. Senior, A.F. Sarofirm, T. Zeng, J.J. Helble, and R. Mamani-Paco, *Fuel Process. Technol.* 2000, 63, 197-213.

[12] O. Lindqvist, J. Wang, and Z. Xiao, *Water, Air and Soil Pollution* 1995, 80, 1217-1226.

[13] B. Hall, Thesis, Department of Inorganic Chemistry, Chalmers University of Technology of Gorteborg, Sweden ISBN 91-7932-682-7, 1992.

[14] J.C. Kramlich, R.N. Sliger and D.J. Going, Reduction of Inherent Mercury Emissions in PC Combustion, Semi-Annual Technical Report, August 1999.

[15] C.L. Senior, D. Lignell, B. Shiley, Z. Chen, and A. Sarofim, Kinetic Models for Predicting the Behavior of Mercury in Coal-Fired Power Plants, ACERC Annual Conference, Salt Lake City, UT, Feb 19-20, 2003.

[16] J. F. Thomas, Mercury Control Technology Field Phase II Testing Status Report, NETL, US DOE, 2005.

[17] M.B. Puchakayala, Ph.D. Dissertation, Mercury Emission Behavior during Isolated Coal Particle Combustion. Mechanical Engineering, Texas A&M University, College Station, TX, 2006.

# Simulations of the Claus Furnace under HiTAC Conditions for Enhanced Sulfur Recovery

**Ashwani K. Gupta**
Department of Mechanical Engineering, University of Maryland, College Park, MD 20742, USA

Sulfur-bearing compounds are very detrimental to the environment and to industrial process equipment. They are often obtained or formed as a by-product of separation and thermal processing of fuels containing sulfur, such as coal, crude oil and natural gas. The Claus process for sulfur recovery has been known and used in the industry for over 100 years. It involves thermal oxidation of hydrogen sulfide and its reaction with sulfur dioxide to form sulfur and water vapor. This process is equilibrium-limited and usually achieves efficiencies in the range of 94-97%, which have been regarded as acceptable in the past years. Nowadays strict air pollution regulations regarding hydrogen sulfide and sulfur dioxide emissions call for nearly 100% efficiency, which can only be achieved with process modifications. A detailed literature survey revealed that most of the early studies have focused on the Claus catalytic stages and/or the tail gas treatment unit. While very little work has been devoted to explore the possible improvements to the combustion process in the Claus furnace. In this study we examine detailed chemical kinetics as well as CFD simulation results of the combustion and thermal stage of the Claus furnace furnace. High temperature air combustion (called HiTAC) is shown to reveal good behavior for enhanced sulfur recovery from the process.

We specifically explore the improvements that could be made on the process by high temperature and low oxygen concentration conditions that are used in the HiTAC conditions. hiTAC conditions results in uniform thermal field in the reactor that is attractive for the Claus process. The colorless combustion offered by HiTAC as well as the mode of fuel injection into the combustion chamber can result in significant improvement for enhanced sulfur recovery in the Claus process.

# Measurements in ISO Room Scale Fires

**Andrew Lock, Gwon Hyun Ko, Matthew Bundy, Erik Johnsson and Anthony Hamins**
Fire Research Division, Building and Fire Research Laboratory, National Institute of Standards and Technology, 100 Bureau Drive, Gaithersburg, MD 20899

## INTRODUCTION

Fire safety research at the National Institute of Standards and Technology's (NIST) is focused on reducing fire losses through several critical programs, including fire fighting technology, advanced measurements and predictive methods, reduced risk of fire spread, the safety of threatened buildings, fires at the wildland-urban interface and hydrogen fire safety. Through the authority of the National Construction Safety Team Act, NIST has investigated the World Trade Center disaster and the Rhode Island Station Night Club fire. These programs interconnect and complement each other to provide a coherent research structure. Within these programs there is also significant opportunity for fundamental research as well. One of the main focuses of NIST fire research is the development of predictive methods such as the Fire Dynamics Simulator (FDS) [1]. To further develop this program new and better flow, chemical species, and temperature data always needs to be provided and new and improved ways of collecting that data are developed as well. Subsequently FDS is then used in higher level research to reconstruct fire events such as the Rhode Island station night club fire as well as in the development of performance criteria for firefighter equipment. Performance metrics and standards in firefighting equipment then help to protect firefighters and provide them with better tools to do their jobs and protect the general population.

The development of field models for the prediction of fire growth and smoke transport is an area of special interest to fire protection engineers well as to design engineers and forensic analysts in arson investigations. FDS, developed at NIST, is a mature tool for these types of models which is continually being improved to enhance its predictive capability. Past versions of FDS had used a mixture fraction based chemistry model which did not predict minor, intermediate production products such as CO and soot. Instead CO and soot production was prescribed from literature data or bench scale experimental investigations [2]. The newest version of FDS (FDS v.5) includes a model for CO production. Experiments such as the full scale compartment fires presented here are designed to provide experimental data with well defined boundary conditions to facilitate the further development of FDS by providing more detailed gas species measurements including CO, $H_2$, and higher hydrocarbon species in addition to temperature data for a full-scale ISO room fire.

A primary goal of these experiments was to provide data on the realistic evolution of a full-scale fire as the heat release rate is increased until the fire reaches the underventilated or vitiated condition. Many previous experimental investigations have been conducted into the behavior of uncontrolled burning of fires in a confined space. The full-scale standard room as defined by the ISO 9705 [3] standard was used for these experiments because it is a large enclosure representative of a room in a residential structure and in order to be consistent with other investigations in full-scale room burning [4-6].

Pitts used the Global Equivalence Ratio (GER) concept as one of the first predictive methods for the production of CO in enclosure fires [5]. This method works by considering the ratio of the total fuel consumption rate to the total air supply rate to the enclosure to calculate the global equivalence ratio and predict the quantity of CO and other species produced. It has been shown that this method works well in predicting CO production in some fires and not in others. Specifically, lean fires are well predicted while most underventilated and extremely hot fires (>900K) are not well predicted. Pitts experiments considered only natural gas as a fuel. More recently Andersson [4] considered the GER as it applied to four more complicated fuels with similar findings that the GER generally works well only for relatively cool fires. The current investigation endeavors to provide more detailed species and will include soot measurements to facilitate future development of effective predictive models for FDS.

The mixture fraction, $Z$, is currently the method of fire product prediction in FDS [7]. The mixture fraction relates the total mass of material originating in the fuel stream present at any given location. This concept was developed by Bilger to analyze flame data [8]. It was adapted to predict turbulent combustion in field models by Peters [9] and has been used to analyze the structure of laminar counterflow and coflow flames [10, 11]. The mixture fraction has been used as a sort of single step chemistry for FDS in the past and has been implemented in such a way as to predict the production of CO in underventilated fires providing a type of two-step combustion chemistry in FDS 5 [7]. Despite this improvement the model is still insufficient to produce highly accurate predictions of CO production, and does not predict soot or other minor species production. In fact it has been shown that the mixture fraction is very poor for the prediction of soot in fires [12].

Another issue surrounding the prediction of CO and other species production is scalability. Good scaling relationships allow a reduced scale or even a bench scale fire to be used as a basis for predicting the behavior of full scale fires without the expense of creating a full scale experiment. On this basis many of the GER experiments as well as other enclosure experiments have been conducted in reduced scale enclosures with the intent of using the results to predict full scale fires. Pitts, Bryner, and others have used the $Ah^{1/2}$ scaling factor, where $A$ is area and $h$ is height of an enclosure vent [5, 13, 14]. The $Ah^{1/2}$ quantity is proportional to the area of the vent and the air flux into the enclosure [15-18]. Using this quantity the ventilation air flow into an enclosure is scaled as the square of the scaling factor. For example, a $2/5$ scale enclosure would have an $Ah^{1/2}$ proportional to $(2/5)^2$ of the original enclosure. A $2/5$ scale ISO 9705 room was used in the earlier component of this investigation and will be referred to here as the reduced scale enclosure (RSE) and used for comparison of results and analysis of scaling [19].

This paper summarizes the current progress in the experimental measurement of the behavior of fires in full scale room fires. Natural gas and heptane fires are considered and experimental data on gas species, heat release rate and temperature are presented. The presented results are compared to a mixture fraction model and with the results obtained previously in reduced scale compartment fires. This experimental data is an improvement of previous data in the sense that it features well defined and steady wall boundary conditions and the gas species measurements are more detailed than previous measurements.

## EXPERIMENTAL CONFIGURATION

The ISO 9705 standard test stipulates a room with internal dimensions of $2.4m \times 3.6m \times 2.4m$ with a door for ventilation that has dimensions of $2m \times 0.8m$ [3]. All dimensions in the room are allowed a tolerance of $\pm 4cm$ and the as constructed dimensions were within that tolerance. The room used in these experiments is shown in Figure 1. The room was lined with a 2.5cm thick noncombustible alumina-silica-zirconia ceramic blanket, K-Lite™ HTZ. Previous investigations in the reduced scale enclosure had shown that ceramic blanket was the most stabile and reliable lining material to be used for these experiments [19]. This finding was further verified here. The structural stability and chemically non-reactive nature of the ceramic blanket also it a much easier boundary condition to prescribe for numerical modeling.

A water cooled pool burner was placed in the center of the room on the floor. The pool burner is a 1m square pan with a total external height of 15cm. The internal surface of the burner is inclined so that different liquid fuel pool surface areas can be achieved by varying the fuel level in the burner. House water was flowed through the burner at a variable rate up to 57 liters per minute to control the burner temperature. Liquid heptane was pumped at a specified continuous rate to control the total heat release rate from the burner. For natural gas the burner was filled with gravel and utility natural gas flow rate was metered to get the desired heat release rate from the burner.

Water cooled, gas sampling probes were inserted into the structure at points A and B, as illustrated in Figure 1. The locations of these points, relative to the origin indicated in Figure 1, are listed in Table 1 as well. These points were chosen to scale up geometrically from the original measurement points in the reduced scale enclosure [19]. Temperature and soot measurement probes were also included at these locations in an attempt to capture a more complete picture of the internal environment of the room. Temperature measurements were made at various other locations inside the room which are not presented here.

The gas samples were pulled through two cooled sample lines to freeze the chemistry and then to a heated line to prevent condensation then to a Nafion polymer membrane which selectively removes water from the gas stream [20]. The dried gas streams then flowed to a Servomex 4100 paramagnetic oxygen analyzer, a Siemens Ultramat 6E non-dispersive infrared (NDIR) CO and $CO_2$ analyzer, a Baseline-Mocon 8800H total unburnt hydrocarbon analyzer using a flame ionization detector (FID) and finally to a micro gas chromatograph (micro-GC). The sample flow path was nearly identical to that reported for the RSE [19]. Each gas analyzer was calibrated before each test, the uncertainty introduced by the span and calibration gases is 1% [19]. The type B uncertainty [21] in gas analyzer measurement is evaluated as the standard deviation of a series of measurements at a steady state condition. This uncertainty comes from the frequency of variations in the room being much faster than the time response of the analyzers. Each gas sampling location had identical and separate equipment except for the micro-GC. Instead of a second micro-GC, a gas

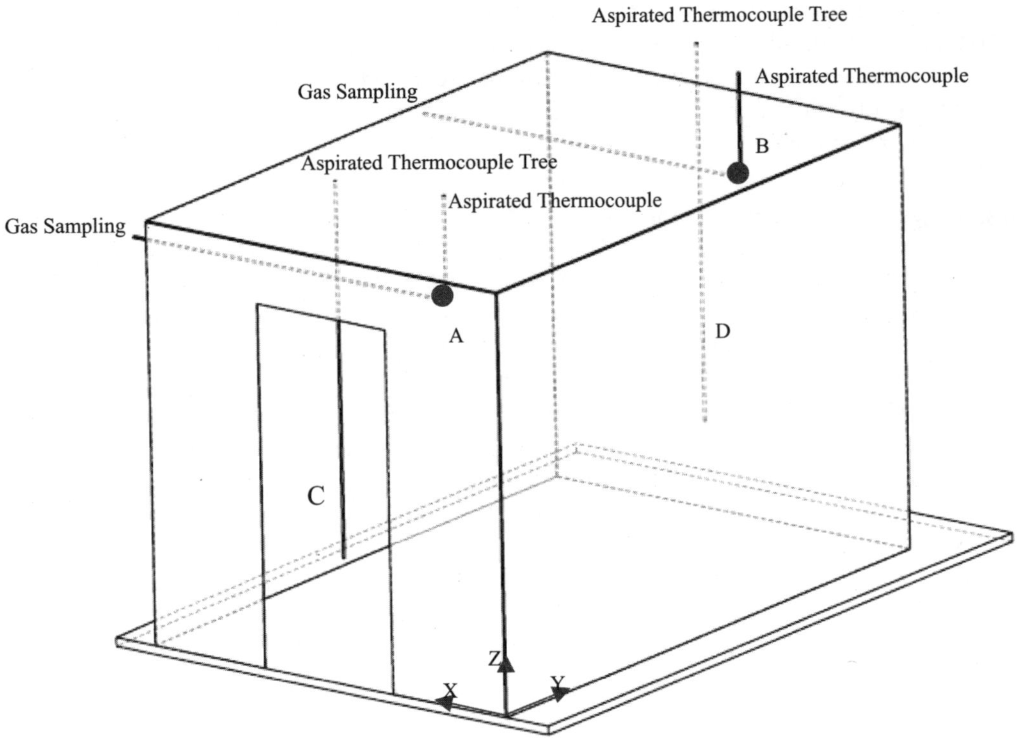


**Figure 1** Schematic drawing of ISO 9705 with internal sampling locations indicated

**Table 1** Location of sampling points inside the ISO Room

| *Sample Point* | *X (cm)* | *Y (cm)* | *Z (cm)* | *Measurements* |
|---|---|---|---|---|
| A | 189 | 25 | 208 | Gas Species, Temperature,Soot |
| B | 189 | 286 | 208 | Gas Species, Temperature,Soot |
| C | 120 | 311 | 180,120,60 | Temperature |
| D | 120 | 20 | 180,120,60 | Temperature |

sampling and storage system was constructed which captured and stored up to sixteen *10ml* samples of gas extracted from the room and stored them to be analyzed after the experiment. The micro-GC was connected directly to the other gas sample line and continually sampled and analyzed the gas flow every *2.5 minutes* throughout each test. The uncertainty in GC measurement is less than 1% based on the calibration gas and the GC method. All of the exhaust gasses from the room fires were captured by the *6 meter* hood at the NIST large fire laboratory (LFL) where oxygen-consumption calorimetry was conducted to determine the heat release rate of each fire [22, 23]. The LFL *6 meter* hood calorimetry has been shown to have an uncertainty of 14%.

Aspirated thermocouples were used for internal room temperature measurements. These thermocouples work by pulling combustion gasses directly from the room over a thermocouple enclosed by two radiation shields at a high velocity [24]. The aspirated thermocouples sacrifice temporal and thermal resolution but result in a significantly reduced uncertainty when compared to bare bead thermocouples in an environment where gas temperatures and radiative environment temperatures are significantly different. Since most of the measurements in this investigation are in the upper layer of the room the uncertainty in temperature measurements is predicted to be 7% [24].

## RESULTS AND DISCUSSION

### Structural Performance

As a continuation of the RSE work, a significant focus of this investigation is to develop a robust structure that is reusable and provides boundary conditions which are easy to mimic in numerical simulations. For this set of experiments one layer of *2.5cm* blanket was used on top of one layer of *1.27cm* gypsum board supported by metal studs and sitting on a plywood base. Internal enclosure temperatures in excess of *1000°C* were expected, and observed cf. Figure 2. Based on this assumption a simple heat transfer calculation was performed, based on a 1-D conduction model of the wall with internal and external convection acting on the surfaces. This model which made many assumptions including that the wall was impermeable to the hot gasses and air predicted that the external wall temperature would not exceed *100°C*. However, due to the permeability of the insulation material and the existence of seams in the insulation, which opened up as the insulation was heated and shrunk, this temperature was exceeded and the paper on the gypsum board burned and the integrity of the structure was compromised. Ultimately the gypsum material was destroyed while the ceramic insulating blanket remained intact. The plywood floor of the enclosure, also covered with ceramic insulating blanket, burned as well. This indicates that for fires of this scale the ceramic blanket had excellent performance while the drywall material were not sufficiently protected from the fire. The solution to this problem is that for the next set of tests the drywall walls and plywood floor of the structure have been replaced with steel.

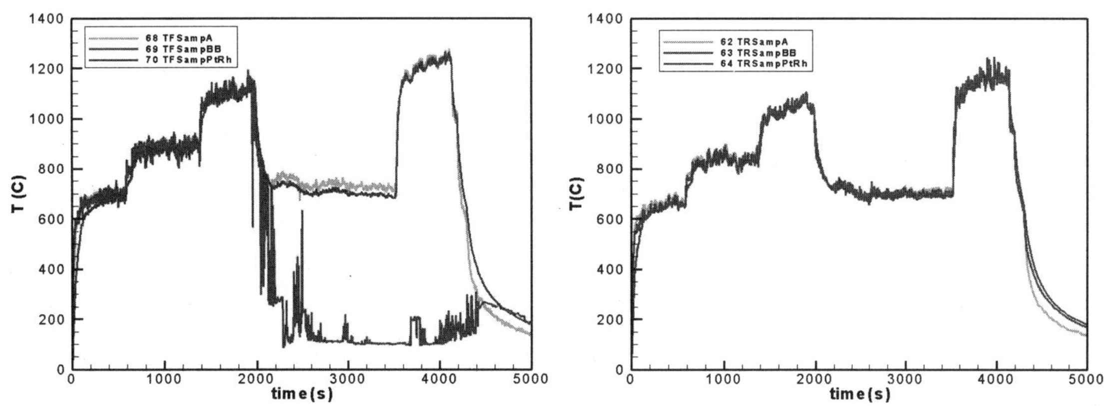


**Figure 2** Thermocouple temperature comparison for an aspirated thermocouple (T_SampA), bare bead type *K* thermocouple (T_SampBB), and a bare bead type *R* thermocouple (T_SampPtRh) for sample position A (left) and sample position B (right)

A robust array of temperature measurement probes were used for these fire tests. Bare bead type *K*, bare bead type *R*, and aspirated thermocouples constructed from inconel were all used for temperature measurements. The aspirated thermocouples are desirable in these situations due to their ability to overcome radiative losses from the thermocouple bead [24]. The three thermocouple types were positioned at the same locations in the box (within *2cm* in each direction), namely positions A and B from TAB. This way all of the thermocouples should be measuring the same temperature and allowing for comparison. Since these thermocouples were located in the upper layer it was expected that as a two layer fire formed the radiative losses from the thermocouples would be minimal, especially since neither of the measurement locations was located near to the door of the enclosure, and thus was unlikely to 'see' a cool radiative sink. Temperature measurements of the three thermocouple probes are compared in Figure 2 for the natural gas test. For both the front and rear positions the three different types of thermocouples show a close agreement in measured temperature. The dramatic drop off in the bare bead type *K* thermocouple temperature was due to failure of the bare bead thermocouple midway through the test. Besides that failure the thermocouple measured temperatures all remained within *50°C* of each other. This indicates that so long as there is not a significant radiation source or sink which can interact with the thermocouple at a drastically different temperature than the surrounding hot gas then an aspirated thermocouple is not necessary. This observation is of special importance for these tests because all of the inconel aspirated thermocouples used in this experiment failed and were destroyed in the fire. Only the type *R* thermocouple consistently provided temperature data without failing.

## General Experimental Observations

Experimental runs were completed at a variety of fire sizes with natural gas and heptane fuels. Here only one experimental run of natural gas, shown in Figure 3, and one experimental of heptane, shown in Figure 4, fueled fires are presented. Each test started at a low heat release rate (HRR) which was gradually increased in steps to a maximum heat release rate of *2.7MW* for the natural gas case and *1.5MW* for the heptane case, as seen in Figures 3 and 4 respectively. Each step in the HRR was allowed to stabilize so that steady state data could be gathered at that fire size. Nominal fire sizes were *0.9*, *1.25*, *2.4*, and *2.7MW* for natural gas and *0.75, 1.25, and 1.5MW* for heptane. Quantitative differences between the gas analyzers and the GC gas sample analysis were generally less than 1%. Figure 3 shows the HRR, temperature, and gas species data for the natural gas test. Figure 4 shows the same information for the Heptane test. From these figures several qualities are apparent. The variation in the gas species concentrations qualitatively follows the expected trend relative to the heat release rate. That is, as the heat release rate increases the temperature in the room increases, oxygen in the room is consumed and more product species, e.g. $CO_2$, and intermediate species, e.g CO, are present. Steady state is reached within 2 minutes after a step change in fuel delivery rate with natural gas burning in the room. The heptane fires approached a steady state more slowly than the corresponding natural gas fires. The heptane fires are inherently more unsteady than the corresponding natural gas fires so that the measured quantities from the heptane fire were significantly more unsteady than the corresponding natural gas fires for the same HRR.

Intermediate species are not observed in appreciable quantities until the HRR reaches *2.7MW* in the natural gas fire and *1.5MW* in the heptane fire. This indicates that in a global sense until this

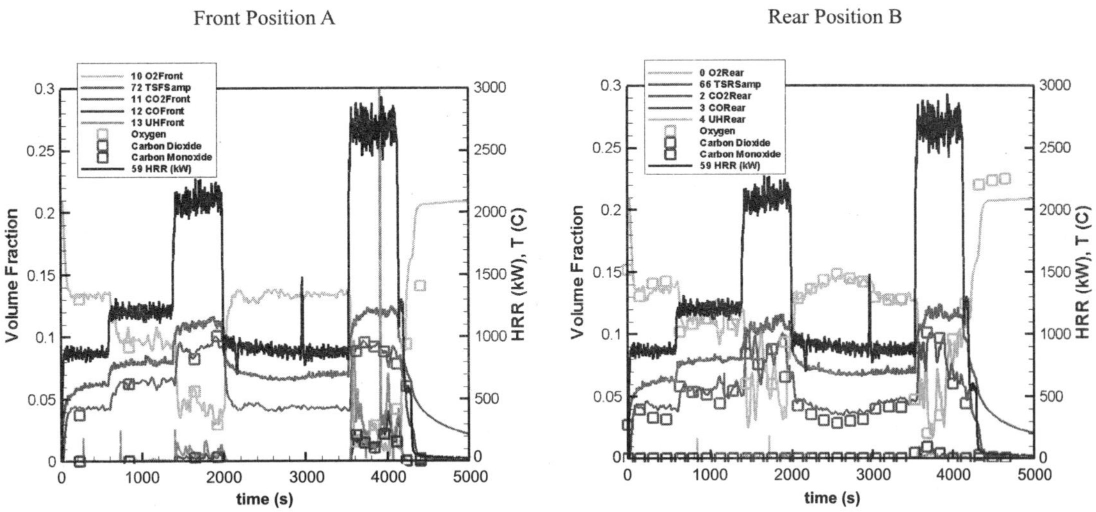


**Figure 3** Experimental heat release rate (HRR), temperature, gas analyzer, and selected GC data at sample position A (left) and sample position B (right) for natural gas fire.

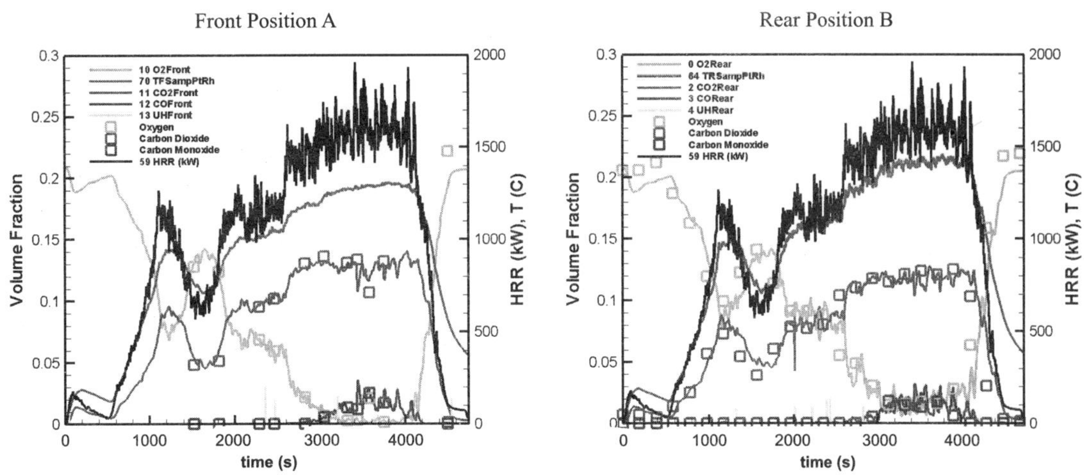


**Figure 4** Heat release rate, temperatures, and species mass fractions as a function of time for position A, front, and position B, rear, in the ISO room for Heptane fire

point is reached, the room is still burning slightly lean and the underventilated condition is not observed. This observation is further validated in the next section. For both fires at their peak heat release rate, less oxygen and more CO and UHC are observed at point A, near the front of the box, while slightly more oxygen and less CO or UHC are observed at point B, in the rear of the box. This indicates that as air enters the room it moves quickly to the rear of the compartment and is only vitiated as it approaches the front of the box, just prior to being exhausted.

## Mixture Fraction Analysis

As mentioned previously the mixture fraction is an important indicator of the local chemical composition and is used as a predictive tool for FDS. A correlation between the global heat release rate and the mixture fraction is not meaningful because the mixture fraction is defined as a local quantity based on the specific gas composition at that point in space and time. Instead local mixture fraction data is presented for the natural gas and heptane fires in Figures 5 and 6, respectively. The stoichiometric mixture fraction, $Z_{st}$, is the mixture fraction at which the ideal ratio of fuel to air is present for complete combustion. For natural gas the stoichiometric mixture fraction is approximately $Z_{st} = 0.0554 \pm 0.0002$ [19]. At larger mixture fractions the theory predicts that the fire burns rich, all oxygen should be consumed and only complete combustion products are formed. At smaller mixture fractions the fire should burn lean and all of the fuel should be consumed producing only complete combustion products. In practice this theory works well at predicting local chemical compositions for lean fires, however it has been well documented that above $Z_{st}$ the measured mixture fraction deviates significantly from the theoretically predicted mixture fraction [19].

The local species mass fractions for the natural gas fire test are plotted as a function of the mixture fraction in Figure 5. For both positions A and B the previous comments about mixture fraction hold true in that the measured local chemical composition follows the theoretical mixture fraction lines well for $Z < Z_{st}$ and then deviates after $Z > Z_{st}$. For the front position, position A, even though the volume fraction of oxygen did not drop to zero the local chemical system does transition to a rich-burning regime where more partial combustion products are produced. Specifically, there is an increase in total unburnt hydrocarbons, mostly methane, and an increase in CO which is not predicted by the mixture fraction theory since the theory assumes only complete combustion occurs. In the rear of the compartment, position B, the oxygen volume fraction is larger and the mixture fraction reaches the point of $Z_{st}$ but does not progress to burn much richer than $Z_{st}$. Here again a strong agreement is observed between the experimentally measured gas species

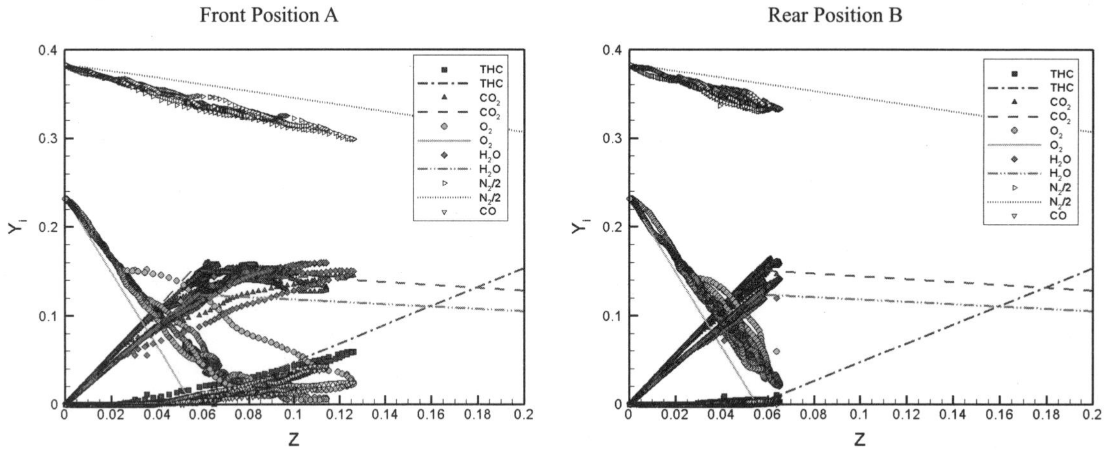


**Figure 5** Measured mass fractions (symbols) and theoretical mass fractions (lines) of important reactant, product, and intermediate species plotted as a function of mixture fraction for point A (left) and point B (right) for natural gas fire

concentrations and the theoretically predicted mixture fraction composition of the gas for $Z < Z_{st}$. However, in position B, the high level of oxygen is unexpected since the sampling location is remote from the doorway vent. This occurs despite locally similar temperatures between the front and rear positions, A and B respectively. Instead it is reasonable to assume that the ventilation profile is such that as fresh air enters the front of the enclosure it flows along the floor and rises in the rear of the enclosure. The local species mass fractions plotted as a function of mixture fraction for heptane fires are presented in Figure 6. Here similar trends to those observed for natural gas fires are observed. Specifically, the mixture fraction theory predicts the fuel oxygen and product mass fractions well for $Z < Z_{st}$ and then the prediction is less successful at predicting the composition for $Z > Z_{st}$. Further, the heptane fires also seem to fall short of achieving the underventilated condition, showing very little data for $Z > Z_{st}$. However, it should be noted that the difference between the measured gas composition and the theoretically predicted gas compositions is smaller for the heptane fire than it was for the natural gas fire.

## Scaling

In the previous investigation with the RSE underventilated fires were easily achieved ($O_2$ fraction dropped to zero in upper layer) for natural gas fires with an HRR of *250kW* [19]. The full scale ISO room was just becoming underventilated at *2.7MW*, cf. Fig. 2. According to the ventilation scaling mentioned earlier [5, 13] the full scale ISO room should be underventilated at *0.75 MW*. This indicates that the $Ah^{0.5}$ scaling is not working as expected. In fact this has been previously reported and is therefore not a surprise. The GER concept on which the ventilation scaling is based has been shown to only be valid for certain ventilation conditions and generally for temperatures less than *900°C* [4, 5]. As seen in Figure 2 the temperature in the full scale room is often above this threshold, with *1000°C* observed even at *1MW*.

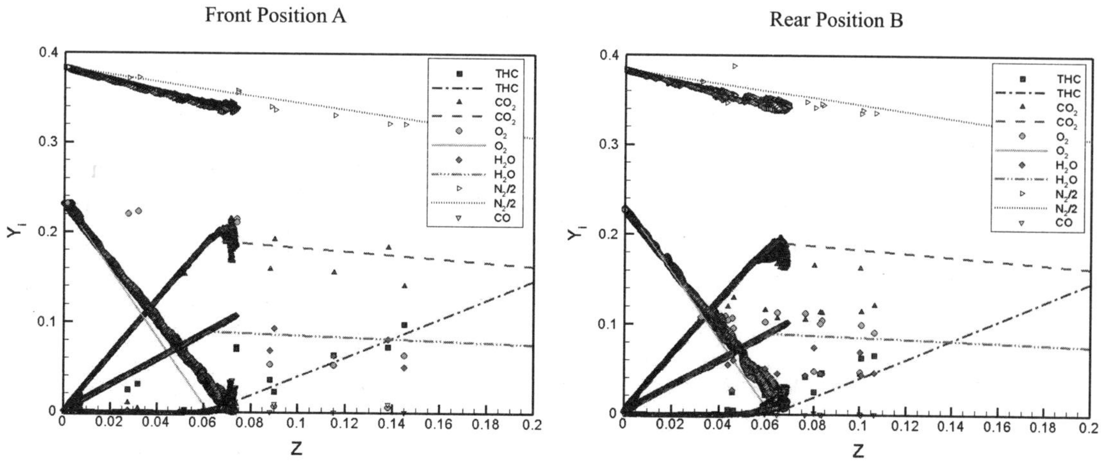


**Figure 6** Experimentally measured and theoretically predicted mass fractions of species plotted verses the mixture fraction at position A, front, and position B, rear, for Heptane fire

## Conclusions

Full-scale room fire tests were done based on the ISO 9705 standard room. Both natural gas and heptane were burned in a pool fire configuration in the center of the room. Heat release rates up to 2.7MW were produced in the room. Instrumentation included oxygen-loss calorimetry for heat release rate measurements, aspirated and bare bead thermocouples for temperature measurements, and gas species measurements.

1. In order to provide a reusable structure and boundary conditions that are easy to reproduce numerically, the structural robustness of the room was investigated. It was found that *2.5cm* of ceramic blanket insulation is insufficient to protect an outer skin of gypsum from the temperatures experienced inside the enclosure. The next evolution of the ISO room will be constructed completely of non-combustable components. The walls and floor will be constructed from steel and *5cm* of insulation with staggered seams will be used to protect the structure and maintain uniform boundary conditions.
2. Fires with a heat release rate of greater than *2.4MW* (and temperatures greater than *1200°C*) were found to destroy most of the temperature measurement probes that were tested. This included inconel sheated type *K* aspirated thermocouples. However, platinum-rhodium type *R* thermocouples mounted in a ceramic support were shown to survive the extreme temperatures inside the enclosure. Additionally, only small temperature differences between the aspirated and bare bead type R thermocouples were observed in the upper layer. This suggests that for many of the temperature measurements in fires of this type the aspirated thermocouples may not be necessary.
3. Oxygen levels were only observed to drop to near zero levels occasionally in the experiments. None of the heat release rates tested produced a steady underventilated fire. This was also evidenced by very low levels of CO, $H_2$, and other unburnt hydrocarbons detected in the fire.
4. The mixture fraction calculations were based on the fuel, oxygen, and $CO_2$ mass fractions measured in the fire. It was observed that for lean natural gas and heptane fires ($Z < Z_{st}$) the mixture fraction does a reasonable job of predicting the mass fractions of unburnt hydrocarbons (fuel), oxygen, and $CO_2$. The heptane fires adhered to the theoretical mixture fraction better than natural gas fires for $Z < Z_{st}$. For $Z > Z_{st}$ the experimental measurements departed from the theoretical predictions significantly. The natural gas fire measurements also showed a large departure from the theoretically predicted mixture fraction for $Z > Z_{st}$
5. The full scale ISO room was only observed to approach an underventilated condition at *2.4MW* while the previous reduced scale enclosure fires became underventilated at *250kW*. This is a factor of 10 difference in required heat release rate which is not consistent with the ventilation scaled prediction for the *2/5* scale reduced scale enclosure. However, it has been established that the ventilation scaling does not work well when the enclosure upper layer temperature is above *900°C*. Enclosure upper layer temperatures well in excess of *1000°C* were observed in the experiments suggesting that this type of scaling will not work for these enclosure fires.

## References


[1] K. B. McGrattan *Fire Dynamics Simulator (Version 4): Technical Reference Guide*; NIST SP 1018; 2004.

[2] A. National Fire Protection; E. Society of Fire Protection, "*SFPE handbook of fire protection engineering.*" National Fire Protection Association: Quincy, Mass, 2002.

[3] ISO9705 *Fire Tests - Full-Scale Room Test for Surface Products First Edition*; International Ogranization for Standardization: Geneva, Switzerland, 1993.

[4] B. Andersson; F. Markert; G. Holmstedt, "Combustion products generated by hetero-organic fuels on four different fire test scales" *Fire Safety Journal* **2005,** 40, (5), 439-465.

[5] W. M. Pitts, "The Global Equivalence Ratio Concept and the Formation Mechanisms of Carbon-Monoxide in Enclosure Fires" *Progress in Energy and Combustion Science* **1995,** 21, (3), 197-237.

[6] W. M. Pitts; E. L. Johnsson; N. P. Bryner, "Carbon Monoxide Formation in Fires By High-Temperature Anaerobic Wood Pyrolysis" *Twenty-Fifth Symposium (International) on Combustion* **1994**, 1445-1462.

[7] K. B. McGrattan; H. Baum; R. Rehm; S. Hostikka; J. Floyd *Fire Dynamics Simulator (Version 5): Technical Reference Guide*; NIST SP 1018-5; National Institute of Standards and Technology: 2007.

[8] R. W. Bilger, "Reaction-Rates in Diffusion Flames" *Combustion and Flame* **1977,** 30, (3), 277-284.

[9] N. Peters, "Laminar Diffusion Flamelet Models in Non-Premixed Turbulent Combustion" *Progress in Energy and Combustion Science* **1984,** 10, (3), 319-339.

[10] A. Hamins; K. Seshadri, "The Structure of Diffusion Flames Burning Pure, Binary, and Ternary Solutions of Methanol, Hepatane, and Toluene" *Combustion and Flame* **1987,** 68, (3), 295-307.

[11] Y. R. Sivathanu; G. M. Faeth, "Generalized State Relationships for Scalar Properties in Nonpremixed Hydrocarbon Air Flames" *Combustion and Flame* **1990,** 82, (2), 211-230.

[12] Y. R. Sivathanu; G. M. Faeth, "Soot Volume Fractions in the Overfire Region of Turbulent-Diffusion Flames" *Combustion and Flame* **1990,** 81, (2), 133-149.

[13] N. P. Bryner; E. L. Johnson; W. M. Pitts *Carbon Monoxide Production in Compartment Fires - Reduced-Scale Enclosure Test Facility*; NIST IR 5568; National Institute of Standards and Technology: 1994.

[14] T. R. Hull; J. M. Carman; D. A. Purser, "Prediction of CO evolution from small-scale polymer fires" *Polymer International* **2000,** 49, (10), 1259-1265.

[15] D. Gross; A. F. Roberston "*Experimental fires in enclosures*", Tenth International Symposium on Combustion, 1965; 1965; pp 931-942.

[16] G. Heskestad "*Modeling of enclosure fires*", Fourteenth International Combustion Symposium, 1972; 1972; pp 1021-1030.

[17] K. Kawagoe *Fire behavior in rooms*; 27; Building Research Institute of Japan: September, 1958.

[18] J. G. Quintiere, "Scaling Applications in Fire Research" *Fire Safety Journal* **1989,** 15, (1), 3-29.

[19] M. Bundy; A. Hamins; E. L. Johnsson; S. C. Kim; G. H. Ko; D. B. Lenhert *Measurements of Heat and Combustion Products in Reduced-Scale Ventilation-Limited Compartment Fires*; TN1483; National Institute of Standrads and Technology: 2007.

[20] K. A. Mauritz; R. B. Moore, "State of understanding of Nafion" *Chemical Reviews* **2004,** 104, (10), 4535-4585.

[21] B. N. Taylor; C. E. Kuyatt *Guidelines for Evaluating and Expressing the Uncertainty of NIST Measurement Results*; NIST TN 1297; National Institute of Standards and Technology: Gaithersburg, MD, 1994.

[22] S. Brohez, "Uncertainty analysis of heat release rate measurement from oxygen consumption calorimetry" *Fire and Materials* **2005,** 29, (6), 383-394.

[23] R. A. Bryant; T. J. Ohlemiller; E. L. Johnsson; A. Hamins; B. S. Grove; A. Maranghides; G. W. Mulholland; W. F. Guthrie *The NIST 3 Megawatt Quantitative Heat Release Rate Facility - Description and Proceedure*; NIST IR 7052; 2004.

[24] L. G. Blevins; W. M. Pitts, "Modeling of bare and aspirated thermocouples in compartment fires" *Fire Safety Journal* **1999,** 33, (4), 239-259.

# Design and Development of Combustion Systems for Aero Gas Turbine Engine

**Venkataraman Shankar*** **and G. Sivaramakrishna#**
*Scientist 'G', Additional Director & Technical Director, E-mail: drshankarv@gtre.org
#Scientist 'E', CFD Group, E-mail: sivaramakrishnag@mail.gtre.org
Gas Turbine Research Establishment, C V Raman Nagar, Bangalore, India

## 1. INTRODUCTION

GTRE (Gas Turbine Research Establishment) of DRDO (Defence Research & Development Organisation) is engaged in design, development, manufacture, testing & supply of airworthiness certified gas turbine engine for the Indian combat aircraft. This paper focuses on the technological issues faced in the development of combustor and afterburner and how successfully the hardware has been realized to meet the specified design targets and performance objectives.

## 2. COMBUSTOR

Figure 1 shows the 3-D CAD model of the state-of-the-art combustor that has been realized. It comprises of pre diffuser, dump diffuser, flame tube and atomizers, the other important components being swirler, flare / heat shield, etc.

### 2.1 Technological Issues Addressed During Development

#### *2.1.1 Erosion of heat shield*

During the developmental testing of one of the versions of combustor, at high pressures and temperatures, it has been observed that the flare / heat shield got eroded. Fig. 2 shows the photograph of the damaged flare / heat shield [1].

Redesigning the flare, by the usage of a better temperature resistant alloy and by altering the manufacturing process, has circumvented this problem. The original flare has been made out of Nimonic C-263 alloy whereas the modified flare has been fabricated out of Haynes – 188 alloy. This alloy has about 150° higher temperature resistance capability and about 90 MPa higher proof

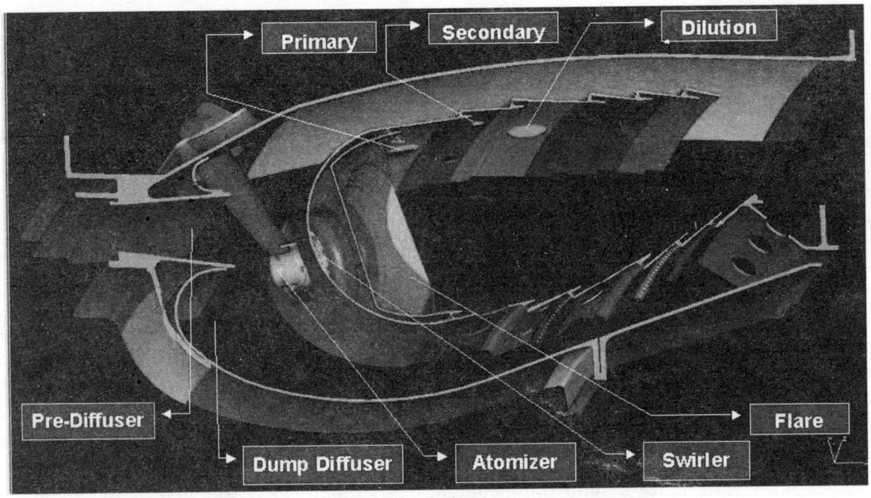


**Figure 1** A 20° Sector CAD Model of Combustor

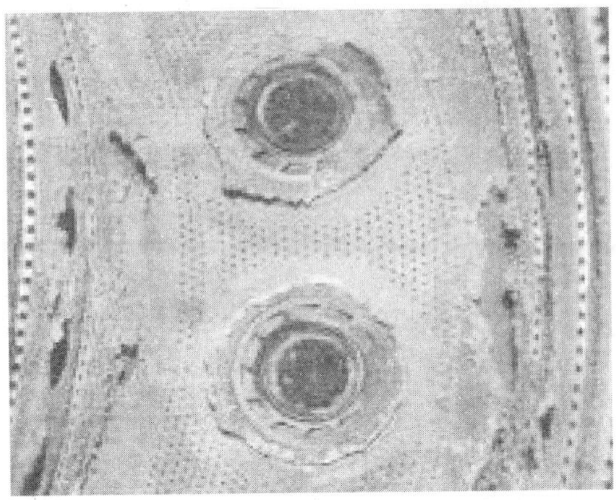

<u>*Test Conditions:*</u>
Inlet Total Pressure: 1600 kPa
Inlet Total Temperature: 800 K

**Figure 2** Photograph of the Damaged Flare / Heat Shield

strength at 1000° C in comparison to Nimonic C-263. Figure 3 shows the drawings of both the original and the modified flares. Combustor tests with this modified flare have confirmed that there is no erosion.

### *2.1.2 Performance improvement of air blast atomizer*

Fuel injection system plays an important role in the development of combustor. The aero version of combustor at GTRE employs an air-blast atomizer. The baseline version of the air blast atomizer has been tested and various parameters viz., spray cone angle, circumferential distribution of fuel concentration and SMD have been measured. Concentration distribution of a typical atomiser has been shown in Fig. 4. From these tests, it was concluded that there is a large non-uniformity between the chosen atomisers, which would be resulting in a bad exit temperature quality.

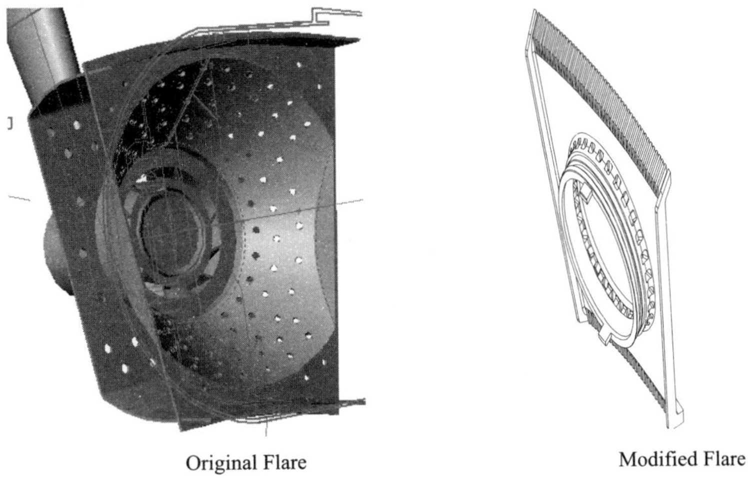


**Figure 3** 3-D Views of the Original and Modified Flares

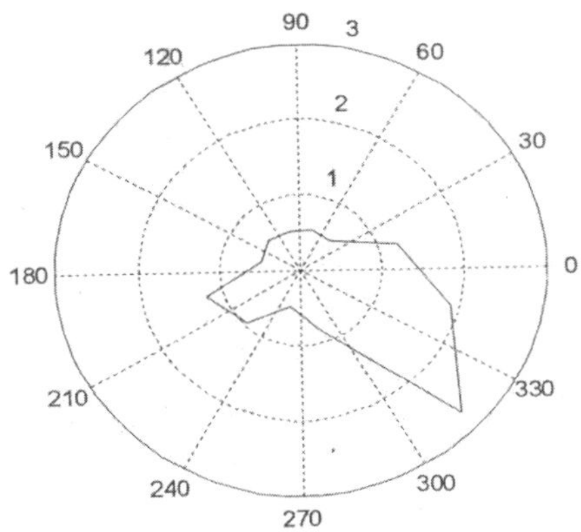


**Figure 4** Circumferential Distribution of Fuel Concentration for the Baseline Atomizer

In order to improve the fuel atomization quality, the atomizer has been modified and the air flow and fuel flow coefficients have been improved. The SMD of the atomizer has been brought down from 30 – 104 to 15 –38 μm for a pressure differential of 2.7 to 10 kPa [2]. 2-D cross sectional drawings of both the baseline and modified atomizers have been shown in Fig. 5. The modified version has swirlers on the pintles, both on the inner and outer surfaces, for better atomization and for better altitude relight capability.

Stand-alone tests of both the original and modified atomizers using optical diagnostics has been carried out at Indian Institute of Technology, Madras. The cone angle and the cross-sectional qualitative spray pattern were obtained by imaging the Mie scattering from the spray of a appropriately oriented laser light sheet, and the axial and cross-sectional droplet size distributions were obtained by means of a Phase-Doppler Particle Analyzer (PDPA).

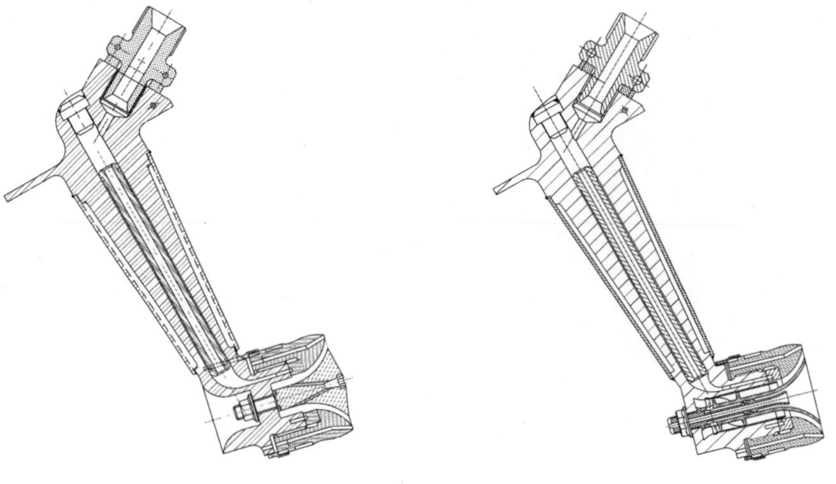

Original Atomiser Modified Atomiser

**Figure 5** 2-D Cross Sectional Views of the Original and Modified Atomizers

The results indicated that the modified atomizer exhibited a greater spread and azimuthal uniformity of the spray close to the lip of the atomizer, and a more uniform spray size distribution in a cross-sectional plane close to the lip of the atomizer when compared to the unmodified atomizer. A comparison of the radial variation of Sauter Mean Diameter (SMD) and Arithmetic Mean Diameter (AMD) for the original and modified Atomisers has been shown in Fig. 6. It can be seen from the figure that while the variation of AMD shows the presence of much finer droplets with the modified atomizer than the original one, the Sauter mean diameter distribution shows entrainment of larger particles in the hollow cone region with the modified atomizer. Further, the atomization is observed to complete within 30 mm of the exit plane of the modified atomizer [3].

Stand-alone atomizer tests, one-burner sector combustor tests and the combustor altitude tests at CIAM facility have demonstrated the above improvements. Figure7 shows the altitude relight characteristic of the combustor with the modified atomizer [4].

### *2.1.3 Combustor behavior in wind mill and spool down test conditions*

Fig. 8 and Fig. 9 show the combustor light up characteristics during wind milling and spool down tests as a part of the core engine testing at CIAM.

It can be seen from the figures that the combustor performed satisfactorily w.r.t. light-up at all the conditions and light up occurred in less than 5 seconds.

### *2.1.4 Manufacture of dome*

One of the critical components in manufacture has been the dome and shown in Fig. 10. It has been realized through spinning process of Nimonic C-263 sheet and at various stages, controlled heat treatment cycles had to be followed to maintain the thickness and to avoid cracking.

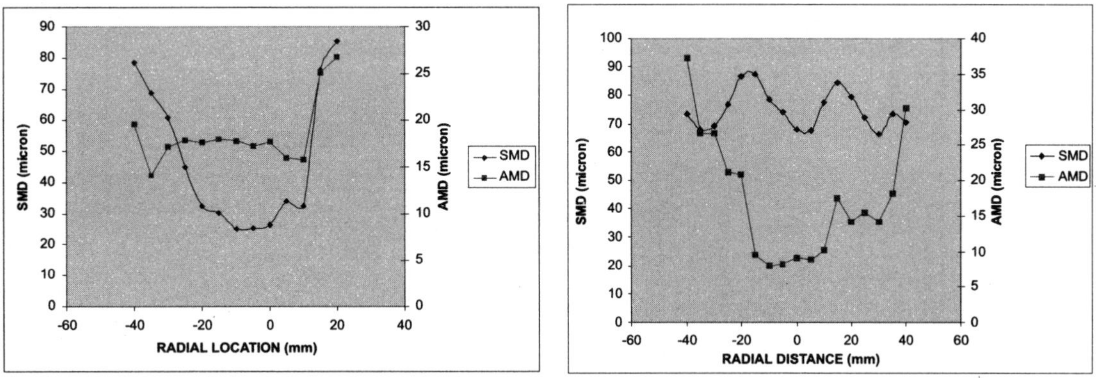


**Figure 6** Comparison of the Radial Variation of SMD and AMD for the Original and Modified Atomisers

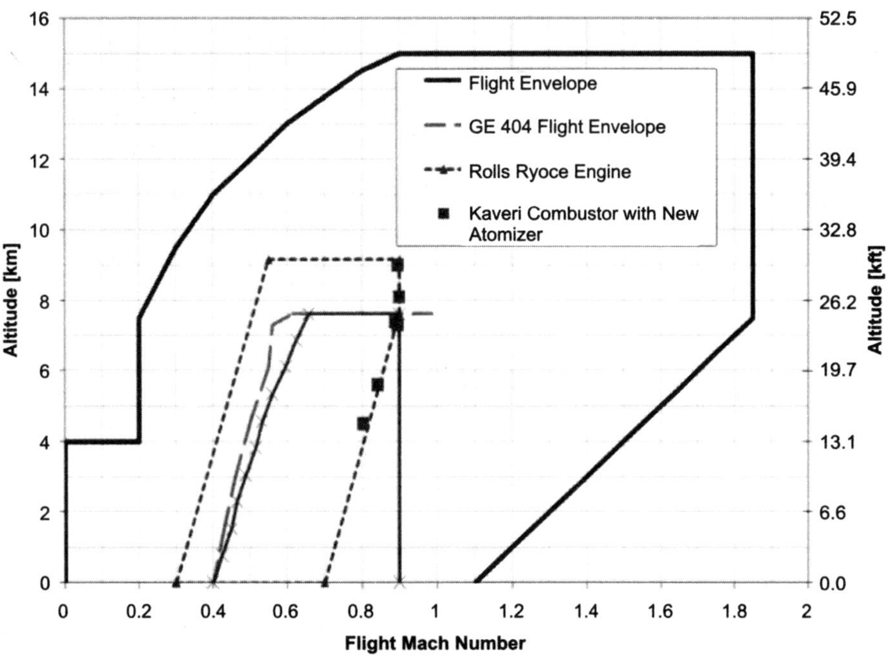


**Figure 7** Combustor Altitude Relight Characteristic obtained at CIAM using the Modified Atomiser

### *2.1.5 Pressure testing of combustion chamber outer casing*

The CCOC (Combustion Chamber Outer Casing) is a pressure vessel and has been successfully made out of Inconel - 718 alloy. During the aircraft flight, this component is subjected to both cyclic pressure and temperature loading and is a highly stressed component. The life limiting features in diffuser casing are the axial and circumferential welds on the casing and the fillet welds around instrumentation bosses. It has been tested to validate the LCF (Low Cycle Fatigue) life and weld

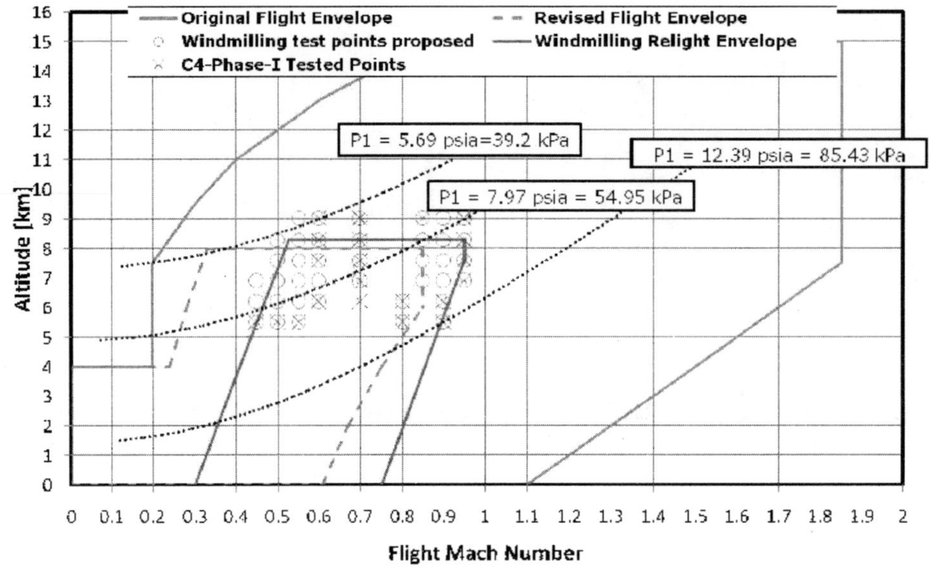


**Figure 8** Combustor Wind Milling Test Results

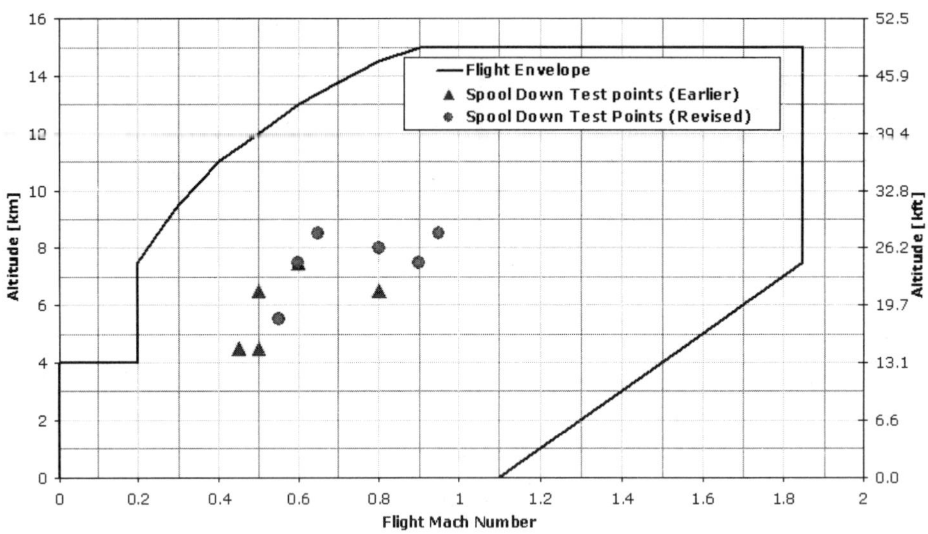


**Fig. 9** Combustor Spool Down Test Results

integrity. The test component has been suitably adapted in the test facility as shown in Fig. 11 and a leak check was performed to check the connectivity of the hydraulic connections and sealing mechanisms.

When the pressure reached the peak value during the leak check, the housing drain valve of the diffuser casing bursted due to insufficient weld penetration. Further, weld crack has been observed

**Figure 10** Photograph of the Modified Dome Assembly

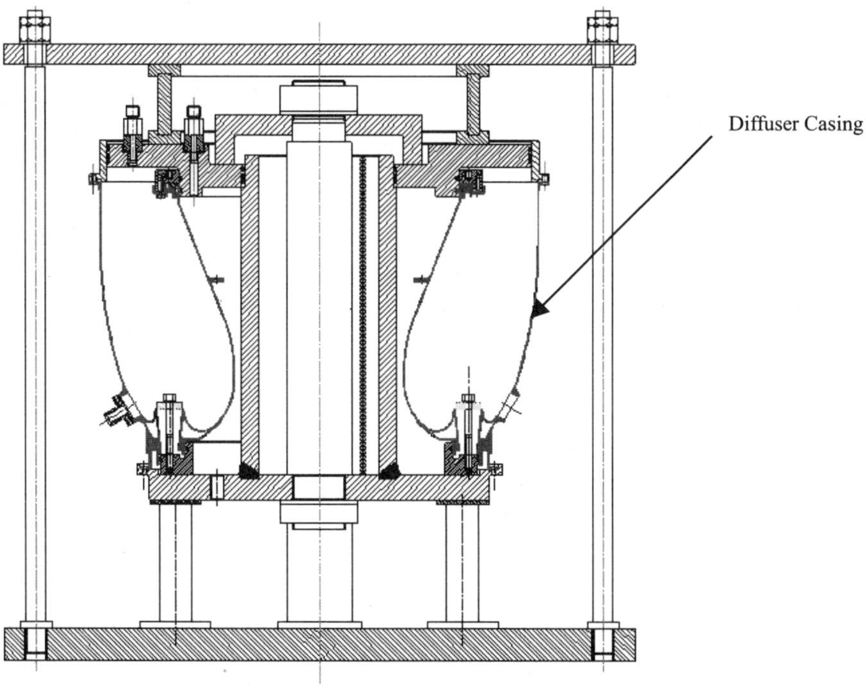


**Figure 11** Arrangement of Diffuser Casing in the Pressure Test Facility

near the boss ignitor, which could be due to the presence of a close proximity weld adjacent to the boss. Modifying the manufacturing process and the type of welding has circumvented both these problems. The modified casing after repeating the test has been cleared without any deformations [5].

#### *2.1.6 Pattern factor improvement*

The initial version of combustor, when tested at high pressures and temperatures, has given a Radial Pattern Factor (RPF) of 0.24 and a Circumferential Pattern Factor (CPF) of 0.6 as against the design intent values of 0.12 and 0.30 respectively. In order to improve the RPF and CPF values and to have an effective liner cooling, the liner open hole configuration has been modified by carrying out a series of CFD simulations. Fig. 12 gives a comparison of the original and modified liner hole configurations. In addition, the configurations of dome and cowl have also been modified. Fig. 13 shows the improved RPF and CPF profiles obtained for the modified configuration.

It can be seen from the figure that through CFD simulations, the RPF and CPF values have been brought down to 0.10 and 0.26 respectively [6].

### 2.2 Tools Employed in Development

#### *2.2.1 CFD codes*

In GTRE, several variants of annular combustors have been designed and developed for aero engine applications. Until the end of the 80's, the design of combustor was based on a one-dimensional basis and empirical approach requiring a large number of tests. This design was supported by CFD TEACH code, axisymmetric inviscid-boundary layer code for the diffuser and later PHOENICS code on a PC. By the end of 90's, ANSWER code was procured and the CFD analysis of the combustor was carried out Pentium Computers. The ANSWER code was validated with extensive measurements carried out on a model combustor and pressure loss and exit temperature profiles obtained on an actual combustor in the rig. This code became a part of the design cycle [7].

In early 2000, GTRE procured two high-end 32 node parallel machines. A 16 node parallel license of Fluent was procured and this became the workhorse for CFD analysis of all new design/ modifications of combustors. Fluent code has also been validated extensively with the rig measurements before using it in the design cycles.

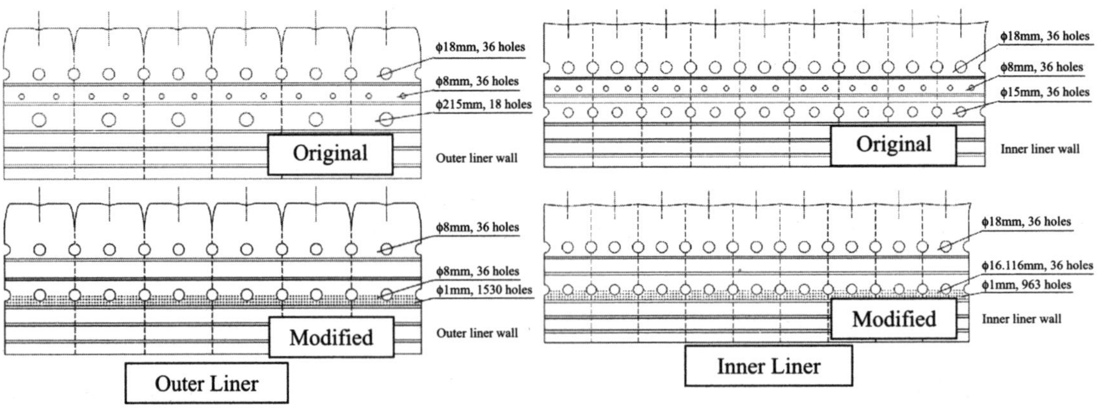


**Figure 12** Liner Hole Configurations of the Original and Modified Combustors

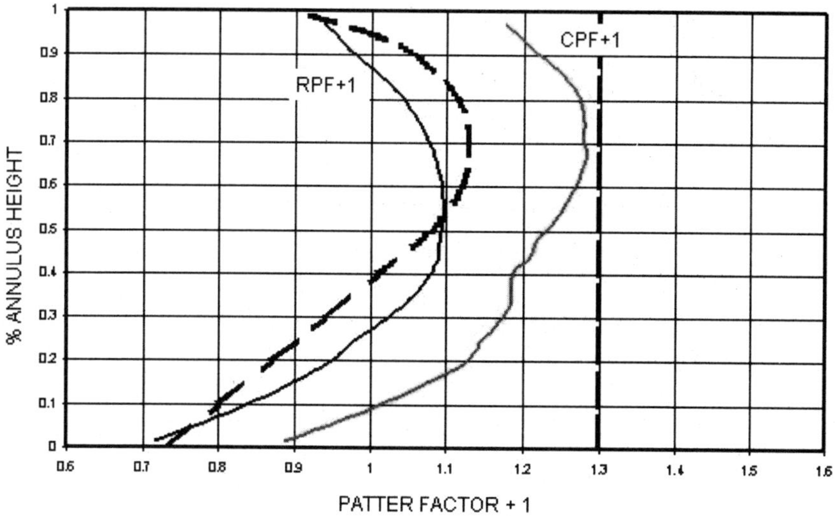


**Figure 13** Comparison of RPF and CPF Profiles obtained for the Modified Combustor and the Design Intent

#### 2.2.1.1 Validation of CFD codes

CFD analysis has been carried out using the CFD code 'ANSWER' for the cold & hot flow experiments carried out at GTRE as well as for the cold flow experiments carried out at Indian Institute of Science (IISc), Bangalore.

Cold flow experiments were carried out on a 120-degree sector Rapid Prototype (RP) model of the combustor to validate the ANSWER code. Tests were carried out at the low speed wind tunnel in the turbomachines lab of the mechanical engineering department, IISc near ambient conditions. Static pressure variation along the combustor walls and the velocity distribution at a few axial locations has been measured in the experiments. Fig. 14 shows the instrumented model mounted in the wind tunnel.

CFD analysis has been carried out on a 40-degree sector of the same combustor using ANSWER code by imposing the experimental conditions as boundary conditions. Fig. 15 shows a typical comparison of the wall static pressure recovery coefficient (Cp) along the outer casing obtained in the experiment and predicted using CFD. A close match is seen between experiments and CFD [8].

With regard to the validation of Fluent code for reacting flow simulations, as the combustor is an annular one with 18 atomizers equally spaced along the circumference, analysis has been carried out for the flow in a 20° sector of combustor with periodic boundary conditions being imposed on the two sides of the extreme boundaries.

A 3-D, hybrid grid having approximately 2 million nodes has been generated using Gambit, the pre-processor of Fluent code. A good quality of grid in terms of aspect ratio and deviation on orthogonality has been generated based on the guidelines given in the Fluent manual. All the features viz., swirler vanes, liner holes, holes on dome, shield, cooling rings and air blast atomizer air passages have been modeled. The overall view of grid generated is shown in Fig. 16.

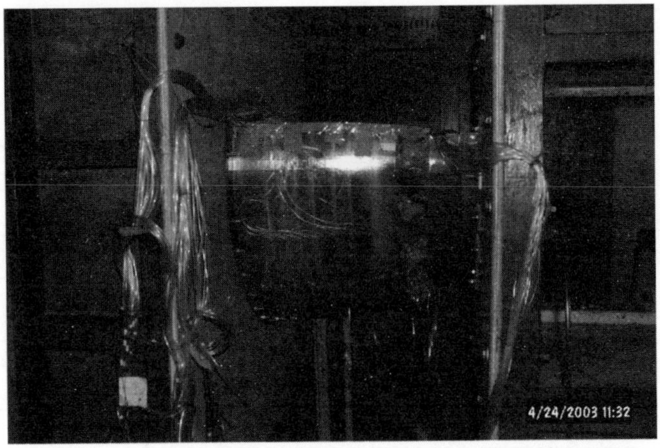


**Figure 14** Instrumented Model Mounted in the Wind Tunnel

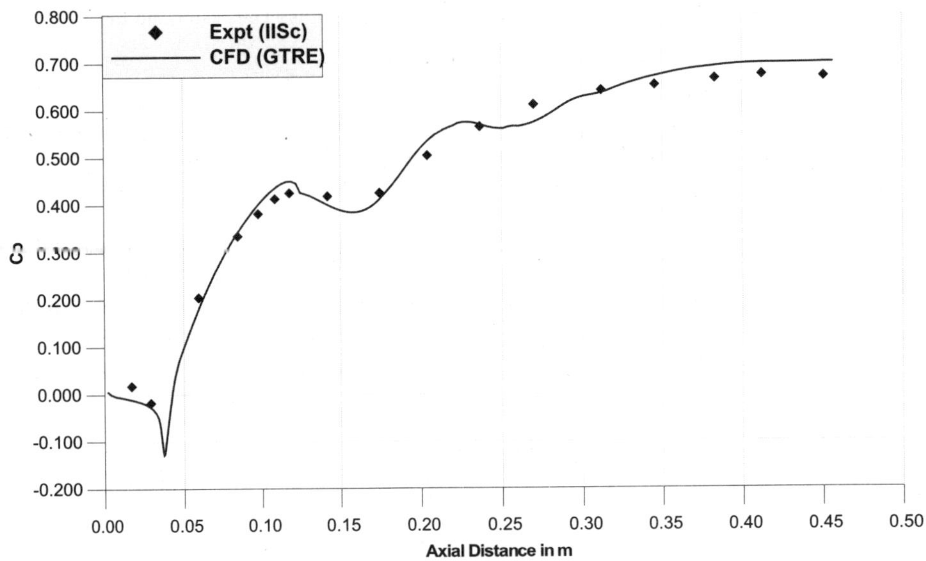


**Figure 15** Comparison of Cp Variation along the Combustor Outer Casing Wall

The simulations have been carried out on 4 processors of IBM P690 parallel system. The solutions are obtained by solving the compressible Reynolds-Averaged Navier-Stokes equations using a finite volume method. Realizable κ-ε model with enhanced wall treatment has been used to model the turbulence. Non-premixed combustion model has been used for the simulation of chemical reactions. Non-adiabatic Flame-let model has been used to model combustion. A total of 17 species and 28 numbers of reactions were considered for simulating the combustion reactions. Aviation kerosene fuel ($C_{12}H_{23}$) has been injected as a discrete phase using cone injection model. The relative velocity of the fuel is adjusted so as to achieve a mean particle diameter of 20 microns.

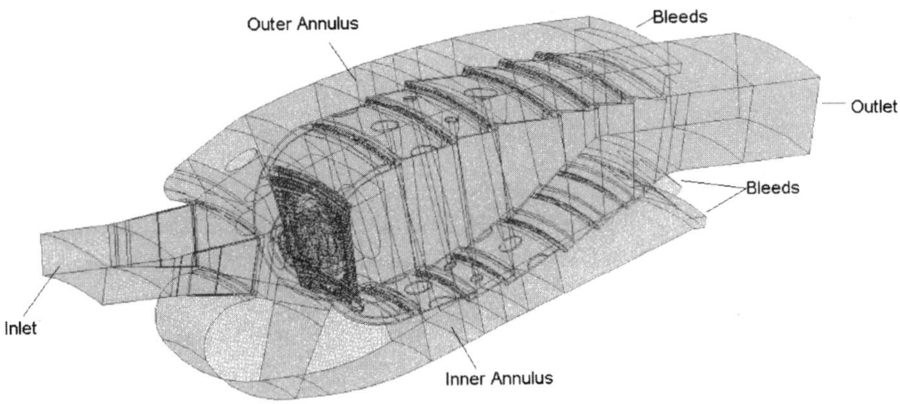


**Figure 16** Overall Grid Generated

Grid independence study has been carried out with three sets of grids for the baseline combustor configuration having grid densities of 1.5, 2 and 2.5 million cells and based on the variation of % total pressure loss between the grids, the 2 million size grid has been used for further analyses.

The converged solution has been post processed to estimate the value of % total pressure loss, RPF and CPF. The comparisons of predicted and measured RPF profiles along the % annulus height at combustor exit have been shown in Fig. 17. From the figure, barring the region close to the hub section, a close agreement can be seen between the predictions and measurements, the maximum deviation being 0.3.

With regard to the % total pressure loss, a maximum deviation of 0.29 has been observed between the measurements and predictions, which is acceptable. With regard to the RPF & CPF values, a maximum deviation of 0.03 between the measurements and predictions in the case of RPF and a maximum deviation of 0.02 in the case of CPF has been observed [9].

### *2.2.2 Thermal & stress analysis codes*

Conjugate Heat Transfer (CHT) and stress analysis are more than necessary in combustor development. CHT analysis has been carried out using Fluent to estimate the liner metal temperature. Fig. 18 shows the contours of liner metal temperature obtained for a typical condition.

A detailed stress analysis has been carried out for the most critical operating condition in order to meet the structural and life requirements using MSC/NASTRAN version 66 on a SPARC-I workstation. An integrated model of the engine static structure has been developed and the stress analysis has been carried out for limit and ultimate loading conditions i.e. pressure, temperature, bearing loads, gyroscopic, maneuver, blade-off, vane loads, unbalance load cases. Fig. 19 shows the finite element model of the engine static structure (180 degree) symmetric model.

This integrated finite element model of casings and frames has been used to estimate growths and stresses in the structure. The estimated Von-Mises stresses are compared with the allowable values of the material at the operating temperature and design objectives for normal and ultimate

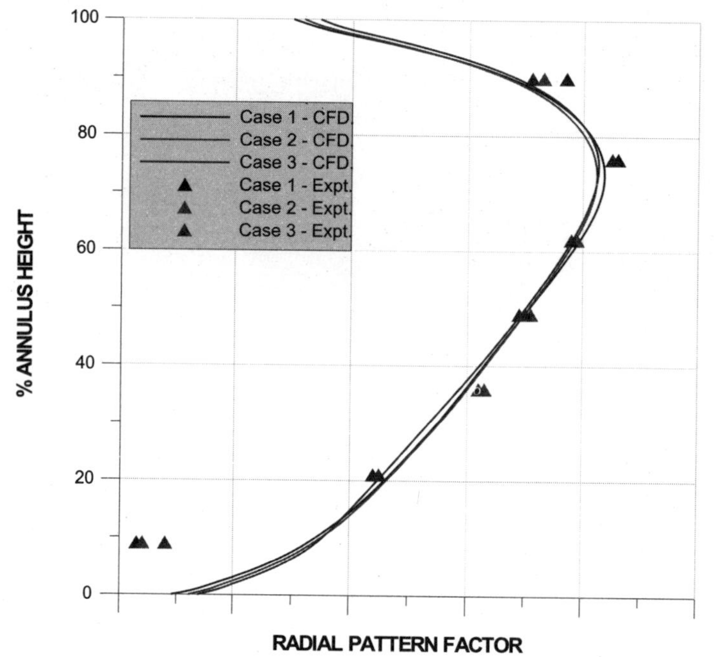


**Figure 17** Comparison between the Measured and Predicted RPF Profiles

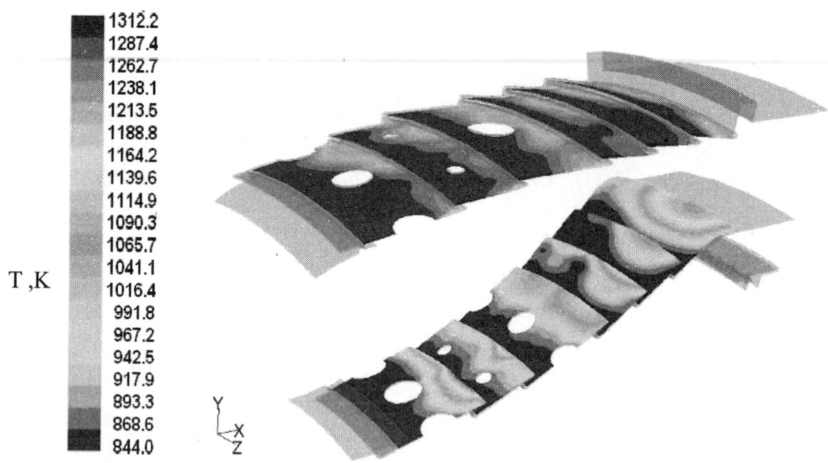


**Figure 18** Liner Metal Temperature Contours obtained through a CHT Analysis

load cases were met. Figs. 20 and 21 show the estimated Von Mises stress distribution for the casing and liner respectively.

Also, the operating stresses have been compared with the properties of Inconnel SU-718 material and it has been found that the CCOC meets twice of design pressure condition with wall

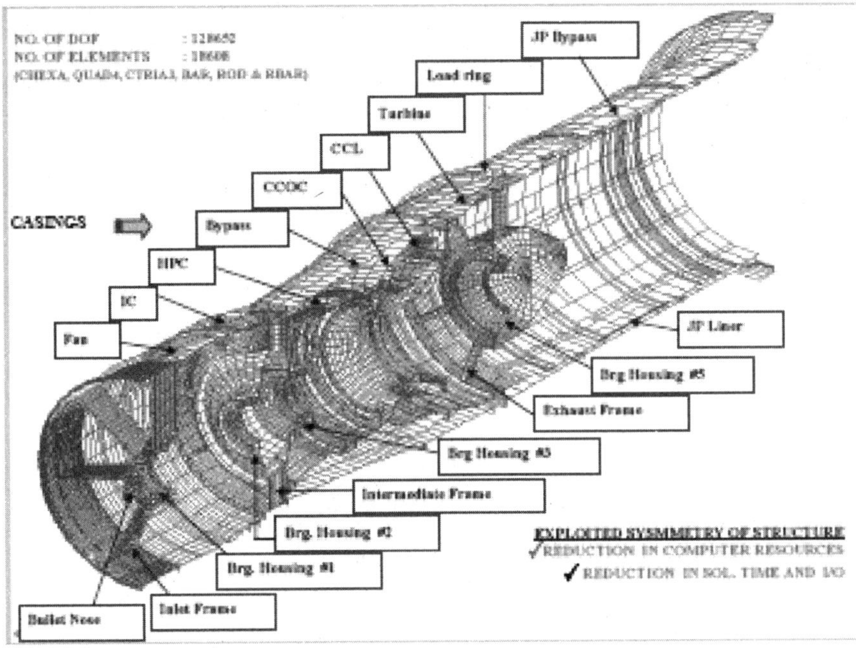


**Figure 19** Finite Element Model of Engine Static Structures

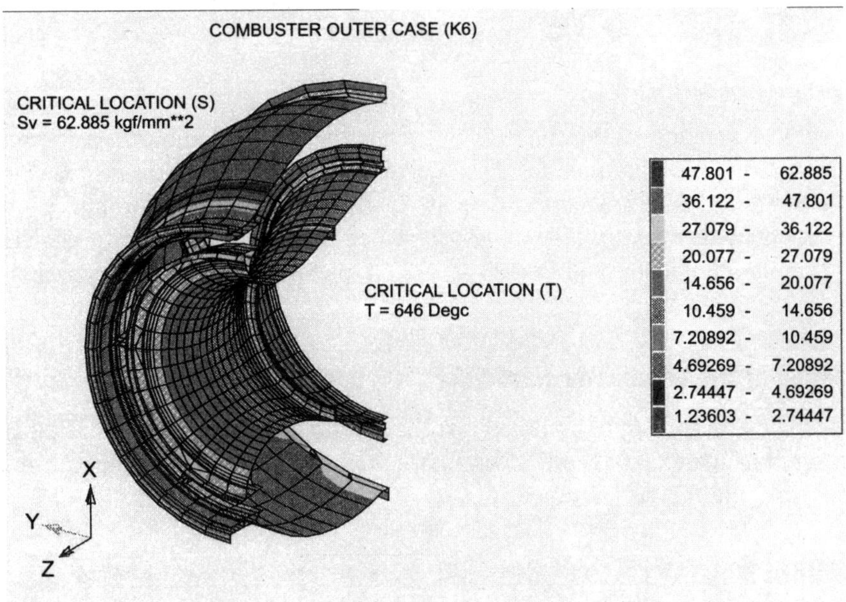


**Figure 20** Von Mises Stress Field for Pressure and Thermal Load Case - Combustor Casing

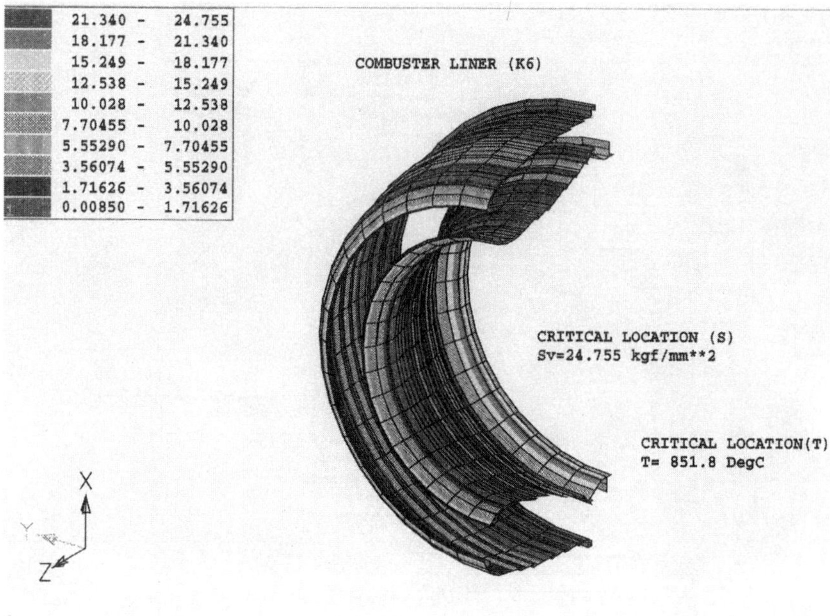


**Figure 21** Von Mises Stress Field for Pressure and Thermal Load Case - Liner

thickness provided. Using a detailed sub-model of CCOC, the casings life has been estimated to be greater than design target i.e., 7000 '0 – max – 0' cycles [10]

#### *2.2.3 Flow visualization technique*

3-D Water flow visualisation studies have been carried out on the combustor to estimate the mass flow distribution as well as total pressure loss. Quality of the flow patterns has been studied using air bubbles as tracers. Fig. 22 shows a typical flow pattern obtained through such study. The pressure losses obtained through this study are scaled using the aero-thermodynamic test data obtained on a similar combustor and Fig. 23 shows a typical total pressure loss variation obtained. [11]

### 2.3 Realization of the Combustor Hardware

Fig. 24 shows the latest version of combustor realized which is planned to be tested at the Central Institute of Aviation Motors (CIAM), Russia for high altitude, high temperature & pressure conditions for verifying the design objectives.

## 3. AFTERBURNER

A state-of-the-art afterburner has been successfully developed and tested in stand-alone mode at CIAM test facility, which is shown in Fig. 25, in order to estimate its performance in terms of light

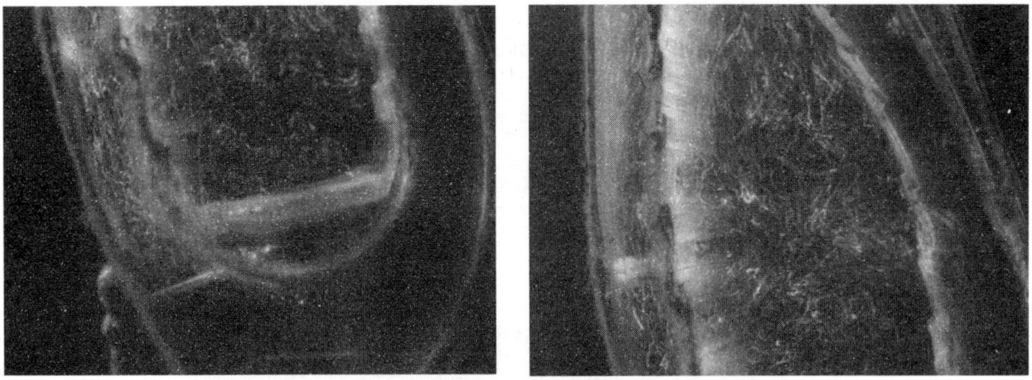

**Figure 22** A Typical Flow Patterns obtained through Water Flow Visualisation Study

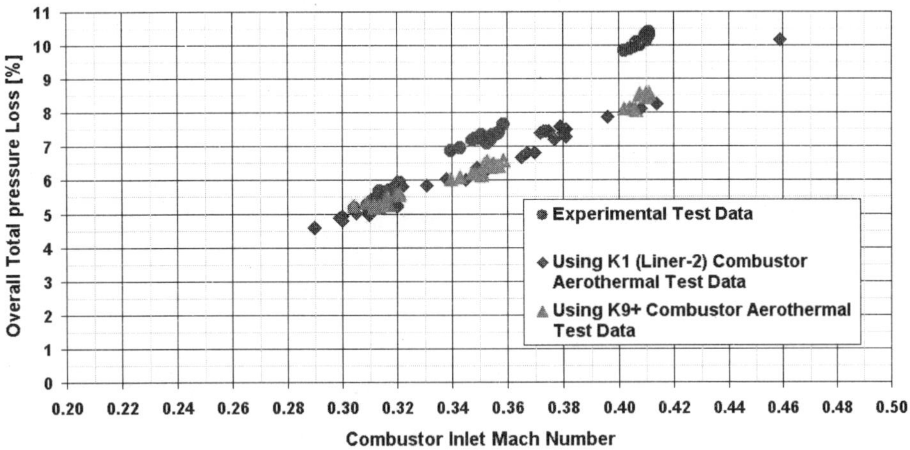


**Figure 23** Comparison of Experimental and Scaled Overall Total Pressure Loss

up, combustion stability, combustion efficiency, total pressure loss and metal temperatures. In these tests, extensive instrumentation has been carried out to characterize the afterburner completely.

Fig. 26 shows the afterburner assembly with convergent-divergent nozzle that has been tested.

## 3.1 Technological Issues Addressed During Development

### *3.1.1 Total pressure loss in exhaust cone*

CFD analysis of the baseline afterburner system indicated that wakes have formed behind the struts and flow separation exists. This lead to a large increase in total pressure loss that resulted in a reduction in the overall engine thrust. Hence, the exhaust cone struts and exhaust cone configuration have been modified through CFD and a significant performance improvement has been obtained. Figure 27 shows the velocity vectors in the afterburner exhaust cone before and after modification of the exhaust cone. [12]

**Figure 24** Photograph of the Modified Combustor Liner

**Figure 25** Altitude Test Facility at CIAM used for Afterburner Testing

### *3.1.2 Improvement of flame stabilization system*

In order to improve the mixing in afterburner thereby improve the combustion efficiency, the baseline V gutter has been modified. Both have one ring and 12 × 6 radial struts. The V gutter blockage has been marginally increased by 2 % due to the additional top gutters. Even though this has resulted in an increase in total pressure loss by 1.5%, the combustion efficiency has increased by 7 %. Fig. 28 shows the baseline and modified V gutter configurations [13].

**Figure 26** Afterburner Assembly with C-D Nozzle

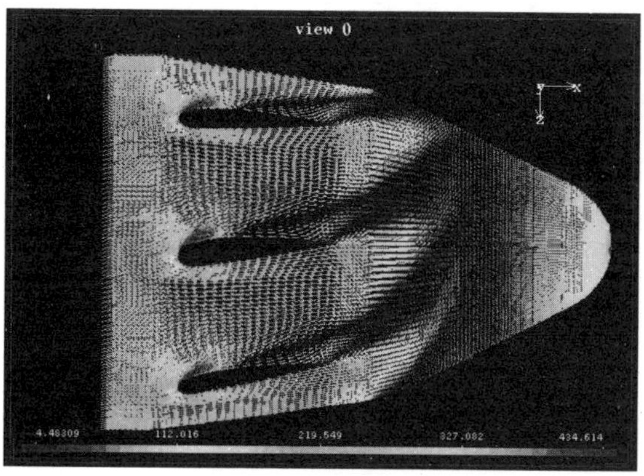


(b) Modified Exhaust Cone

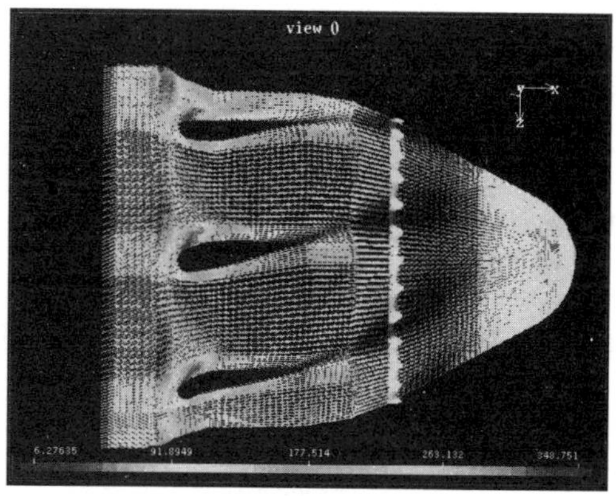


(a) Baseline Exhaust Cone

**Figure 27** Velocity Vectors on the Exhaust Cone Surface

### *3.1.3 Improvement of fuel injection system*

The baseline afterburner had four manifolds, each manifold using fuel in the transverse direction through several 0.5mm orifices. Performance analysis revealed that the afterburner combustion efficiency has been found to be about 68 %. In order to improve the quality of the fuel air mixture and thereby increase the combustion efficiency, the manifolds system has been converted into a spray bar system by keeping the same flow number and different fuel injection configurations have been studied. Fig. 29 shows the original and modified fuel injection systems.

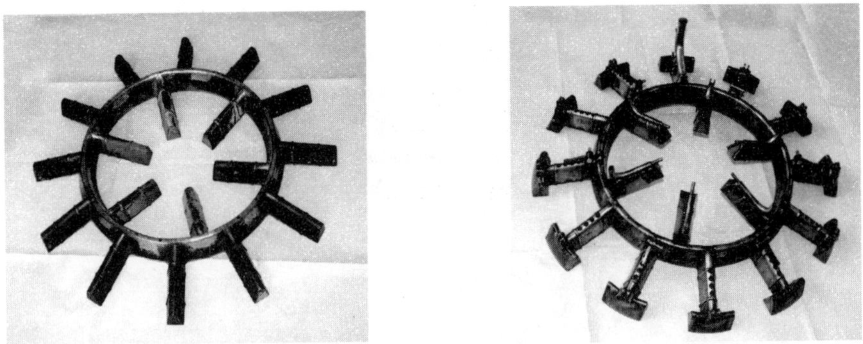

(a) Baseline (b) Modified

**Figure 28** V-Gutter Configurations

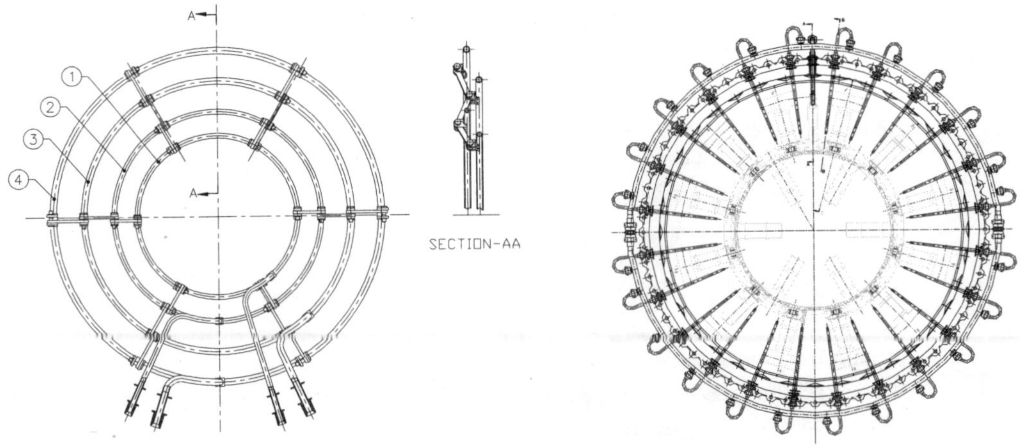


Manifolds Spray Bars

**Figure 29** Original and Modified Fuel Injection Systems

### *3.1.4 Improvement of ignition system*

The ignition system for such an afterburner should have a high reliability as it leads to single point mission failure. A Pilot Ignition System (PIS) has been successfully developed and is shown in Fig. 30.

The light up at various altitudes and Mach number conditions using this has been demonstrated as shown in Fig. 31.

### *3.1.5 Afterburner performance improvement*

The afterburner performance in terms of combustion efficiency, core total pressure and bypass total pressure along the afterburner length have been shown in Fig. 32. It can be seen from the figure that the maximum combustion efficiency is ~ 88 % and the core total pressure loss is ~ 17 % [14].

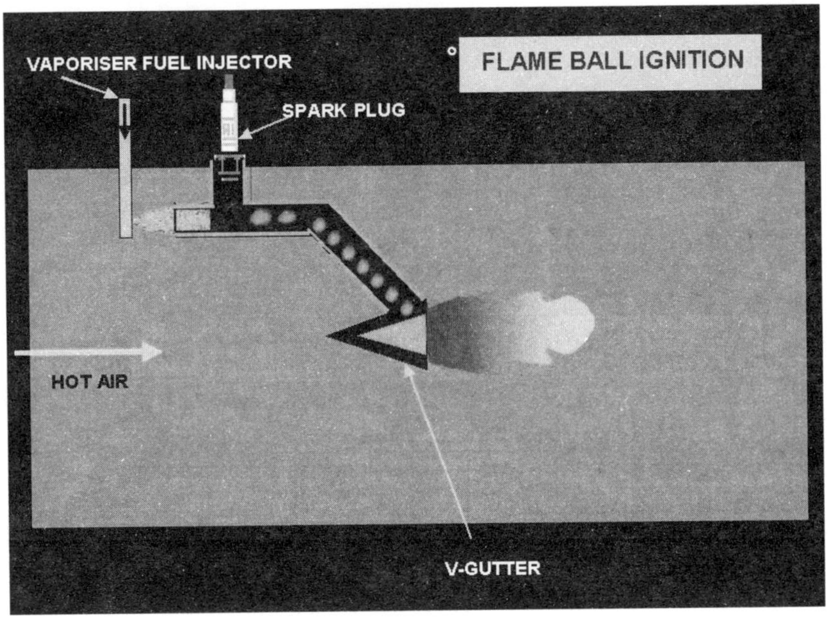


**Figure 30** Pilot Ignition System

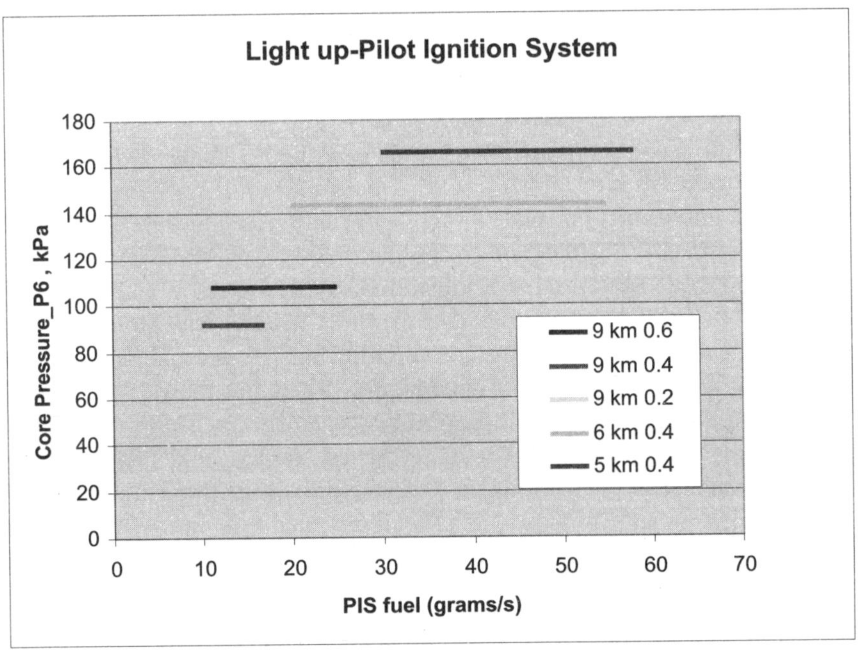


**Figure 31** Light up Characteristics of PIS

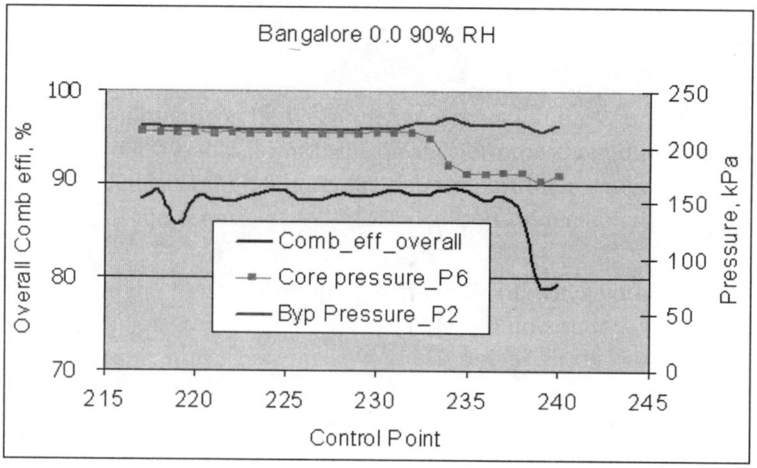


**Figure 32** Measured Values of Combustion Efficiency; Core and By-pass Total Pressures

## 3.2 Tools Employed in Development

### *3.2.1 CFD tools*

At GTRE, commercial code PHOENICS, FLUIDYN-NS and STAR-CD have been used to understand/estimate and improve the afterburner performance. The PHOENICS code had been used during early stage of afterburner analysis without resolving small holes like screech and cooling ring holes. But in the later stage, both FLUIYN-NS and STAR-CD codes have been used to carry out cold and hot flow CFD analysis resolving chute, screech, cooling ring holes and linkages. Both FLUIYN-NS and STAR-CD have been validated extensively with the rig measurements and engine data before using them as design tools. A number of modifications in the baseline afterburner have been carryout to improve the afterburner performance viz., pressure loss reduction, combustion efficiency improvement and thrust boost.

The afterburner has been analyzed by an integrated approach namely a CAD to MESH to CFD approach. The CAD model incorporated all the complex geometrical details of the afterburner starting from low-pressure turbine exit to the C-D nozzle exit. The flow in the afterburner system is treated to be steady, turbulent, compressible and reacting. The CFD analysis is accomplished using 120° sector model afterburner with 2 million grid points as shown in Fig. 33. The solutions are obtained by solving the compressible Reynolds-Averaged Navier-Stokes equations using a finite volume method. Turbulence has been taken into account using the standard $\kappa$-$\varepsilon$ / non-linear quadratic $\kappa$-$\varepsilon$ turbulence model. Combustion has been modeled using a two-step chemical reaction mechanism and the reaction rate has been calculated using a combined Arrhenius and Eddy Dissipation Model.

Figures 34 and 35 show the predicted total pressure and total temperature variation along with the experimental data at the core exit (i.e. at the nozzle entry plane) under cold flow condition. It can be observed that there is a good agreement between them. In addition, the comparison of total thrust developed by the afterburner under cold and hot flow conditions differs by 3.8% [15].

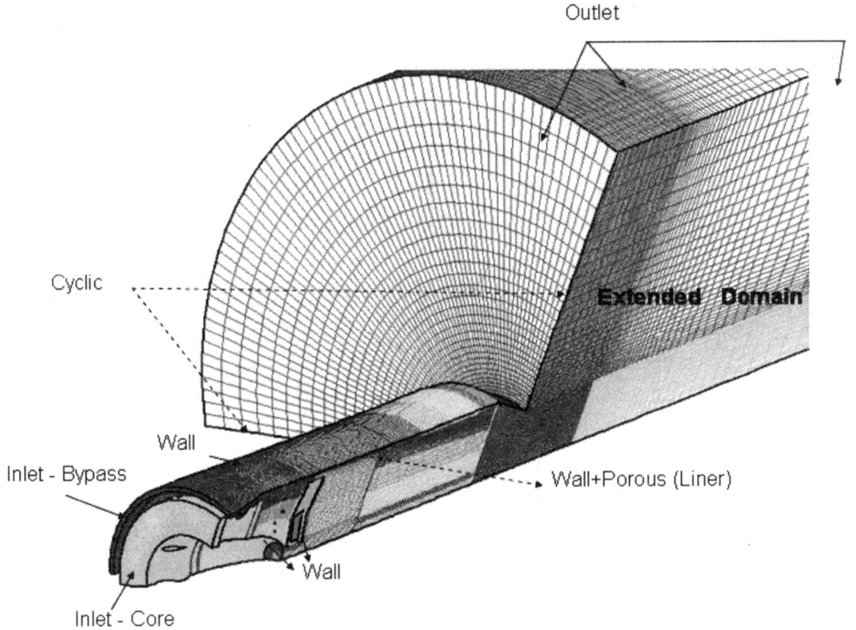


**Figure 33** Computational Grid and Boundary Regions

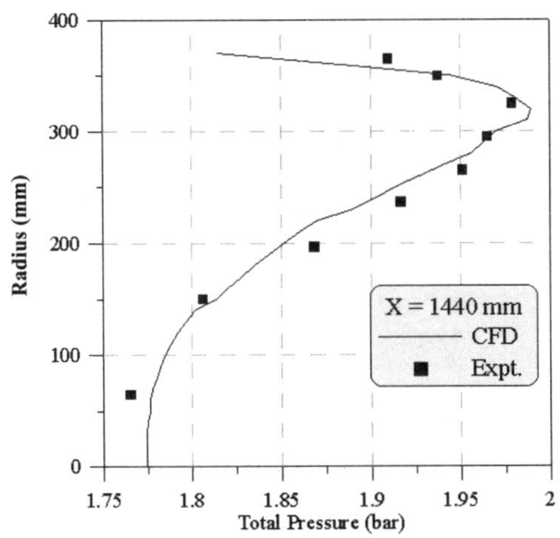


**Figure 34** Radial variation of total pressure at the nozzle entry

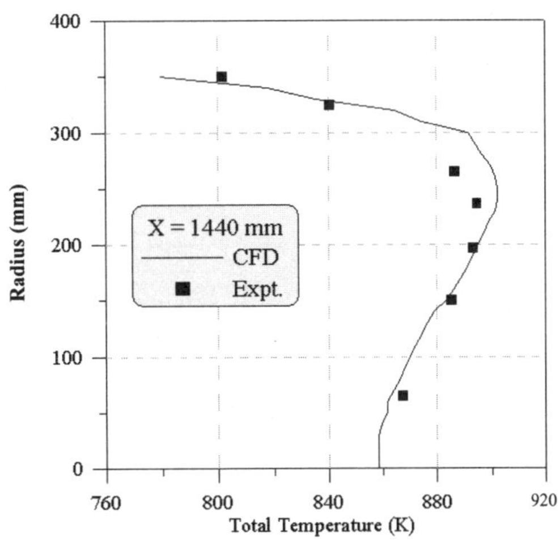


**Figure 35** Radial variation of total temperature at the nozzle entry

### *3.2.2 Thermal and stress analysis codes*

The jet pipe liner and jet pipe casing temperatures have been predicted using a 1-dimensional heat transfer model, which accounts for radiation and convection from hot gas. Empirical correlations have been used for the heat transfer coefficients and adiabatic film temperature estimation. Using this model, liner metal temperature has been predicted for life calculations at various engine-operating conditions. The predicted metal temperatures and measured temperatures have been found to be in good agreement at all engine speeds.

Fig. 36 shows the comparison between the measured and the estimated skin temperature at all steady state points at one typical point each on the jet pipe liner during a particular engine test condition with partial reheat. In this figure, the variation of liner temperature has been plotted with gas temperature and all the temperatures have been non-dimensionalised with the maximum gas temperature [16].

It can be seen from the figure that the predicted data matches within an accuracy of ± 4% with the measured data. This discrepancy could be due to the uncertainty in estimation of some of the parameters like gas temperature, gas emissivity and coolant air distribution.

The buckling strength of the Jetpipe Liner has been improved by addition of circumferential stiffened rings, pin supports and corrugations and a best configuration, which meets the buckling and weight requirements for the worst load condition, has been achieved. This has been achieved through an Eigen value-buckling analysis. A Finite element shell model of the baseline configuration as shown in Fig. 37 has been generated using MSC/PATRAN. The Liner has been subjected to aerodynamic pressure and thermal loads. As a part of optimization study, various design iterations have been carried out wherein the buckling strength has been chosen as the primary objective function along with weight as the other important design aspect.

Initially, the baseline configuration of the jetpipe liner has been analyzed for the buckling requirement and the buckling factor was found to be 0.485, which does not satisfy the buckling strength criteria. Hence, an exhaustive iterative approach has been adopted to improve the structure

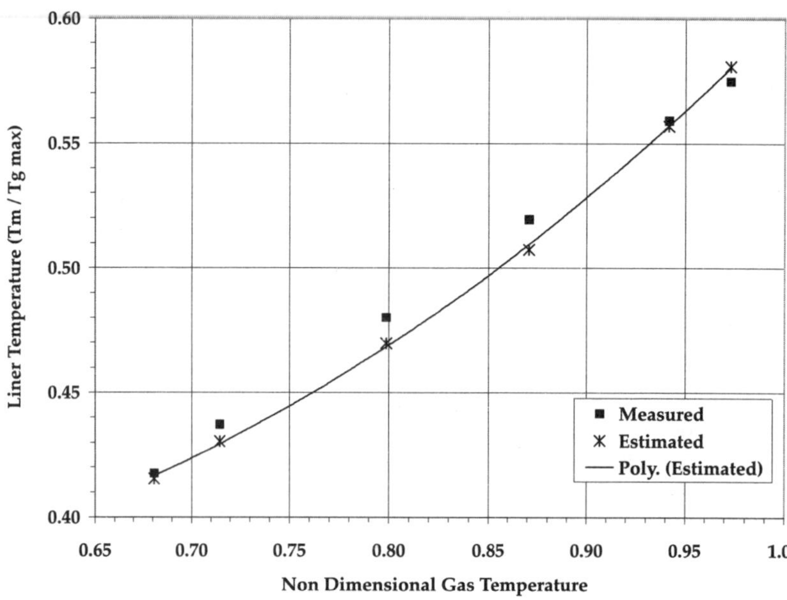


**Figure 36** Comparison between Estimated and Measured Skin Temperature for Jet Pipe Liner

**Figure 37** A typical finite element model of the Jetpipe Liner

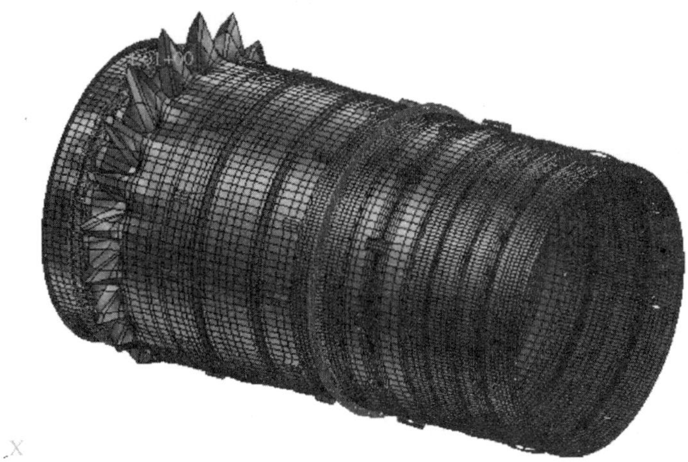

**Figure 38** Buckling Pattern for Jetpipe in Fundamental Mode

and Fig. 38 shows the buckling pattern of the optimized configuration for the flameout case. It has been seen that the buckling factor is 1.29 for this configuration [17].

### *3.2.3 Screech & buzz codes*

Afterburner system is prone for combustion instability, in the form of screech and buzz, which is detrimental not only to the performance, but also to the hardware. In order to avoid their occurrence, a large number of 3 mm diameter holes have been drilled on the jet pipe liner over a length of 350 mm. Afterburner tests at CIAM tests at full reheat conditions (Mn = 0.7; Sea Level) showed acoustic instability in one of tangential mode with the frequency in the range 3500-3600 Hz. Further, eight transducers demonstrated spectrum peaks in the range 557-576 Hz which are less steep than those at 3500-3600 Hz, which can be interpreted as afterburner be an acoustic resonator which selects from acoustic noise and amplifies the oscillations of one of Eigen frequency [18,19].

In the operation mode, at an altitude of 0.9 km and 0 Mach number, no screech excitation has been observed up to 110% reheat strength. Only the tendency is seen in the spectrum to the elevation of oscillations in the range 3500-4000 Hz. In addition, narrow band oscillations of the first tangential mode (557-567 Hz) are indicated once again by most of the transducers. At an altitude of 8 kms, weak longitudinal oscillations with 256 Hz are discovered in full reheat conditions. [14]

#### 3.2.4 Gas analyser to measure species concentration

During the afterburner tests at CIAM, species concentration has been measured using a gas analyzer system. A Data Acquisition System (DAS) records CO2, CO, HC, 02 volumetric concentrations (either % or ppm) and calculated values for Fuel / Air Ratio (FAR) and combustion efficiency. CO2 and CO concentrations are measured and recorded on "dry" basis and in the process of FAR/ETA calculations they are corrected to a "wet" sample. HC concentrations are measured in "wet" samples and no correction is needed. $O_2$ concentrations do not participate in FAR or ETA calculations. As $H_2$ concentrations have not been recorded in the DAS, the calculated efficiency values have been amended to take into account measured $H_2$ using the following relation.

$$\text{Efficiency} = 100 - 0.0236\ \text{EI(CO)} - 0.1166\ \text{EI(HC)} - 0.2783\ \text{EI}(H_2),\ \%$$

(here EI represents so called "emission index", grams of the substance per kg of fuel).

These values of efficiency have been compared with the CFD predictions and a good agreement has been obtained between CFD and measurements. [14]

### 3.3 Realization of the Afterburner Hardware

The state of the art afterburner has been realized and successfully tested in stand-alone conditions and in engine conditions. This gives thrust boost of about 45%.

## 4. FURTHERING COMBUSTOR AND AFTERBURNER TECHNOLOGIES

As far as the combustor is concerned, we are just in the process of initiating a program targeting a total pressure loss of the order of 4% at an entry Mach number of 0.35 and entry pressure of 21 bar. Further, it is necessary to initiate work on pattern factor control through active fuel flow modulation to realize a CPF of 0.3 and RPF of 0.12 at all operating conditions.

As far as the afterburner is concerned, we are presently focusing our attention to improve the hardware further to obtain 10% more thrust boost, 3% more reheat efficiency and 3% reduction in pressure loss. The present afterburner has been designed for a maximum fuel throughput of 4.5 kg/s resulting in a maximum temperature to about 1400K. We are planning to increase it to 1800 K. Action has already been initiated for adopting a fully corrugated liner and ISO grid design for the jet pipe for reducing the weight by about 30 to 40 %.

## Acknowledgements

The authors would like to thank Director, GTRE for his support and encouragement in preparing the paper. The authors would also like to thank Dr. S. Kishore Kumar, Dr. C. Badarinath, N.

Muthuveerappan, Dr. S.K. Patel, V. Kesavan, P. Ramesh, Raju Navindgi, G.N. Jayaprakash, Dr. K. Ashirvadam, T. Krishna Murthy, Batchu Suresh, N.R. Vishwanathan, Arjun Rathod, Dr. S. Ganesan, N. Sarangi, V. Kumarappan, Akhilesh Kumar Yadav, V. Srishanthi, Dalton Maurya, K. Hari, Ramakrishna Narayana Bhat, M. Srinivasa Rao, Nirmal Shankar, Mukund Kumar Pandey, K. Saravanan, D. Maria Christu Raja, S. Manikandan and other officers and staff of combustion group, afterburner group, heat transfer group, CFD group, structural integrity and mechanical analysis group and structural mechanics group – 2 of GTRE for providing the necessary information in preparing the paper. The efforts of Dr. V.A. Skibin, S.A. Volkov, Dr. A.A. Gorbatko, V.I. Furletov and V.M. Velichkovskiy in the developmental testing of combustor and afterburner at CIAM are duly acknowledged.

## References

1. Skibin, V. A., 2004, "Completion of testing as per appendices B, E of the scope of testing (on the stand C 5-2)", CIAM Report. (Classified)
2. Skibin, V. A., 2005, "Kaveri Engine Main Combustor Tests and Tests on Kaveri Engine Main Combustor Components at CAIM Test Facility (WP No. 4; Stage 2; Stage 2)", CIAM Report. (Classified)
3. Chakravarthy, S.R. and Sujith, R.I., 2006, "Atomisation Characteristics of the Modified GTRE Atomiser", Department of Aerospace Engineering, IIT, Madras.
4. Skibin, V. A., 2007, "Technical Report – Altitude Tests of Kaveri Engine Main Combustor Version – 4 (WP No. 4; Phase 3)", CIAM Report. (Classified)
5. Sarangi Niranjan, Ramakrishna Narayana Bhat and Ramesh, P., 2004, "Analysis and Fatigue Life Validation of Diffuser Casing of a Turbofan Engine", Proceedings of an International Conference on Total Engineering Analysis and Manufacturing Technologies.
6. Skibin, V. A., 2006, "Testing for improvement of Kaveri engine main combustor pattern factors", CIAM Report. (Classified)
7. Kishore Kumar, S., Venkataraman Shankar, Sivaramakrishna, G. and Srinivasa Rao, M., 2007, "Application of CFD Tools in the Design & Analysis of a Gas Turbine Combustor", Proceedings of 7$^{th}$ Asian CFD Conference.
8. Sivaramakrishna, G., Muthuveerappan, N., Kishore Kumar, S. and Venkataraman Shankar., 2006, "3-D CFD Analysis of An Aero Gas Turbine Combustor and Validation With Lab Scale Experiments", Proceedings of 8$^{th}$ National Conference on Air Breathing Engines and Aerospace Propulsion.
9. Srinivasa Rao, M., Sivaramakrishna, G., Muthuveerappan, N., Kishore Kumar, S. and Venkataraman Shankar., 2007, "CFD Analysis of an Aero Gas Turbine Combustor –Validation and Experiments", Proceedings of International Conference & XX National Conference on IC Engines and Combustion.
10. Patel, S. K., 2002, "Structural analysis of Kaveri (K6) engine Static Structure", GTRE Internal Technical Report No. GTRE/KAVERI/SWE/STR/116/02.
11. Dalton Maurya and Badarinath, C., 2004, "Estimation of Overall Total Pressure Loss of an Annular Combustor by Water Tunnel Experiments", Proceedings of 7$^{th}$ National Conference on Air Breathing Engines and Aerospace Propulsion.
12. Ganesan, S., Kishore Kumar, S., Venkataraman Shankar, Ganesan, V. and Soundranayagam, S., 2005, "Numerical Study of Exhaust Cone Struts Effect on Gas Turbine Afterburner Performance", Proceedings of 8$^{th}$ Annual CFD Symposium.

13. Ganesan, S., Kishore Kumar, S., Venkataraman Shankar and Ganesan, V., 2007, "Mixing and Combustion of Cold & Hot Flows in a Gas Turbine Afterburner", Proceedings of 7th Asian CFD Conference.
14. Skibin, V. A., 2007, "Kaveri Engine After Burner tests and After Burner detail tests", Draft copy of the CIAM Report. (Classified)
15. Ganesan, S., Kishore Kumar, S., Venkataraman Shankar and Ganesan, V., 2006, "CFD Studies on V-gutter configurations for Aero Gas Turbine Engine After Burner", Proceedings of 9th Annual CFD Symposium.
16. Suresh Batchu, Arjun Rathod, Kesavan, V. and Venkataraman Shankar, 2006, "Thermal Modelling and Validation of Jet pipe Liner and Casing for an Aero Gas Turbine Engine", Proceedings of 8th National Conference on Air Breathing Engines and Aerospace Propulsion.
17. Srishanti, V. and Patel, S.K., 2005, "Buckling Analysis of Afterburner Liner of Gas Turbine Engine", Proceedings of MSC Users Conference.
18. Ashirvadam, K., Kishore Kumar, S., Haran, A.P. and Paul, P.J., 2001, "Combustion Instabilities in Gas Turbine Afterburners", Proceedings of 15th International Symposium on Air Breathing Engines.
19. Ashirvadam, K., Kishore Kumar, S., Haran, A.P. and Paul, P.J., 2004, "Screech in Gas Turbine Afterburner – a Combustion Instability Problem (Theory & Experiment)", Proceedings of 16th International Symposium on Air Breathing Engines.

# Is Gasoline the Best Fuel for Advanced Diesel Engines? - Fuel Effects in "Premixed-Enough" Compression Ignition (CI) Engines

**Gautam T. Kalghatgi**
Shell Global Solutions, U.K.

## 1. INTRODUCTION

In conventional compression ignition (CI) engines diesel fuel is injected near the end of the compression stroke, near top dead centre (TDC), into the cylinder when the charge pressure and temperature are high. Chemical reactions start and continue as the fuel vaporises and the fuel vapour and air mix with each other. Auto-ignition occurs when this mixture is quite rich, with an equivalence ratio, $\phi \sim 4$; soot is produced during this rich combustion [1]. Following this pre-mixed combustion, the rest of the fuel burns in a diffusion flame and nitrogen oxides (NOx), primarily made up of NO [2], are formed during this phase or in the hot gases that remain after the end of combustion [1]. It is difficult to control particulates and NOx emissions from diesel engines. After-treatment systems to control them are being developed but are expensive, complicated and will reduce fuel economy and thus reduce the primary advantage of diesel engines. Moreover, regulations on emissions from diesel engines are becoming more stringent in all parts of the world. Hence there is a great deal of interest to develop engine combustion systems that offer the efficiency of a diesel engine, but with low smoke and low NOx.

For the purposes of the current work, we define diesel fuels only in terms of their autoignition quality as those with Cetane Number (CN) $> \sim 30$ and gasoline fuels as those with CN $< \sim 30$ or Research Octane Number (RON) $> \sim 60$ [3]. In practice, diesel fuels are significantly less volatile than gasoline fuels [3]. Auto-ignition is determined by chemical kinetics [4], which depend on the pressure and temperature development, mixture strength and chemical composition of the fuel. Typical, practical diesel fuels have CN $> \sim 40$ and will auto-ignite very quickly before the fuel and air are mixed sufficiently. If the mixing is accelerated or if the chemical reaction is slowed down, auto-ignition can be made to occur after the fuel and air are better mixed and soot levels can be reduced. NOx levels can be reduced by reducing combustion temperature by either running lean, pre-mixed or using EGR, exhaust gas recirculation [2].

Several operating strategies have been employed to promote pre-mixed and low-temperature combustion in CI engines. In homogeneous-charge compression-ignition (HCCI) the fuel and air are fully premixed as in a spark ignition engine and compressed till the charge auto-ignites. In modern direct-injection (DI) diesel engines with common rail injection systems, HCCI combustion can be obtained by injecting the diesel fuel very early in the cycle, giving it enough time to mix completely with air. However, with HCCI, there is no in-cycle control over the phasing of the heat release, which will be determined by the initial conditions of the mixture at the start of the compression stroke and the auto-ignition characteristics of the fuel. This is in contrast to conventional SI or diesel engines where combustion phasing is determined by the spark timing or the fuel injection timing. Thus HCCI combustion is very difficult to control. Moreover, because diesel fuel ignites very easily, HCCI combustion can be achieved only with low compression ratios and even then, at low loads. Otherwise heat release will occur during the compression stroke and this will lead to very high pressure-rise rates.

Thus practical CI engines always need fuel injection near TDC to control combustion phasing. Advanced diesel engines try to achieve sufficient premixing of fuel and air before combustion. High injection pressures and high swirl rates increase the mixing rate while chemical reaction is delayed through the use of cooled EGR, late injection and lower compression ratios. Numerous strategies are employed to promote premixed combustion in diesel engines. In the Toyota UNIBUS system a small amount of fuel is injected very early in the cycle and a second fuel pulse near TDC initiates the combustion [5,6]. In the Nissan MK (Modulated Kinetics) system, premixing is promoted using high levels of cooled EGR to delay auto-ignition and reduce temperatures [7,8,9]. High swirl, late (after TDC) fuel injection and high injection pressures also contribute to the promotion of mixing prior to ignition in such systems. Hashizume et al.[10] introduced the MULDIC (Multiple stage Diesel Combustion) system. In this system part of the fuel is injected very early and undergoes premixed lean auto-ignition; fuel is injected again well after TDC into the mixture with high temperature. Multiple injection strategies have also been combined with high EGR rates to reduce NOx and smoke [11,12,13].

Conventional diesel fuels are not suited for premixed CI combustion. In fact, most of the high technology used in advanced diesel engines, which makes them complicated and expensive, is required by the necessity to promote premixed combustion by overcoming the propensity of diesel fuel to autoignite easily. Thus high injection pressures and high swirl are used to increase mixing rates while high levels of cooled EGR are used to delay combustion; this in turn requires higher boost pressures to achieve the required loads. Even then, with conventional diesel fuels, low NOx and low smoke with partially premixed CI combustion is possible only at low loads. Modern engines now require after-treatment systems to further reduce NOx and particulates, making them even more complicated and expensive. If fuels which are more resistant to autoignition than conventional diesel are used, premixed combustion, and hence low NOx and low particulates could be more easily attained. Hashizume et al. [10] did use a fuel in the gasoline auto-ignition range with a CN of 19 - RON ~ 85 according to [3] -, in their experiments. With early injection, diesel fuels, which are very prone to auto-ignition, will ignite very early in the cycle unless the mixture strength is very low [13]. Even with multiple injections, the heat release from the first, early, injection will occur during the compression stroke and reduce efficiency [10,13] for fuels with high CN. Both smoke emission and fuel consumption improved when the CN of the fuel used during the first stage was reduced [10]. Shimazaki et al. [15] have also reported that a lower cetane fuel increases the mixing time before ignition and enables lower smoke levels to be achieved compared to a higher cetane fuel. Their experiments were done in an engine with a CR of 16.5 and they compared two fuels of 19 CN and 61 CN [14].

In this paper we discuss experimental results [15,16] that confirm that fuels with higher resistance to autoignition are indeed much better for partially premixed CI combustion. Indeed gasoline was found to allow significantly higher loads to be achieved with low smoke and low NOx compared to other diesel-like fuels tested.

## 2. EXPERIMENTAL DETAIL

The experiments were done at KTH, Stockholm and have been reported previously in [15] and [16].

### 2.1 The Engine

The engine used is a single cylinder engine based on a heavy duty SCANIA D12 6-cylinder CI-engine. It has a flat cylinder head and a shallow bowl piston. The compression ratio used in the experiments presented in this paper was 14 – see Table 1 for other dimensions. The cylinder pressure was measured with a water-cooled piezo electric AVL pressure transducer. Boost pressure, simulating turbo charging, was supplied by an external compressor. Heat release was calculated from the pressure signal. The intake pressure was measured with a Gems pressure transducer of membrane type. The intake temperature was measured with a K thermocouple at a distance of 30 cm from the intake port. A PID regulator held the temperature within 2 °C of the set temperature. The airflow rate was measured with a rotary piston Aerzener gas meter. The mixture strength is represented by $\lambda$ and was obtained by calculating the ratio between the measured airflow rate and fuel flow rate. It was also measured using a broadband lambda sensor and calculated from the exhaust gas concentrations by the exhaust gas analyzer (Horiba EXSA 1500). The opacity, for soot measurements, was measured with an AVL opacimeter 439.

**Table 1** Engine dimensions

| | |
|---|---|
| Compression Ratio | 14 |
| Bore [mm] | 127 |
| Stroke [mm] | 154 |
| Connecting rod length [mm] | 255 |
| Number of valves | 4 |

EGR was supplied from a spark ignited SAAB B205R engine working at stochiometric or close to stochiometric conditions at a fixed speed and load. This was done instead of recycling the exhaust from the single cylinder engine to obtain a fast and stable system that makes it possible get a fixed EGR composition independent of the test engine operating conditions. These EGR gases went through a three-way catalyst and were then cooled before mixing with the intake air. The EGR backpressure was atmospheric. The amount of external EGR was controlled with the measured $CO_2$ concentration in the intake gases. The EGR level quoted in the text below is the volume percent of the exhaust from this stoichiometric engine in the intake flow of the test engine. Thus "25% EGR" in the present paper means that the intake flow had ~15.8% vol oxygen and 4% vol carbon dioxide. The EGR level, in terms of volume percent of the exhaust from the test engine, which runs leaner, would be higher than these quoted values.

A standard diesel injector with 8 holes with a hole diameter of 0.2 mm was used in all the tests. The injection pressure used was 1300 bar for the diesel fuels and 1100 bar for the gasoline. One of the intake ports was closed during the tests to increase the swirl ratio, which was estimated to be 2.4.

## 2.2 The Fuels

The properties of the four fuels used are summarised in Table 2. Swedish MK1 is a commercial low-sulphur Swedish diesel fuel. Diesel 1 and Diesel 2, with lower CN, are made by blending a diesel kerosene component with an aromatic solvent which has a very high octane number and hence a low CN. The CN for these fuels was determined by using a blending rule established by measuring the cetane numbers of other blends of these components. It can be seen that they are slightly more volatile than Swedish MK1. The fourth fuel used is a commercial, low-sulphur gasoline and its CN is estimated to be about 15 [3]. It is also more volatile than the other three fuels with an initial boiling point (IBP) of 32ºC and a final boiling point (FBP) of 176ºC. All these fuels had sufficient lubricity additive to ensure that no damage was caused to the injectiosystem. All the fuels have very similar lower heating values.

## 2.3 The Experiments

Indicated Mean Effective Pressure (IMEP) and CA50 quoted below are averages based on 100 engine cycles. Emissions and smoke opacity were also measured at each experimental point. Many different injection strategies were also used as described in [15,16,17]. All the experiments are done at 1200 RPM with the intake temperature at 40°C.

**Table 2** Properties of the fuels used in the tests

| *Fuel* | *CN* | *Density g/cc* | *IBP °C* | *T10 °C* | *T50 °C* | *T90 °C* | *FBP °C* | *Aromatics %Vol* | *LHV** MJ/kg* |
|---|---|---|---|---|---|---|---|---|---|
| Swedish MK | 154 | 0.81 | 195 | 208 | 240 | 273 | 297 | ~3 | 43.8 |
| Diesel | 1 39 | 0.81 | 167 | 179 | 196 | 220 | 246 | 34 | 43.5 |
| Diesel | 2 30 | 0.83 | 167 | 179 | 198 | 222 | 246 | 50 | 43.3 |
| Gasoline | ~15* | 0.726 | 32 | 50 | 102 | 144 | 176 | 29 | 43.2 |

.* RON of 94.7 and MON of 85.9

**LHV – lower heating value

# 3. RESULTS

## 3.1 Single Injection Results

With injection near TDC, diesel fuels auto-ignite quickly after injection, before they can mix well with air. The crank angle difference between the start of injection (SOI) and the crank angle when 50% of the total heat release occurs (CA50) is defined [15] as the engine ignition delay (EID); EID = CA50 – SOI.. The higher the EID, the better mixed the fuel is with air at combustion. Diesel fuels of different CN and volatility characteristics were tested by Risberg et al. [17] in the engine described above but with a compression ratio (CR) of 11.4 at 1200 RPM. It was found that for the same combustion phasing, CA50, fuels with lower CN give higher values of EID e.g. compare U.S. diesel of 39.5 CN with EU diesel. In consequence, the measured NOx values were also lower for the lower CN fuels - see Figure 1 in [15].

In the rest of the paper the effect of changing fuel auto-ignition quality on EID and the consequent effects on emissions is further discussed using the four fuels listed in Table 2 with CR fixed at 14. In Figure 1, CA50 is plotted against start of injection, SOI for all four fuels using data from Figure 3 in Ref. 15 at a fixed operating condition - intake pressure, $P_{in}$ of 1.5 bar abs.; no EGR; 1200 RPM; intake temperature, $T_{in}$ of 40 C; single injection of fuel at 0.6 g/s giving $\lambda$ of 3.8.

For the diesel fuels with CN 30 or greater, as expected, CA50 increases – heat release occurs later in the cycle – as SOI increases. This was also true for gasoline for SOI later than about –30 crank angle degrees (CAD). Indeed the engine was very stable on gasoline in this range of SOI. Note that gasoline needs to be injected much earlier than the diesel fuels to get the same CA50. If gasoline is injected earlier in the cycle, there is more time for mixing with air and the combustion heads towards HCCI and with this operating condition, becomes more unstable. In fact if gasoline was injected before 40 CAD BTDC (before TDC), no combustion was possible. Thus it is essential for there to be inhomogeneity or stratification, presumably with rich pockets present, provided by "late" injection to ensure combustion; this allows the control of combustion phasing by changing

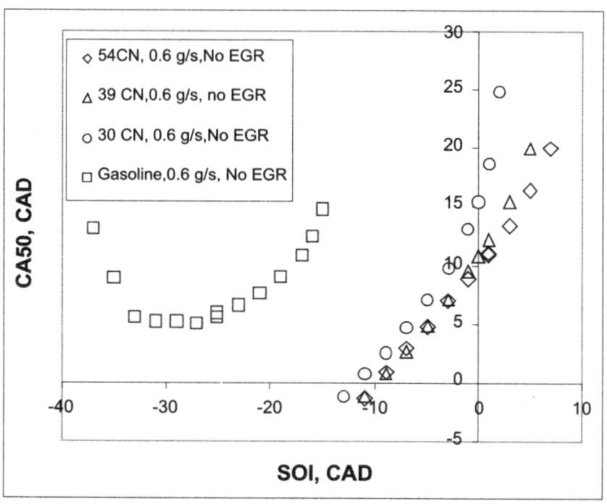


**Figure 1** CA50 vs Start of Injection (SOI) for the four fuels using data from Fig.3 of Ref. 15. $P_{in}$ =1.5 bar abs.; no EGR; 1200 RPM; intake temperature, $T_{in}$ = 40 C; single injection of fuel at 0.6 g/s, $\lambda$ = 3.8

SOI. With more premixing, at the mixture strength and operating conditions used, gasoline is too resistant to autoignition to ensure ignition and combustion. This underlines the importance of not fully premixing fuel and air in CI combustion i.e. not running the engine in HCCI mode in order to ensure combustion and control its phasing. Though the gasoline is not fully premixed with air, it is more premixed i.e. the EID is higher at the time of combustion compared to the diesel fuels as shown in Figure 2 where the data in Figure 1 is replotted as EID vs CA50. In Figure 3, the measured NOx levels are plotted against EID for the four fuels. It can be seen that for gasoline, because of the high ignition delay, NOx is significantly lower than for the diesel fuels. With high EID, there is more mixing of fuel and air before combustion and the local mixture strength is nearer the global mixture strength which is lean i.e. the fuel rich pockets for gasoline are not as fuel rich as with the other fuels. This leads to significantly lower NOX. In Figure 4, NOx is plotted against mean IMEP (based on 100 cycles) and it can be seen that for the same IMEP, gasoline gives significantly lower NOx because of higher EID compared to the diesel fuels. In these experiments, smoke was very low for all the fuels but as the load (IMEP) is increased by injecting more fuel into the cylinder, smoke goes up. We now consider the fuel effects on smoke emissions. In the rest of the paper we consider results with $P_{in}$ = 2 bar and EGR mostly at 25% stoichiometric.

In Figure 5, the AVL smoke meter reading is plotted against IMEP for Swedish MK1 diesel fuel. The fuel was injected in a single pulse starting at TDC and higher IMEP was gained by increasing the amount fuel injected by increasing the injection pulse width. The Filter Smoke Number (FSN), can be obtained by dividing the opacity percentage by 5 [18] over the range encountered. It can be seen that as load is increased, smoke increases because more and more fuel is injected after heat release starts – there is more overlap between the injection and heat release events.

Smoke can be reduced by late injection even with diesel fuel [7,8,9] – Nissan MK style combustion. However CA50 occurs later with later injection and efficiency decreases i.e. IMEP decreases as shown in Figure 6. With the same mass of gasoline injected (1.2 g/s), again, CA50 increases and IMEP decreases with increasing (SOI) but the smoke level is always very low –

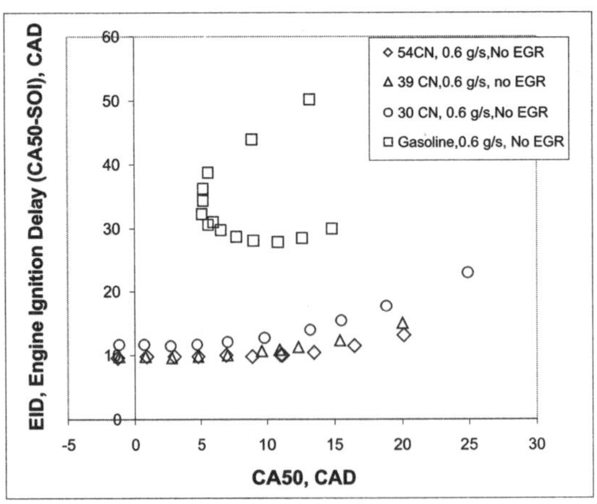


**Figure 2** EID vs CA50 for data in Fig.1

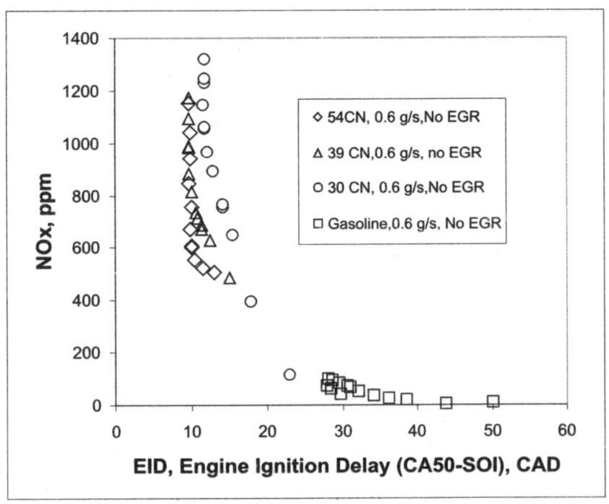


**Figure 3** NOx vs EID for experiments considered in Fig.1

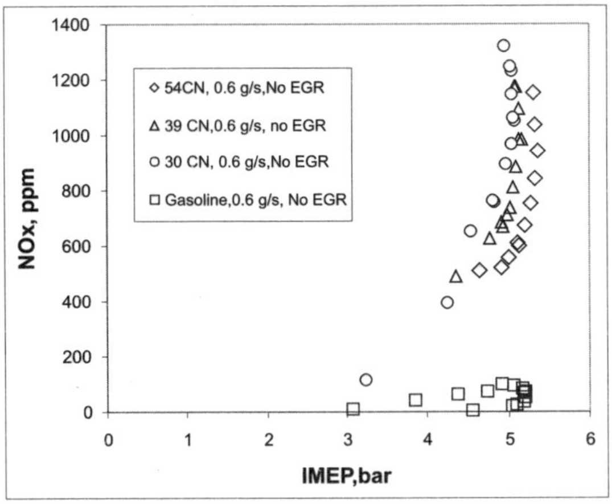


**Figure 4** NOx vs IMEP for experiments considered in Fig.1

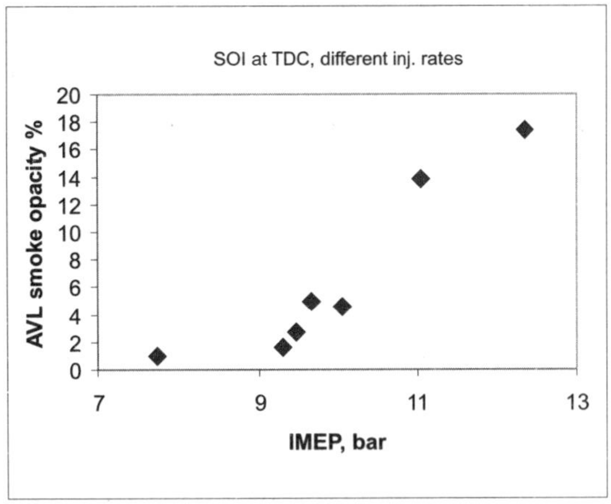


**Figure 5** Smoke vs IMEP for Swedish MK1 diesel. $P_{in}$ = 2 bar abs.; 25% EGR; 1200 RPM; intake temperature, $T_{in}$ = 40°C; single injection of fuel at TDC [16]

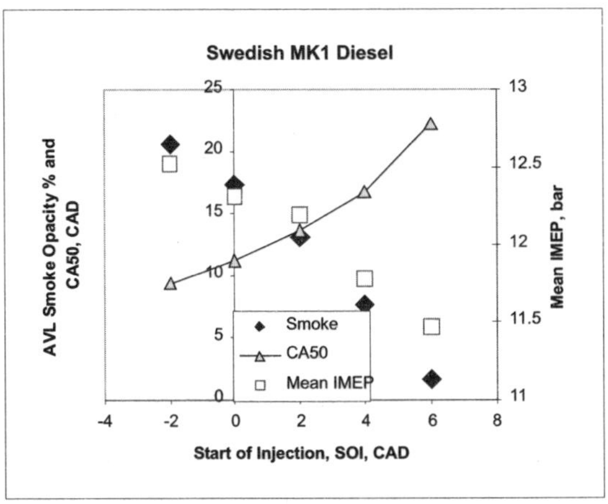


**Figure 6** Smoke, CA50 and ISFC vs SOI for Swedish MK1 diesel. $P_{in}$ = 2 bar abs.; 25% EGR; 1200 RPM; $T_{in}$ = 40°C; single injection of fuel at different SOI. λ = 1.8 [16]

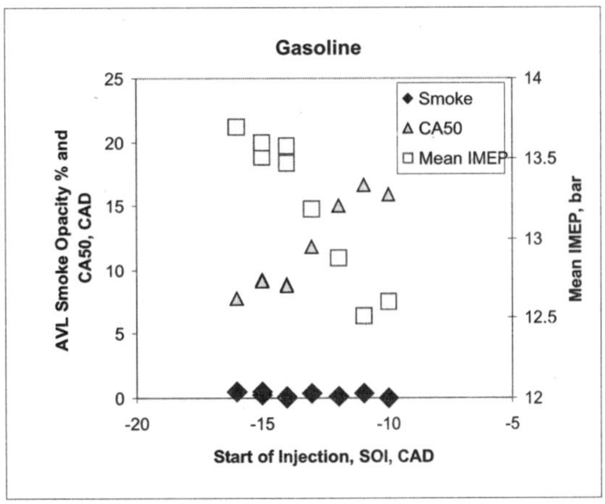


**Figure 7** Smoke, CA50 and ISFC vs SOI for gasoline. $P_{in}$ = 2 bar abs.; 25% EGR; 1200 RPM; $T_{in}$ = 40°C; single injection of fuel at different SOI. λ = 1.8 [16]

Figure 7. Note that gasoline is injected much earlier in the cycle than the diesel fuel to get similar CA50 and thus has larger EID. In Figure 8, the smoke level for the two fuels from the previous two figures is plotted against EID and it can be seen that gasoline has very low smoke because of high EID. In Fig. 9 the injector needle lift and heat release curves are shown for the high load and low load cases in Figure 6 for diesel and a typical case for gasoline from Figure 7. As the diesel fuel is injected later, with SOI after TDC, in Nissan MK style combustion, the overlap between the injection event and the heat release event decreases and smoke decreases. However, for gasoline, the heat release starts after the injection event is over and the smoke level is very low.

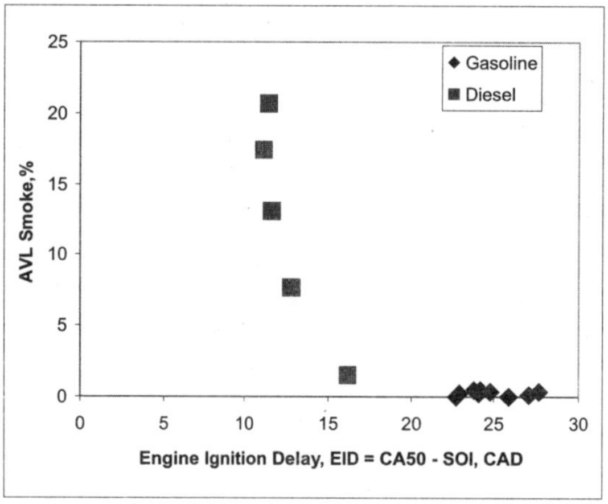


**Figure 8** Smoke from Figs. 6 and 7 vs EID for gasoline and diesel

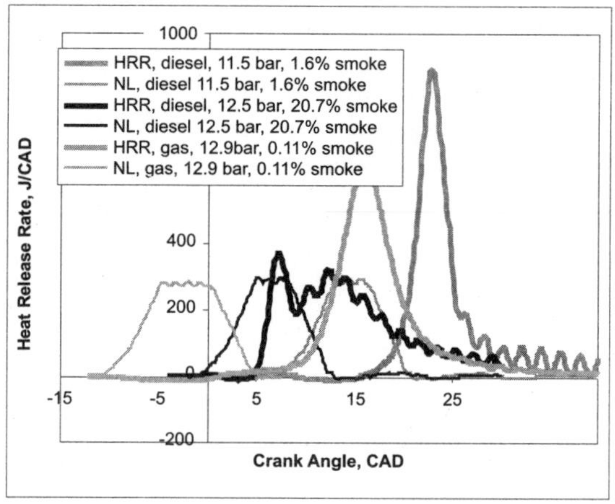


**Figure 9** Needle lift and heat release rate profiles for the high smoke and low smoke cases for diesel in Fig. 6 and a gasoline case from Fig. 7

Along with smoke, NOx also needs to be controlled and EGR is a very effective way of reducing NOx. This is shown in Figure 10 – the data are from Figure 15 of Ref. 15 – where indicated specific NOx, ISNOx, is plotted against IMEP for gasoline at different fuelling levels. Thus it can be seen that with 0.8 g/s of gasoline, an EGR rate of 10% reduces the NOx level significantly for the same IMEP and, with 25% EGR, NOx level can be further reduced. With the EGR level held at 25%, the amount of fuel was gradually increased and as IMEP increased, ISNOx also increased (Figure 10) because the overall mixture strength becomes richer. This was achieved without increasing the smoke level up to the point where mean IMEP was 14.86 bar. However, when the fuel amount was increased slightly, by about 2.5%, at this point, the smoke level increased suddenly from about 1.8% AVL smoke reading to 38%. Figure 11 shows ISFC for the cases considered in Figure 10. Also shown in Figure 11 for comparison is ISFC for the same fuel but run in HCCI mode by injecting the fuel very early in the inlet port of a different single cylinder engine [19]. HC and CO emissions and exhaust temperatures for these experiments are discussed in [15]. At the "best" operating point, with mean IMEP at 14.86 bar, before the smoke increases, ISNOx, ISFC, ISCO and ISHC are 1.21 g/kWh, 178 g/kWh, 3.4 g/ kWh and 3.6 g/kWh respectively. The standard deviation of IMEP, based on 100 cycles was 0.115 bar; thus the engine is very stable. However HC and CO levels might be too high.

Figure 12 shows the average pressure signal based on 100 cycles, the heat release rate from it and the injector needle lift profile for this operating point. There is a small negative heat release immediately after injection, probably caused by fuel vaporization and charge cooling. Ignition and initial heat release appear to start half way through the injection event and almost all the heat release takes place after the injection has stopped. The figure also highlights the fact that if the combustion is more premixed, the peak heat release rate and the pressure rise rate will be higher. Thus with single injection, there is a limit on how much gasoline can be injected into the cylinder to get low smoke and NOx with low ISFC. At this limiting IMEP, HC and CO might be higher than normally encountered in diesel engines but will be significantly lower than in HCCI engines. This,

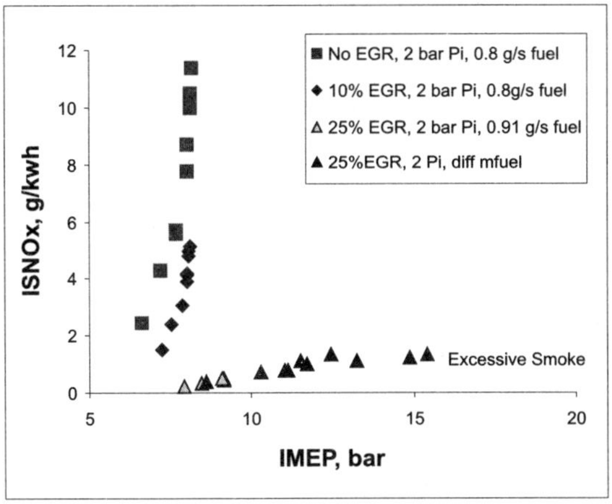


**Figure 10** ISNOx vs IMEP for gasoline [15]. 1200 RPM; $P_{in}$ = 2 bar abs.; $T_{in}$ = 40°C; single injection of fuel, different fuel amounts, SOI, EGR

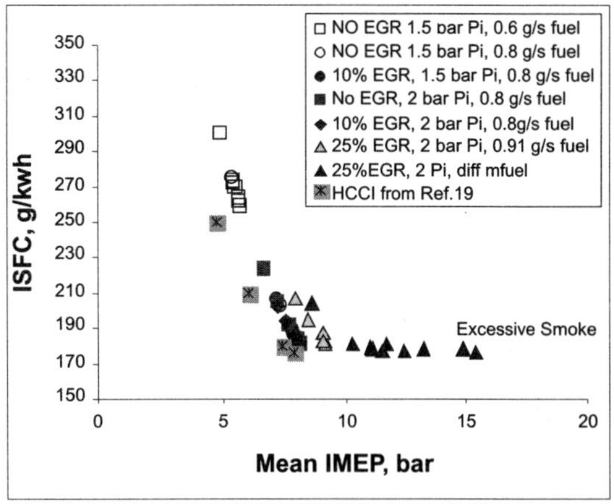


**Figure 11** ISFC vs IMEP for cases in Fig.10 and HCCI for the same gasoline from Ref.19

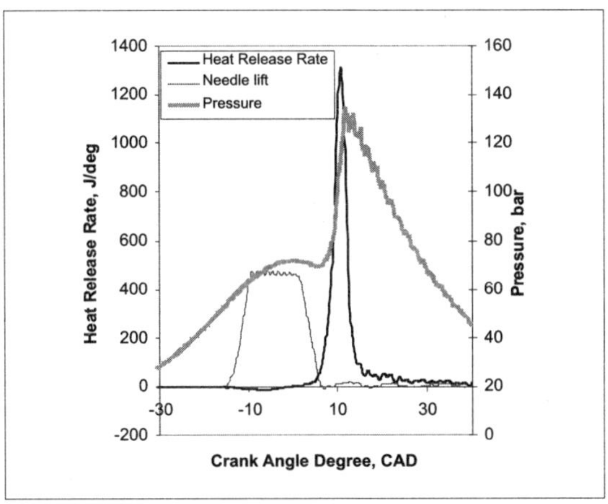


**Figure 12** Pressure, heat release rate and needle lift curves for gasoline with 14.86 bar IMEP, low smoke (FSN < 0.4) and ISNOx , ISFC, ISCO and ISHC of 1.21 g/kWh, 178 g/kWh, 3.4 g/ kWh and 3.6 g/kWh respectively. Needle lift, arbitrary scale

like high heat release rates and high pressure rise rates, is a feature of premixed combustion with an overall lean mixture. Injecting the fuel in two pulses would help reduce the width of the injector pulse near TDC and would also allow some shaping of the heat release curve, helping to moderate pressure rise rates. Such injection strategies were considered in Ref. 16.

## 3.2. DOUBLE INJECTION

Many different double injection strategies were tried and have been described in [16]. Here we present some selected results. Examples of heat release rate (from the mean pressure) and the needle lift are shown in Figures 13a and 13b for diesel and gasoline respectively for dual injection.

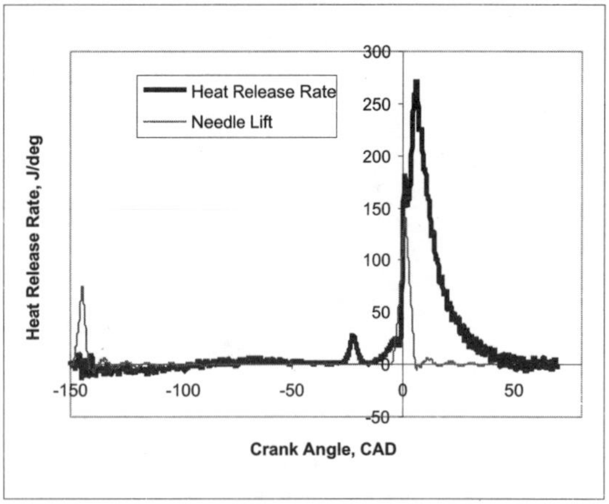


**Figure 13a** Needle lift and heat release rate for Swedish MK1 diesel fuel, 1.2 g/s [16]. Main injection SOI at – 6 CAD. Mean IMEP 11.84 (std 0.115 bar) bar. AVL smoke opacity 8.7%. In g/kWh, ISFC = 183, ISNOx = 0.3, ISHC = 11.3, ISCO = 10

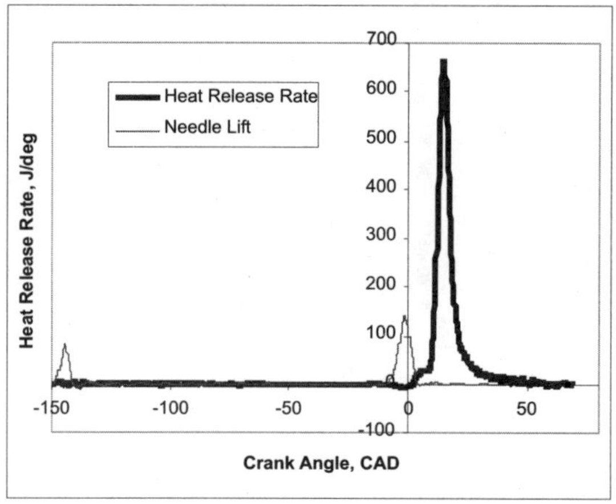


**Figure 13b** Needle lift and heat release rate for gasoline fuel, 1.2 g/s [16]. Main injection SOI at – 9 CAD. Mean IMEP 12.86 (std 0.108 bar) bar. AVL smoke opacity 0.9%. In g/kWh, ISFC = 174, ISNOx = 0.39, ISHC = 6.8, ISCO = 9

In both cases, 30% of the total fuel (1.2 g/s) was injected in a pilot injection starting at 150 CAD before TDC as the valves closed. The rest of the fuel was injected near TDC with SOI at – 6 CAD. The average pressure traces for these two cases are shown in Figure 14. With the diesel fuel (Figure 13a), at the fuelling rates used, the pilot injection leads to some heat release before TDC because of its high reactivity. This reduces the efficiency as noted in [10]. In contrast, with gasoline – Figure 13b -, no heat release is discernible in the compression stroke. Again the heat release is essentially separate from the injection event and this leads to lower smoke. When the main injection occurs with the diesel fuel, fuel is injected into a hotter environment and the ignition delay is reduced so that almost all that fuel will undergo diffusive combustion.

In Figure 15 smoke opacity has been plotted against mean IMEP for both the gasoline and diesel fuels; the fuelling rate was fixed at 1.2 g/s (to within 2%) but the fuel was injected using different injection strategies discussed in [16]. It is clear that with the diesel fuel, at the total fuelling rate used, with dual injection, it was not possible to reduce the smoke level to match the lowest level obtained with single injection – the lowest filled black square in Figure 15. Figures 16 and 17 show the indicated specific fuel consumption (ISFC) plotted against CA50 and IMEP respectively. Figure 16 shows that for the same CA50, dual injection for diesel fuel increases ISFC mostly because of the heat release occurring during compression as discussed above. Figure 17 shows the expected negative correlation between ISFC and IMEP. Thus with double injection, diesel fuel, unlike gasoline, leads to worse smoke and efficiency because of the heat release during the compression stroke. Further experiments described in [16] investigated if higher IMEPs were possible with gasoline using double injection while maintaining low smoke and NOx and ISFC. Some of these results are described below.

In this series of experiments with gasoline, the SOI of the main injection was at 11 CAD before TDC with the fuelling rate at 1.19 g/s; the fuelling rate of the pilot injection with SOI at 150 CAD

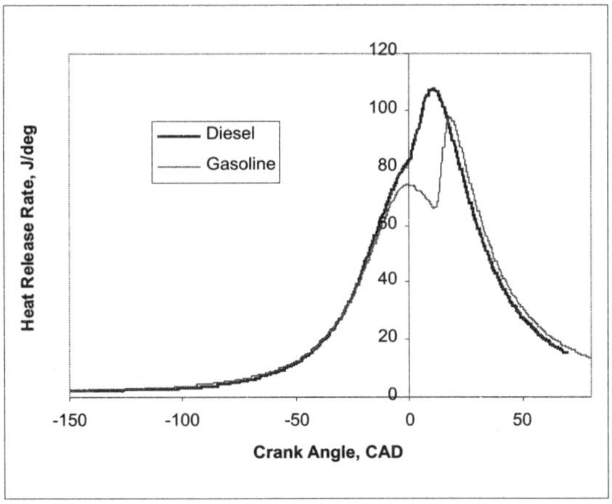


**Figure 14** Average pressure trace corresponding to the cases shown in Figures 13a and 13b

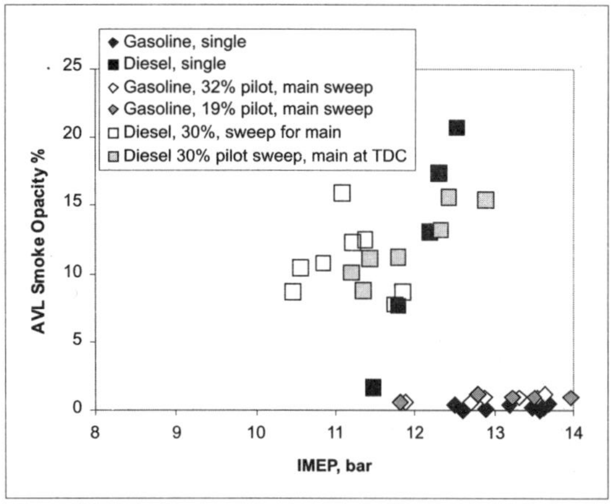


**Figure 15** AVL smoke opacity % vs IMEP for both the diesel and gasoline fuel used. Total fuelling rate 1.2 g/s but different injection strategies [16]

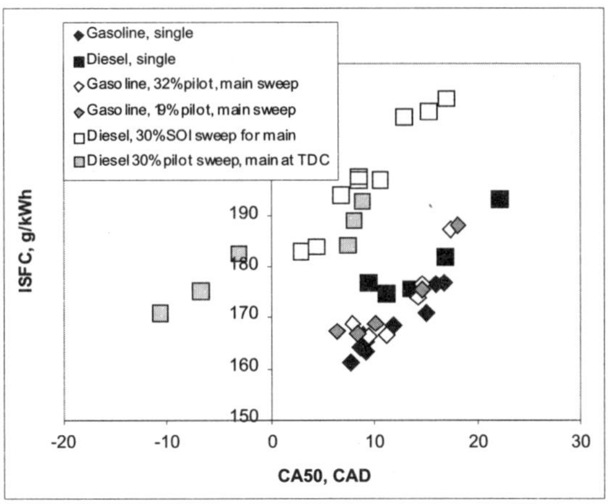


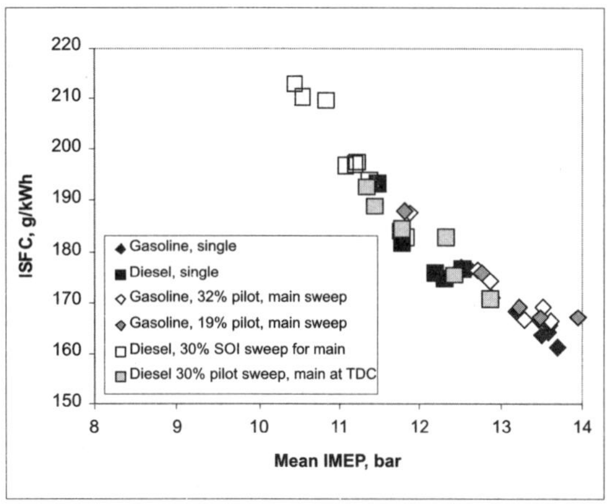


**Figure 16** Indicated specific fuel consumption (ISFC) vs CA50 for both the diesel and gasoline fuel used. Total fuelling rate 1.2 g/s but different injection strategies [16]

**Figure 17** Indicated specific fuel consumption (ISFC) vs Mean IMEP for both the diesel and gasoline fuel used. Total fuelling rate 1.2 g/s but different injection strategies [16]

BTDC was increased to get higher loads. Figure 18 shows the AVL smoke opacity for these tests with gasoline, plotted against mean IMEP – also shown for comparison are the results for gasoline with single injection described in the previous section. As the total amount of fuel injected is increased by increasing the pilot injection, IMEP increases but eventually smoke also increases. However, a mean IMEP over 16 bar could be reached with the AVL smoke opacity less than 0.5 % (FSN < 0.1).

In Figure 19, the average pressure traces are compared for gasoline for three different cases – a) the best operating point with single injection with mean IMEP of 14.86 bar considered above and

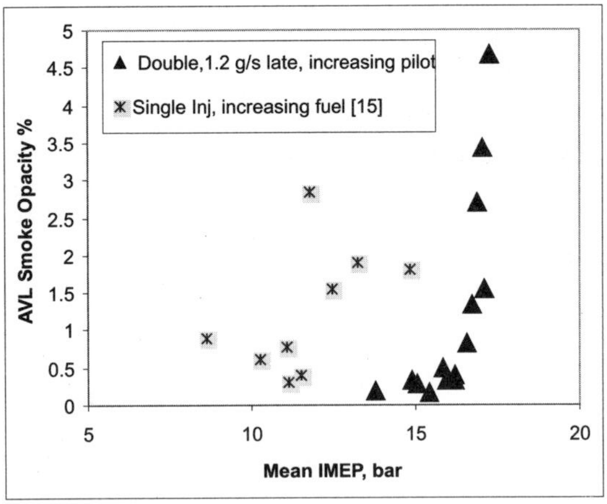


**Figure 18** AVL smoke vs IMEP for gasoline. Single injection and double injection

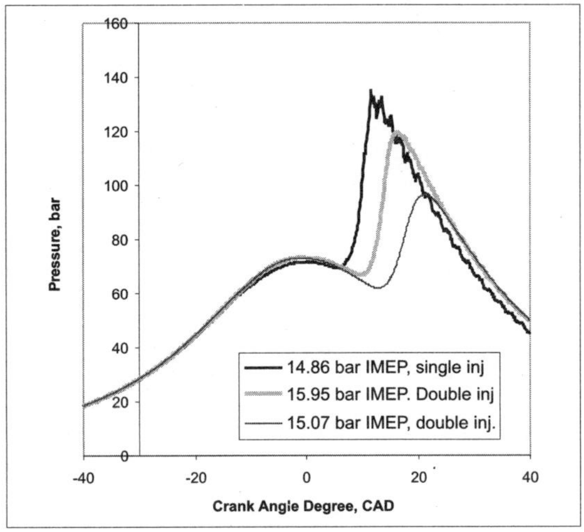


**Figure 19** Average pressure curves for cases a), b) and c) listed in Table 3 for gasoline

two cases with double injection with the main injection pulse rate fixed at 1.19 g/s (SOI –11 CAD) and different pilot fuelling rates, b) mean IMEP of 15.07 bar and c) mean IMEP of 15.95 bar. All these cases were run at 1200 RPM with the same intake pressure of 2 bar abs., the same EGR level as indicated by the intake CO2 concentration (Table 3) and the same intake temperature of 40 C. In all three cases the injection event is completed just as the heat release starts. Figure 20 shows the heat release rate curves calculated from the average pressure for these three cases. It can be seen (Table 3) that with double injection, higher IMEP can be reached with lower smoke, NOx, HC, maximum heat release rate and peak pressure compared to the single injection case while maintaining comparable fuel economy. For the two cases with double injection shown, peak heat release occurs fairly late but interestingly, the cyclic variation on IMEP is still very low. If the fuel and air were fully premixed (HCCI combustion), the cyclic variation would be unacceptably high

**Table 3** Summary data for the three cases considered in Figures 19, and 20

| *Case* | *Injection* | *Fuel Rate Mean g/s* | *Co2 Intake %* | *IMEP* Mean bar* | *IMEP* std bar* | *AVL smoke % opacity* | *ISNOx g/kWhg* | *ISFC g/kWh* | *ISHC g/kWh* | *ISCO g/kW* | *HRR* J/deg* | *Angle of MHRR* CAD* |
|---|---|---|---|---|---|---|---|---|---|---|---|---|
| a | Single** | 1.436 | 4.05 | 14.86 | 0.115 | 1.81 | 1.21 | 178 | 3.6 | 3.4 | 1446 | 10.8 |
| b | Double*** | 1.46 | 4.14 | 15.07 | 0.138 | 0.28 | 0.59 | 179 | 3.0 | 5.8 | 817 | 18.2 |
| c | Double*** | 1.59 | 4.16 | 15.95 | 0.112 | 0.33 | 0.58 | 179 | 2.9 | 6.8 | 1393 | 14.1 |

Mean from 100 cycles

** SOI @ -16 CAD

*** 1.19 g/s @ -11 CAD and rest at -150 CAD

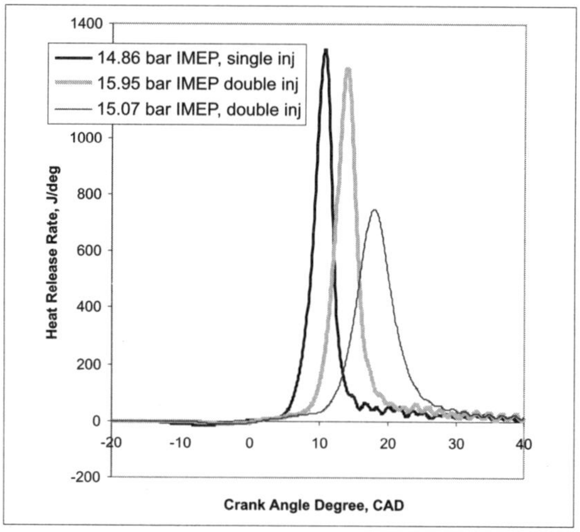


**Figure 20** Average heat release rate curves for cases a), b) and c) listed in Table 3 for gasoline

with such late heat release [19,20]. The later and lower maximum heat release rate with double injection, compared to the single injection case would result in lower maximum pressure rise rate and hence lower noise compared to single injection. As the total fuelling rate is increased, the heat release occurs earlier and the maximum heat release rate also increases. Eventually, some of the fuel injected near TDC will interact with combustion products and smoke increases.

## 4. DISCUSSION

The results discussed demonstrate that it is possible to run a CI engine on gasoline at high load and low smoke, low NOx and with low ISFC. The combustion mode is partially pre-mixed – "pre-mixed enough" – autoignition. In this type of operation, heat release can be controlled using the timing of the main fuel injection near TDC. This makes it much more practical while realising most of the benefits in terms of NOx and particulates compared to HCCI combustion, when fuel is fully premixed with air. In HCCI combustion heat release is completely controlled by autoignition chemistry and once the fuel is injected early, will occur where it will. This makes it more difficult to control and will also restrict the operating regime ; indeed at most of the operating conditions considered in this paper, the engine could not be run in HCCI mode on gasoline. On the other hand HCCI mode will offer lower NOx and particulates. It was argued in [3] that a gasoline-like fuel, because it enables higher loads at a given set of operating conditions, was better than a diesel-like fuel for HCCI combustion. Of course, a gasoline-like fuel has difficulty autoigniting at low loads but this difficulty can be more easily overcome by thermal management such as using large amounts of internal EGR using negative valve overlap. For a smaller engine this might be more difficult, because of the lack of sufficient heat in the exhaust gases, and a lower octane gasoline might be needed to enable low-load operation in the HCCI or partially pre-mixed mode. Such a fuel would still offer benefits over a diesel fuel in the type of combustion discussed in this paper.

The loads achievable in this type of partially premixed autoignition combustion are much higher, while maintaining low smoke and NOx, with gasoline compared to a diesel fuel because of the resistance to autoignition and the consequent higher engine ignition delays with gasoline. Splitting the fuel in two injections per cycle enables higher IMEPs to be reached while alleviating some of the problems associated with premixed combustion such as high heat release rates. In this work we have not attempted to optimise injection and mixing strategies and injector design though there is a lot of scope for this. It is very likely that if the number of injections were increased above two, heat release could be further spread out and the pulse width of the last injection reduced further. This should enable more fuel to be injected into the cylinder and hence higher IMEP to be realised than in the present case while maintaining low smoke, low NOx, low noise and low fuel consumption. However this requires the fuel to have sufficient resistance to autoignition so that there is sufficient premixing of fuel and air before heat release occurs. Otherwise, as discussed for diesel fuel above, smoke levels will go up. In fact with the Swedish MK1 diesel fuel, with the same operating conditions, to get smoke opacity below 0.5% as was possible with the gasoline (at nearly 16 bar IMEP), mean IMEP could not be greater than about 6.5 bar. Even higher fuelling rates and hence higher IMEPs while maintaining low smoke should be possible with gasoline with higher intake pressures and further optimisation of injection and mixing e.g. more injector holes. It might be possible to use much lower injection pressures with gasoline.

A CI engine operating on gasoline through the entire range is conceivable. Such an engine could be started by using spark ignition or a powerful glow plug, run in HCCI mode, with a single injection pulse early in the cycle at low and moderate loads; this would require the use of internal EGR or other thermal management. At high loads the fuelling rate would be increased by additional injection pulses later in the cycle. The engine will need to have an oxidizing catalyst to control HC and CO emissions, a turbocharger or a supercharger and a variable valve train to control the HCCI part of the combustion. Of course further work is needed to understand the effect of engine size and higher engine speeds, which will reduce the time available for autoignition, on partially premixed compression ignition of gasoline. So far we have concentrated on achieving high loads but it will be desirable to establish the lowest loads achievable using gasoline in this type of combustion. As discussed above, it might be more difficult to run smaller engines on gasoline at low load because of the lack of sufficient heat in the exhaust and a gasoline of lower octane number than used in this paper may be required. Oxygenates like ethanol and natural gas will also lead to high ignition delays. Oxygenates might be particularly advantageous since the presence of oxygen in the fuel helps to reduce smoke by itself.

Thus if the aim is to promote premixed combustion i.e. increase engine ignition delay in CI engines in order to control smoke and NOx, the fuel needs to be as much like gasoline as possible. The extent to which this is possible in practice depends on whether other critical requirements can be met. Thus the engine needs to start and run at low loads satisfactorily; this might mean that in smaller engines the octane number of the fuel should be lower than considered in the current paper. It must have acceptable noise which depends on the maximum pressure rise rate and its phasing but these could be controlled using multiple injections. As already stated HC and CO need to be controlled through the use of an oxidising catalyst. At the moment, the practical debate on fuel autoignition quality in CI engines is about whether future engines will require higher cetane in the 40 to 60 range, not whether diesel fuel should be replaced by gasoline. If these future engines want to promote premixed combustion, higher fuel cetane number will certainly not help; it might hinder such operation. Of course, other diesel quality issues such as sulphur and volatility levels and other compositional issues are also important in the debate about future fuel requirements. These will eventually have to be agreed by the different stakeholders such as the auto and oil industries, governments and regulatory bodies and environmental interests. In general the prospect of more pre-mixed combustion in future compression ignition engines is likely to reduce the pressure on the cetane requirement but might increase, to a lesser extent, the pressure on the volatility requirement of fuels for such engines.

This work also highlights the possibility of using fuels of low cetane number to simplify the technology in diesel engines and hence reduce their cost. As discussed in the introduction much of this technology such as high injection pressures is aimed at overcoming the difficulty presented by the easy autoingnition of current diesel fuels.

## Conclusion

1. The engine can be run on gasoline in partially pre-mixed mode even when it cannot be run with early injection in HCCI mode.
2. For the diesel fuel, with single injection, smoke decreases as fuelling rate is decreased for a given injection timing and as injection timing is retarded for a given fuelling rate.
3. For a given condition, there is significantly lower smoke and lower NOx, if the global mixture strength is lean enough, with gasoline compared to the diesel fuel because of higher ignition delay.
4. For a given total fuelling rate, compared to single injection, with pilot injection early in the compression stroke, there can be sufficient heat release for the diesel fuel during the compression stroke to reduce efficiency. This increases specific fuel consumption and makes the lowest smoke level possible much higher compared to single injection.
5. In contrast, with gasoline, pilot injection early in the compression stroke helps reduce the maximum heat release rate for a given IMEP and enables heat release to occur later with low cyclic variation compared to single injection. This enables higher mean IMEP to be reached with lower smoke, NOx and maximum heat release rate compared to single injection.
6. One of the operating points reached with gasoline had mean IMEP of 15.95 bar (Stdev. 0.112 bar), AVL smoke opacity of 0.33% (FSN < 0.07), ISNOx of 0.58 g/kWh, ISFC of 179 g/kWh, ISHC of 2.9 g/kWh, ISCO of 6.8 g/kWh and peak pressure of ~ 120 bar. At the same operating conditions, to get such low level of smoke with Swedish MK1 diesel fuel, IMEP had to be below 6.5 bar.

There is scope for significant improvement by increasing the intake pressure and EGR level and optimising the injection and mixture preparation strategy – e.g. more than two injections, increased number of injector holes, higher injection pressure and optimised injection timings. Lower cetane fuels – gasoline is an extreme example – are better suited for partially premixed CI combustion.

## Acknowledgments

This work was supported by Shell Global Solutions. The experiments were done at KTH Stockholm and were first reported in [15] and [16]

## References

1. Dec, J.E., " A conceptual model of DI diesel combustion based on laser sheet imaging", SAE 970873, 1997.
2. Heywood, J.B., "Internal combustion engine fundamentals", Ch. 11., McGraw Hill Book Co., 1988
3. Kalghatgi, G.T., "Auto-ignition quality of practical fuels and implications for fuel requirements of future SI and HCCI engines" SAE 2005-01-0239, 2005.
4. Westbrook, C.K., "Chemical kinetics of hydrocarbon ignition in practical combustion systems", Proceedings of the Combustion Institute, vol. 28, pp. 1563-1577, 2000.
5. Yanagihara, H., Sato, Y. And Mizuta, J., " A simultaneous reduction in Nox and soot in diesel engines under a new combustion system (Uniform Bulky Combustion System- UNIBUS)", 17$^{th}$ International Vienna Motor Symposium, pp. 303 – 314, 1996.
6. Kitano, K., Nishiumi, R., Tsukasaki, Y., Tanaka, T. and Morinaga, M., " Effects of fuel properties on pre-mixed charge compression ignition combustion in a direct injection diesel engine", JSAE 20030117 and SAE 2003-01-1815, 2003
7. Kimura, S., Osama, A., Ogama, H., Muranaka, S., " New combustion concept for ultra-clean and high-efficiency small DI diesel engines", SAE 1999-01-3681.
8. Kimura, S., Osamu, A., Kitahara, Y. And Aiyoshizawa, E., " Ultra-clean combustion technology combining a low temperature and premixed combustion concept for meeting future emission standards" SAE 2001-01-0200
9. Kimura, S., Ogawa, H., Matsui, Y. And Enomoto, Y., "An experimental analysis of of low-temperature and premixed combustion for simultaneous reduction of of NOx and particulate emissions in direct injection diesel engines", Int. J. Engine res., vol 3, no.4, pp. 249-259, 2002
10. Hashizume, T., Miyamoto, T., Akagawa, H. And Tsujimura, K., " Combustion and emission characteristics of multiple stage diesel combustion", SAE 980505, 1998.
11. Su, W., Lin, T. And Pei, Y., " A compound technology for HCCI combustion in a DI diesel engine based on multi-pulse injection and the BUMP combustion chamber", SAE 2003-01-741, 2003
12. Mueller, C.J., Martin, G.C., Briggs, T.E., Duffy, K.P., " An experimental investigation of in-cylinder processes under dual-injection conditions in a DI diesel engine", SAE 2004-01-1843, 2004
13. Dronniou, N., Lejeune, M, Balloul, I. and Higelin, P., "Combination of high EGR rates and multiple injection strategies to reduce pollutant emissions", SAE 2005-01-3726, 2005.
14. Shimazaki, N., Tsurushima, T. and Nishimura, T., " Dual mode combustion concept with pre-mixed diesel combustion by direct injection near top dead center", SAE 2003-01-0742, 2003
15. Kalghatgi, G. Risberg, P., and Ångström H-E, " Advantages of fuels with high resistance to auto-ignition in late-injection, low-temperature, compression ignition combustion", SAE 2006-01-3385, 2006.
16. Kalghatgi, G. Risberg, P., and Ångström H-E, "Partially pre-mixed auto-ignition of gasoline to attain low smoke and low NOx at high load in a compression ignition engine and comparison with a diesel fuel", SAE 2007-01-006, 2007.
17. Risberg, P., Kalghatgi, G. and Ångström H-E, " Auto-ignition quality of diesel-like fuels in HCCI engines", SAE 2005-01-2127, 2005.

18. Schindler W., Nöst M., Thaller W., Luxbacher T., “Stationäre und Transiente Messtechnische – Erfassung Niedriger Rauchwerte”, MTZ 10 / 2001.
19. Kalghatgi, G.T. and Head, R.A.,” Combustion limits and efficiency in a HCCI engine”, Int. J. Engine Res., vol 7, no. 3, pp 215-236, 2006
20. Kalghatgi, G.T., “ Fuel effects in gasoline HCCI engines”, in “Homogeneous charge compression ignition (HCCI) and controlled auto ignition (CAI) engines for the automotive industry”, Ed. H. Zhao, Woodhead Publishing Ltd., 2007.

# Review and Simulation of Combustion in the Wave Rotor

**M.R. Nalim**[1], **V. Kilchyk**[2] **and P. Akbari**[3]

[1, 2]Indiana University-Purdue University Indianapolis, Indiana, USA & Purdue University, West Lafayette, Indiana, USA

[3]Manhattan College Riverdale, NY, 10471

In recent decades, considerable efforts have been made to exploit non-steady combustion and gasdynamic phenomenon. The performance potential of non-steady flow machines has led to development of various oscillatory flow devices such as pulse detonation engines, wave rotors, pulse jets, and non-steady ejectors. There is now particular interest in devices that can achieve constant-volume combustion for a gas turbine, to create a Humphrey cycle heat engine that combines pressure-gain combustion with full-expansion work extraction. The wave rotor appears to have considerable potential to enhance the performance and operating characteristics of such hybrid gas turbine engines. After earlier attempts were thwarted by mechanical design problems, recent efforts in Switzerland and the US gained from design expertise developed with pressure-exchange wave rotors. This work first briefly reviews the history, potential benefits, and existing challenges in developing these non-steady combustors. Then, the effects of strong channel rotation and pressure waves, and the use of a hot jet igniter under experimental low-pressure test conditions are investigated using two-dimensional numerical simulations of combustion and gasdynamics. The jet of hot gas can be created from a cross-over duct or a separate torch igniter. Strong rotational acceleration equivalent to about 4000g has limited impact on the combustion rates when jet ignition is used. Weak shock waves and vortices produced during the ignition process appear to have a stronger impact on flame dynamics than rotation-induced buoyancy forces.

## I. INTRODUCTION

Combustion engines have improved remarkably in performance, emissions, costs, and operational characteristics over a century. While intermittent-combustion positive-displacement piston engines remain dominant for automobiles, steady-combustion gas turbine engines dominate aircraft propulsion and a growing range of power-generation applications. However, intermittent or pulsed combustion devices such as pulse jets [1] and pulsed detonation engines (PDE) [2] remain an important possibility for propulsion and other applications. Indeed intermittent combustion was used in the earliest gas turbines, but the subsequent improvement of compressor technology obviated the need for pressure-gain combustion. Today, energy costs and greenhouse gas controls are driving a reappraisal of this approach.

Steady combustion involves a loss in stagnation pressure, but non-steady combustors typically produce a pressure gain during combustion process. When properly integrated with turbomachinery, pressure-gain combustors can improve turbofan and other gas-turbine engine efficiency and power output or thrust.[3] For such systems, positive-displacement mechanisms are generally too heavy, while unvalved or check-valved single pulse combustors have limited efficiency improvement in addition to subjecting the system to pulsations. Significant developments in multi-tube PDEs aimed at a "hybrid" engine are discussed briefly below as a prelude to the review of the wave rotor combustor. Wave rotor technology provides an opportunity to generate pressure gain with nearly steady inflow and outflow, for seamless integration with turbomachinery working under steady-flow principles.

## Constant-Volume Combustion for Gas Turbines

Cyclic constant-volume combustion combined with full working-gas expansion creates an efficient thermodynamic cycle [4, 5] known as the Humphrey cycle. This is illustrated in Fig. 1, which compares schematic temperature-entropy (T-s) diagrams of a conventional Brayton engine (1-2-$3_b$-$4_b$) and the replacement Humphrey cycle (1-2-3-4) for fixed turbine inlet temperature and compressor discharge pressure. Significant pressure gain ($p_3 > p_2$), enables greater turbine power output and higher cycle efficiency for fixed combustion energy input. When compared with years of effort to reduce relatively smaller losses in turbomachinery components, constant-volume combustion represents a revolutionary improvement in modern gas turbines.

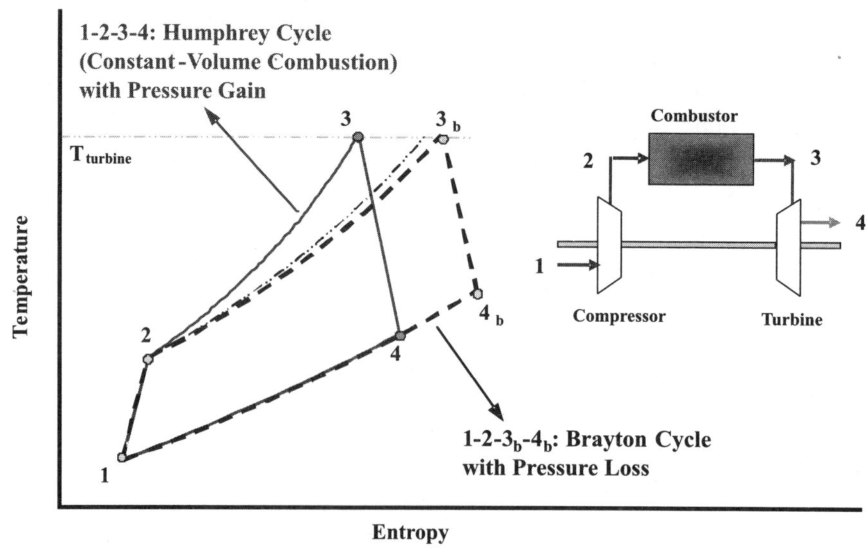


**Figure 1** Superior performance of the Humphrey cycle (1-2-3-4) over the Brayton cycle (1-2-$3_b$-$4_b$)

Ideally, constant-volume combustion can be achieved by both deflagrative and detonative combustion modes. In a pulse detonation engine (PDE), detonable mixtures of fuel and air are admitted into open-ended tube(s) and ignited, generating a detonation that provides a pulse of pressure and thrust. [16, 17]. Multiple-tube PDEs have been considered for turbine engines. Simle analytical [12, 20] and preliminary numerical simulations [21] predict good performance benefits, but detailed multi-dimensional modeling [23] and experimental tests [24, 25] indicate substantial flow losses. Performance can be degraded by flow obstructions for detonation transition, periodically stagnating duct flows, and unsteady flow in turbines and nozzles. The wave rotor is a non-steady-flow machine that can accomplish nearly constant-volume deflagration by mechanical confinement, with relatively simple integration with turbomachinery flows.

## II. WAVE ROTOR APPROACH

A wave rotor consists of a rotating drum as schematically seen in Fig. 2,[12] with multiple, axially oriented gas channels around the circumference. The drum rotates between two stationary endplates, each of which has ports or manifolds at different pressures, controlling the fluid flow through the channels. Through rotation, the channel ends are periodically connected to the ports located on the stationary endplates. The sudden opening or closing of the rotating channel to the stationary ports initiates gasdynamic compression and expansion waves. With proper timing of the port interactions, these waves in the channel effect pressure and energy exchange between fluids, dynamically, i.e. without pressure equilibration.

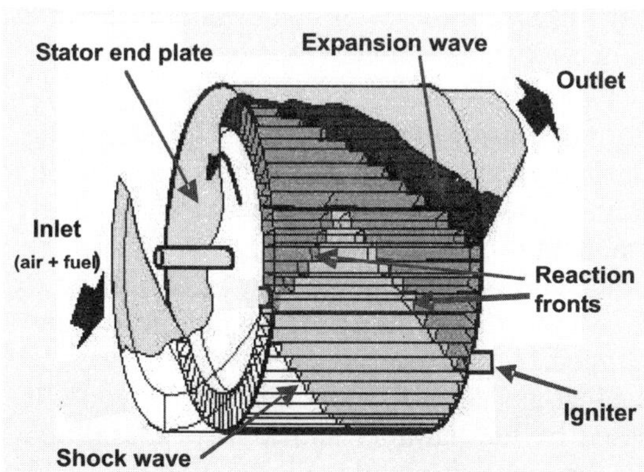


**Figure 2** Schematic configuration of a combustion wave rotor[12]

Wave rotors were initially developed as dynamic pressure exchangers for applications including [13] supercharging of piston engines, air conditioning, and hypersonic wind tunnel gas compression. Significant commercial success was attained by the Comprex® supercharger for diesel car engines. A comprehensive review of pressure-exchange wave rotor is available in Ref. 14. Pressure-exchange wave rotors can enhance the performance of conventional gas turbines by acting as a topping spool to produce a net gain in turbine inlet pressure, while retaining a conventional steady-flow combustor at elevated pressure.[15, 16]

Introducing a combustion process in the rotor channels creates the wave rotor combustor,[17, 18] with pressure rise achieved by the combination of the pressure-wave compression and confined combustion. Each channel is an individual combustion chamber which periodically charges and discharges as it rotates past the partial-annular inlet and outlet ports, with the end walls functioning as the inlet and exit valves. Combustion within the channels requires a method of introducing fuel, and a method of ignition that will ensure rapid combustion of all fuel. For example, the air inlet duct may incorporate fuel nozzles to disperse fuel in controlled increments, and igniters may be located at the opposite endplate. One inlet and one outlet port are minimally required, but some designs [19] may use two outlet ports at different pressures for complete evacuation of the channels. Combustion is initiated in all channels by continuously-firing stationary igniters at one of the endplates.

From both mechanical and aerodynamic points of view, there are several advantages of the wave rotor configuration. With a large number of rotating channels receiving and discharging flow at any moment, the wave rotor has essentially steady inlet and exit flows, continuous ignition system, and simple valveless filling and purging processes with low losses. The compressor and turbine experience normal continuous flow. Mixture stratification along the rotor channels for improved flame propagation and emissions control can be achieved by dividing the inlet port into several partitions.

## Internal Working Principle

Figure 3 is a developed (unwrapped) view of the wave rotor, where the circumferential motion of the rotor channels is represented on paper by a vertical upward translatory motion, while the horizontal direction represents axial distance along the rotor channels. The complex inflow-outflow gasdynamic processes occurring inside the rotor channels are shown as a simplified wave pattern without showing the combustion details, reflected waves, and flow patterns. Snapshot images are shown of three sample channels.

The simplest wave rotor combustor cycle involves three steps: the filling of the device with a combustible mixture, the initiation and completion of the combustion, and finally, the exhaust of high-pressure products into the turbine. The filling process starts when the channels are exposed to the inlet port bringing combustible mixture (state 2). Meanwhile the purging process has been already begun from the other end of the channels by opening the exhaust port and generating expansion waves (state 3). While the filling process continues, the scavenging of burned gas through the exit port is stopped by closing the exhaust port, generating a compressive wave or "hammer shock". Some residual gas and fresh air-fuel mixture (state A) are trapped in the channel and are favorably compressed by the hammer shock wave that propagates backward toward the inlet wall. At the moment it arrives to the inlet wall, the inlet port is closed. With both ends of the channel closed, the combustion is initiated by an igniter (not shown). The combustion process under constant-volume condition increases the pressure and temperature in the channel significantly to state B. The presence of residual gas means that the combustion process in the working gas is not purely constant volume. This can be avoided by using a second exhaust port at lower pressure to completely evacuate the channel. Opening the exhaust port creates expansion waves which cause the burned gas to leave the channels at state 3, and another operating cycle starts. The features of the cycle results in a relatively uniform inflow and outflow, significant pressure gain, and overall temperature ratio that is compatible with the gas turbine engine requirements. The wave rotor is a self-cooled machine as the rotor channels are swept alternately by hot and cold fluids. Therefore, the average wall temperature of each channel is lower than the combustion peak temperature.

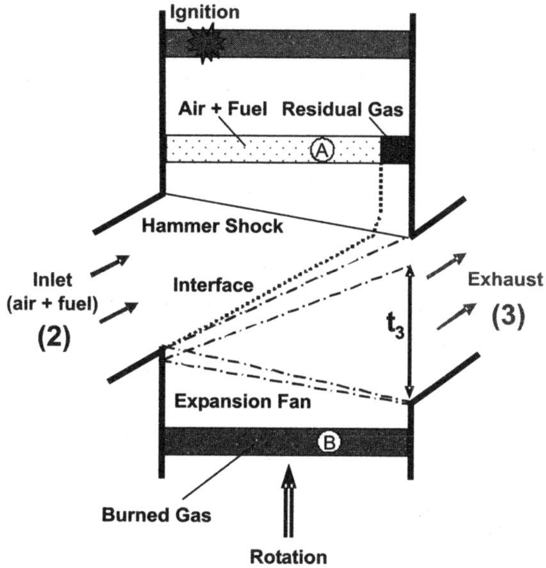


**Figure 3** Wave Pattern in a developed view of the combustion wave rotor

There is limited knowledge of the possible combustion processes in a wave rotor. In principle, both deflagrative and detonative combustion modes can provide confined combustion for wave rotors, creating pressure rise across the channels. The deflagration combustion mode has been studied both theoretically [12, 20, 21] and experimentally,[19] but the detonation mode only theoretically.[22] Detonation requires particular fuels in limited ranges of stoichiometry and channel width, and generates concentrated pressure and momentum in the outflow. A rapid deflagrative combustion process can be more reliable, yet minimize the duration of high pressures and temperatures within the channels.

The combustion initiation is achieved by employing one or more igniters. The igniter(s) can be placed at either the forward or aft end of the channels, depending on whether forward or backward flame propagation is desired. A cross-fire tube or an end-wall cavity that injects or recirculates hot gas makes the combustion process essentially continuous, improves ignition reliability and burning rate which can lead to a self-sustaining engine.[23-25]

## III. BRIEF HISTORICAL REVIEW

Following ideas proposed in the 1950's,[26] the first attempts to build and test a combustion wave rotor encountered mechanical difficulties, not surprisingly for novel technology.[27] With the success of the Comprex® supercharger, ABB commenced research on gas turbine engine application.[19] In the 1990's, tests in a rotary-valved single-channel experiment suggest good combustion with low emissions, and was followed by a combustion wave rotor with 36 channels.[28, 29] Ignition was generated by spark plug for starting, and by a self-sustaining jet injection of already-burned gas from a neighboring channel. During the tests, various fuels were tried and fuel mixture was stratified via four injection nozzles. Operating in two cycles per revolution, the engine had high-pressure and low-pressure outflows per cycle, enabling it to scavenge the exhaust gas in one cycle. ABB encountered specific challenges: (i) inhomogeneous mixture in the cell, causing a slow diffusion flame, (ii) leakage at higher chamber pressures, causing premature ignition and misfiring, (iii) thermal stresses on the ignition ring end-plate, (iv) inadequate support of cantilever single bearing, (v) over-sensitive electromechanical leakage gap controller. Major remedies recommended were (a) lead away duct for leakage gas removal, (b) rotor cooling by air, (c) two-sided rotor support, and (d) mechanical control for thermal expansion.

Meanwhile, a wave rotor research program at NASA quantified loss mechanisms and developed modeling tools for pressure-exchange wave rotors, especially a quasi-one-dimensional (Q1D) code with comprehensive loss models.[30] The feasibility on-rotor combustion was assessed,[20] and combustion prediction capability was added to the Q1D code.[12] A single reaction progress variable was used for uniform mixtures subject to detonation, while multiple species were represented for stratified mixtures in deflagration mode. A simple eddy diffusivity model, ignition-temperature kinetics, and total-energy based flammability limit indicated that longitudinal fuel stratification could be accomplished over a reasonable time in wave rotors. The code's one-dimensional detonation prediction capability later became widely useful for study of PDE.[8, 33]

In recent years, a joint effort by Rolls Royce and IUPUI is examining detonative and near detonative internal combustion. A preliminary design method based on a sequence of computational models [31, 32] were used to design combustion processes in an experimental test rig. New solutions to mechanical design challenges such as clearance control and sealing issues have been developed and tested.[16, 34, 35]

## IV. CHALLENGES AND RESEARCH NEEDS

Combustion in a wave rotor presents some unique challenges. Basic combustion phenomena, ignition, and control methods remain poorly understood at the fundamental level. Non-steady combustion in the unusual wave rotor geometry involves complex flame-wave interactions fundamentally different from familiar internal combustion engines and steady-flow gas turbine combustion. The interaction of a flame front with strong pressure waves is not commonly present or is avoided in other conventional combustion systems.

For most gas turbine engine combustors, detonative propagation appears unlikely, and premixed turbulent combustion processes must be demonstrated in basic experiments. Deflagrative combustion is strongly influenced by turbulence,[19, 36] originating in the upstream compressor, ducting, fuel spray, the opening/closing processes, and shock compression processes, and likely to persist over the intake period of a wave cycle.[20] Pressure waves interacting with flames or other density gradients or discontinuities produce vorticity by baroclinic forces (Richtmyer-Meshkov instability), accelerating or sometimes quenching combustion locally.[37] Energetic and rapid ignition methods, such as hot gas injection at high speed and the use of partitions for thermal management and pilot ignition,[36] introduce additional complexity.

## V. EFFECT OF ROTATION ON TURBULENT FLAME PROPAGATION

A significant feature that distinguishes the wave rotor combustor from other existing combustors is the presence of strong rotational forces. Strong spinning of a tube is known to notably change overall combustion reaction rates in it. For instance, axial flame propagation in a rotating tube has been reported to speed up five times.[38]

In the wave rotor combustor, rotation about a longitudinal axis parallel to the channel causes the development of a pressure gradient across in the channel, given sufficient time after charging. For typical wave rotor operating conditions, the resulting buoyancy force is equivalent to several thousand times the standard gravitational acceleration. It is important to know whether the conditions within a wave rotor would support flame speed amplification similar to the spinning tube, as this may strongly influence its design. Knowledge of effective flame speed is crucial for the development of the wave rotor combustor.

In practice, fast filling of the working volume and the ignition process are accompanied by shocks and expansion waves that also strongly influence the flame speed.[39, 40] The flame-shock interaction results in a Richtmyer-Meshkov instability (RM) that may accelerate the combustion rate. RM instability is produced when a shock wave travels through an interface separating regions of fluid with different densities. Gas compression and interface deformation produced by the shock are relatively well understood for non-reacting fluid when the angle between the shock wave and density interface is small. The interaction is more complex for larger angles [41] and in reacting fluids. An indirect effect of channel rotation may involve tilting of the flame to provide a more RM-sensitive orientation to the roughly planar waves. The presence of close boundaries adds further complexity through boundary layer effects and introduction of multiple reflected shock and expansion waves.However, little is known about flame acceleration effects of rotation in the presence of weak shock waves. This is among the features of wave rotor combustion that can be studied by computational simulation, where the effect of a particular feature or process can be isolated.

## VI. COMPUTER SIMULATION OF TURBULENT FLAME PROPAGATION

A second-order, time-accurate density-based solver of the Fluent computational fluid dynamics code was used in the study of combustion in the rotating channel with hot jet ignition. Combustion simulations for a propane-air mixture used an eddy-diffusion model with finite-rate chemistry and a single-step global reaction mechanism to complete combustion. Standard reaction coefficients from the Fluent database were selected for the calculations. The standard *k-E* turbulence model was employed for the simulations.

For jet-ignited combustion, the two-dimensional model of a channel of length 0.8 m and radial height 64 mm (see Fig. 4) uses a.grid resolution varying from 0.25 to 2.7 mm. Mesh resolution at the left end of the channel and along the injected gas trajectory was increased. No-slip adiabatic walls have been assumed. Along the wall grid resolution was made fine enough for the enhanced wall treatment function of Fluent to be used. Initially the channel is filled with a quiescent stoichiometric propane-air mixture at 300 K and 0.1 MPa. Rotation at 4000 rpm was simulated with a uniform acceleration of 40,000 m/s$^2$ applied in the y direction.

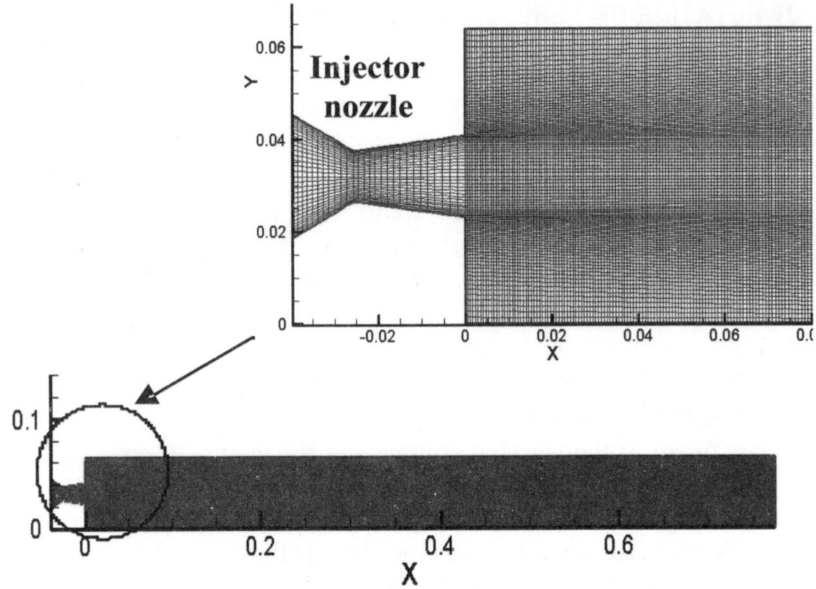


**Figure 4** Channel geometry. Dimensions are shown in meters

The ignition nozzle inlet boundary was assigned 0.202 MPa static pressure and 2200° C, with the composition corresponding to the product of the complete combustion of stoichiometric propane-air mixture. Jet geometry and parameters were selected to insure a stable ignition source that consumed no more than 1% of the total wave rotor combustor fuel consumption. The injector is stationary in the modeled plane. To simulate finite injection duration due to rotation, the nozzle inlet boundary was changed to a wall boundary after 0.75 ms.

Turbulence in the channel is produced from complex effects, including the filling process. Upstream duct wall, fuel-injection struts and nozzles, and turning vanes all contribute, in addition to the process of channel opening and then closing to the flow. The turbulence decay rate and effects of rotational forces are difficult to estimate. Based on typical duct flow experience, the residual level of turbulence before ignition has been chosen to correspond to the flow in the wave rotor combustor channel at 10 m/s average velocity with 10% turbulence intensity.

The combustion process without and with the effect of rotation is illustrated in Figs. 5 and 6, respectively, in terms of pressure and temperature history. Gas exiting the nozzle produces a shock wave its total pressure 101 kPa higher than the total pressure of the gas in the channel. In both cases, this shock wave travels through the channel and is then reflected back and forth several times. The total cycle duration of the wave rotor combustor is expected not to exceed 15 milliseconds (ms), with combustion processes taking place in only a portion of the channel length. Therefore, computations have been carried out to approximately 9 ms, during which the flame was observed to propagate to approximately two-thirds of the entire channel length. Due to the confined combustion, the overall pressure level increases during the combustion process.

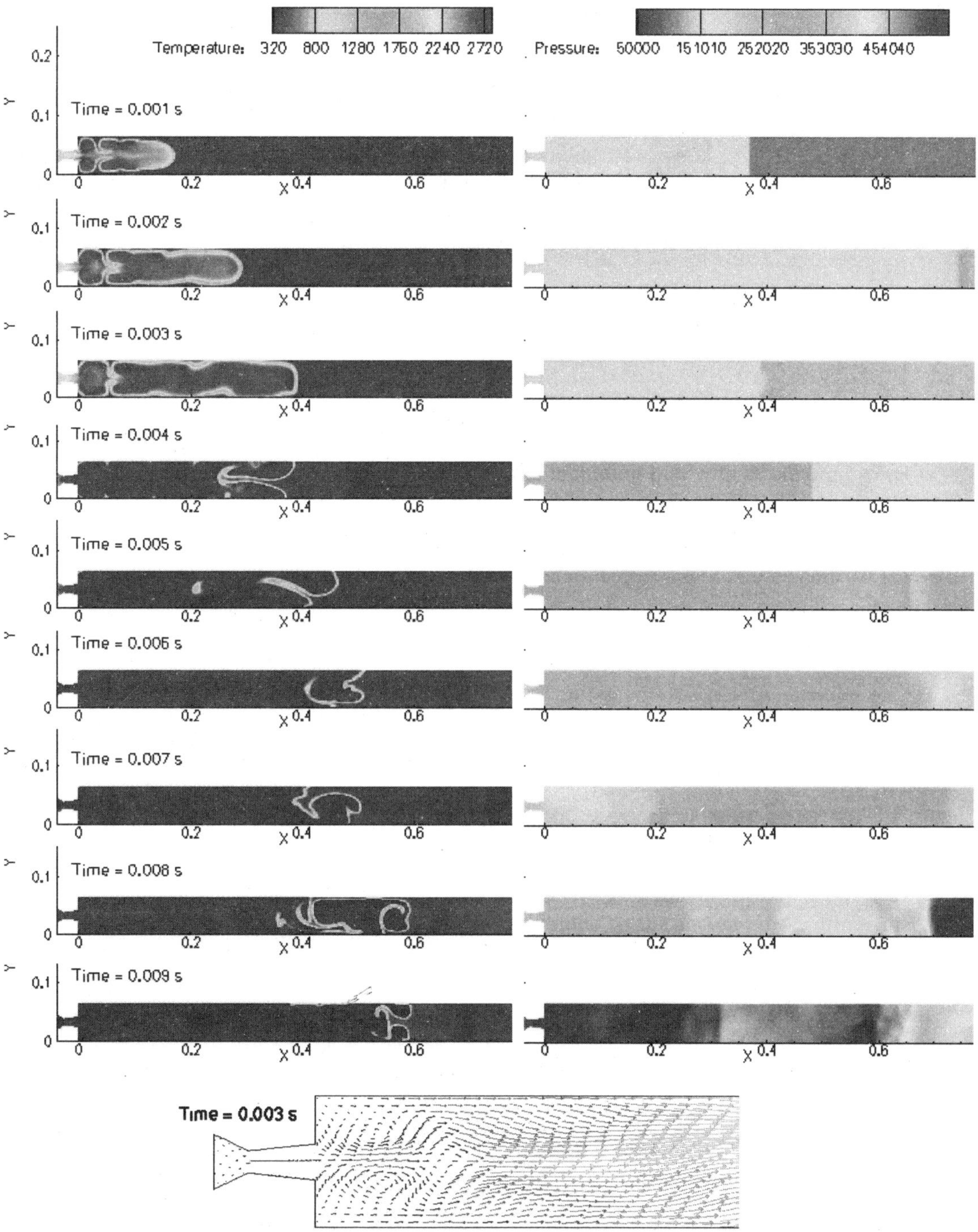


**Figure 5** Temperature (K) and pressure (Pa) history, and ignition jet flow field in the non-rotating channel

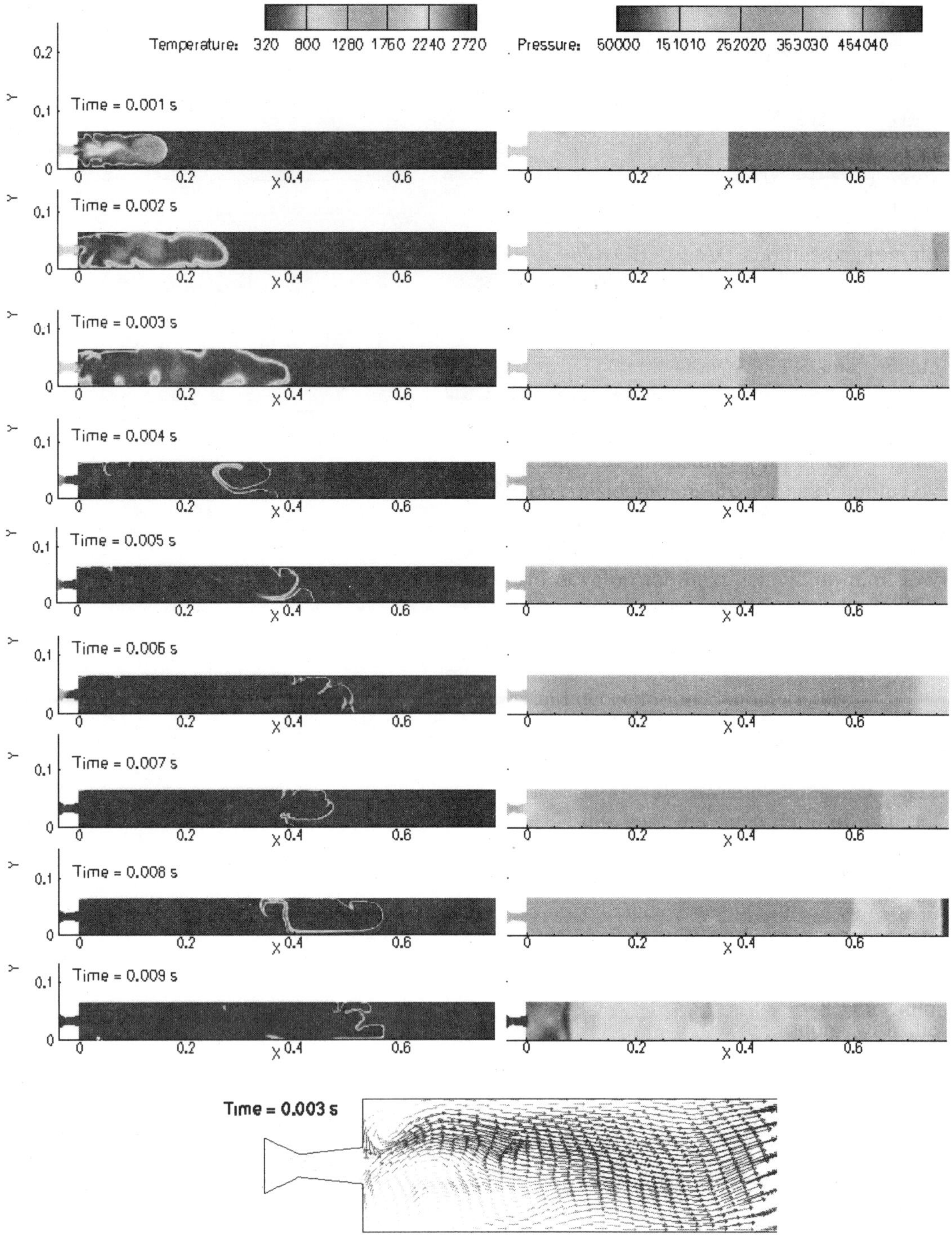


**Figure 6** Temperature (K) and pressure (Pa) history, and ignition jet flow field in the rotating channel

In the case of no rotation as illustrated in Fig. 5, it is observed that the flame developed almost axisymmetrically until it collided with a reflected shock wave. RM instabilities then deformed the flame, increasing the flame surface area. The overall reaction rate in the channel depends on unburned gas temperature and flame surface. The increased flame surface area following the shock resulted in increased overall reaction rate, as shown by Fig. 7. The process of shock-flame interaction was repeated again with a successively reflected shock at approximately 6.6 milliseconds. Overall, the shock–flame interactions have been observed to produce a highly corrugated flame surface, which leads to a significant increase in the overall rate of reaction as each successive pressure wave that interacts with the flame. There is a lag between flame surface increase and the corresponding growth in reaction rate (see Fig. 7).

The flame dynamics early in the simulation (0 – 3 ms) may be characterized through examination of the ratio of inertial forces to buoyancy forces acting on the jet. It may be seen from Fig. 6 that buoyancy forces affected the jet shape only slightly. Soon after ignition the flame occupies the full cross-sectional area of the channel and the resulting overall combustion rate was similar to the non-rotating case (Fig. 7). At the later stages of the combustion process, jet momentum dissipates, and shock–flame interaction and resulting RM instability dominated flame dynamics.

Comparison of combustion in the rotating and non-rotating channel revealed that the effect of the rotation-produced buoyancy forces is not as significant as other effects. Jet-induced motion that is shown at the bottom of Fig. 6 along with shock–flame interactions prevented any persistent radial stratification. The jet-produced motion and pressure waves appear to have greater impact on the flame dynamics than buoyancy forces. This can be explained by the fact that the fully developed pressure gradient due to rotation produced only ~3.7 kPa pressure difference across the channel, versus ~100 kPa gage pressure difference across shock waves from the jet ignition. The former might be too small to produce noticeable effects in the relatively short time necessary for flame

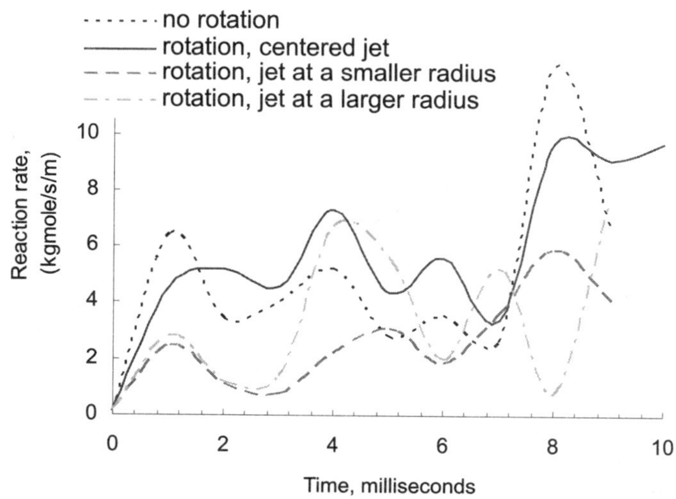


**Figure 7** Overall reaction rates as functions of time

propagation. In a practical wave rotor, even stronger shock waves may be present from the opening and closing of main flow ports.

## Conclusion

The wave rotor combustor is a non-steady pressure-gain combustion device where the combustion process occurs within the rotor channels, creating deflagrative or detonative combustion modes. It is expected to offer significant fuel savings and greenhouse gas emission reduction for gas turbines. A brief review of prior and recent work reveals that, while the only sustained effort was technically successful, important challenges remain.

Further careful experimental efforts are needed on well-instrumented test rigs for adequate understanding of the basic combustion processes. Continued research on sealing and thermal expansion control are also needed to ensure practically useful operation. Computer simulation has been an important tool in predicting performance and designing experiments. Simulations with gradually increasing complexity, from thermodynamic and aerothermodynamic models, to quasi-one-dimensional and eventually multi-dimensional models have allowed preliminary optimization and design.

The effect of rotation and pressure waves on the overall flame propagation rate has been studied in the rotating channel under the conditions close to design conditions of an experimental rig. It was found that the effect of rotation on the overall fuel consumption rate is minor when ignition is provided with a hot gas jet. Gas motion produced during ignition, along with shock waves, had a stronger impact of the flame dynamics than buoyancy forces. The main increase in the overall reactions rate is a result of flame-shock interactions. The larger flame surface area produced by Richtmyer-Meshkov instabilities and temperature increase due to the shock compression resulted in a significant increase of the global reaction rate. This phenomenon must be confirmed by experiments, and the overall combustion rate must be verified.

If a high rate of combustion and a sound mechanical and thermal design is achieved, the wave rotor combustor will be an attractive candidate for revolutionary redesign of gas turbine engines to reduce fuel consumption and greenhouse gas emissions significantly.

## References

1. Kentfield, J. A. C., *Nonsteady, One-Dimensional, Internal, Compressible Flows*, Oxford University Press, Oxford, 1993.
2. Roy, G. D., Frolov, S. M., Borisov, A. A., and Netzar, D. W., "Pulse Detonation Propulsion: Challenges, Current Status, and Future Perspective," *Progress in Energy and Combustion Science*, Vol. 30, No. 6, 2004, pp. 545-672.
3. Kentfield, J. A. C., "On the Feasibility of Gas-Turbine Pressure-Gain Combustors" *International Journal of Turbo and Jet Engines*, Vol. 12, No. 1, 1995, pp. 29-36.
4. Nalim, M. R., "Thermodynamic Limits of Work and Pressure Gain in Combustion and Evaporation Processes," *Journal of Propulsion and Power*, Vol. 18, No. 6, 2002, pp. 1176-1182.
5. Heiser, W. H., and Pratt, D., "Thermodynamic Cycle Analysis of Pulse Detonation Engines," *Journal of Propulsion and Power*, Vol. 18, No. 1, 2002, pp. 68-76.

6. Kailasanath, K., "Recent Developments in the Research on Pulse Detonation Engines," *AIAA Journal*, Vol. 41, No. 2, 2003, pp. 145-159.
7. Wu, Y., Ma, F., and Yang., V, "System Performance and Thermodynamic Cycle Analysis of Air-Breathing Pulse Detonation Engines," *Journal of Propulsion and Power*, Vol. 19, No.4, 2003, pp. 556-567.
8. Paxson, D. E., "Performance Evaluation Method for Ideal Airbreathing Pulse Detonation Engines," *Journal of Propulsion and Power*, Vol. 20, No. 5, 2004, pp. 945-947.
9. Xia, G., Li, D., and Merkle, C., "Modeling of Pulsed Detonation Tubes in Turbine Systems," AIAA Paper 2005-225, 2005.
10. Baptista, A., Rasheed, A., Badding, B., Velagandula, O., and Dean, A. J., "Mechanical Response in a Multi-Tube Pulsed Detonation Combustor-Turbine Hybrid System," AIAA Paper 2006-1234, 2006.
11. Glaser, A. J., Caldwell, N., and Gutmark, E., "Performance Measurements of a Pulse Detonation Combustor Array Integrated with an axial Flow Turbine," AIAA Paper 2006-1232, 2006.
12. Nalim, M. R., and Paxson, D. E., "A Numerical Investigation of Premixed Combustion in Wave Rotors," *Journal of Engineering for Gas Turbines and Power*, Vol. 119, No. 3, 1997, pp. 668-675.
13. Shreeve, R. P., and Mathur, A., *Proceeding ONR/NAVAIR Wave Rotor Research and Technology Workshop*, Report NPS-67-85-008, Naval Postgraduate School, Monterey, CA, 1985.
14. Akbari, P., Nalim, M. R., and Müller, N., "A Review of Wave Rotor Technology and Its Applications," *Journal of Engineering for Gas Turbines and Power*, Vol. 128, No. 4, 2006, pp. 717-735.
15. Welch, G. E., Jones, S. M., and Paxson, D. E., "Wave Rotor-Enhanced Gas Turbine Engines," *Journal of Engineering for Gas Turbines and Power*, Vol. 119, No. 2, 1997, pp. 469-477.
16. Wilson, J., Welch, G. E., and Paxson, D. E., "Experimental Results of Performance Tests on a Four-Port Wave Rotor," AIAA Paper 2007-1250, 2007.
17. Nalim, M. R., and Pekkan, K., "Internal Combustion Wave Rotors for Gas Turbine Engine Enhancement," ASME Paper IGTC-2003-FR-303, 2003.
18. Nalim, M. R., and Pekkan, K., "A Review of Rotary Pressure-Gain Combustion Systems for Gas Turbine Applications," ASME Paper GT-2003-38349, 2003.
19. Anonymous, "A Pressure-Wave Machine with Integrated Constant-Volume Combustion," NEFF Funding of Swiss Energy Research 1977-1997, Project No. 426, 1997, pp. 142-153.
20. Nalim, M. R., "Assessment of Combustion Modes for Internal Combustion Wave Rotors," *Journal of Engineering for Gas Turbines and Power*, Vol. 121, No. 2, 1999, pp. 265-271
21. Nalim, M. R., "Longitudinally Stratified Combustion in Wave Rotors," *Journal of Propulsion and Power*, Vol. 16, No. 6, 2000, pp. 1060-1068.
22. Snyder, P. H., Alparslan, B., and Nalim, M. R., "Gas Dynamic Analysis of the Constant Volume Combustor, A Novel Detonation Cycle," AIAA Paper 2002-4069, 2002.
23. Keller, J. J., "Method for Preparing the Working Gas in a Gas Turbine Installation," United States Patent No. 5,197,276, 1993.
24. Bilgin, M., Keller, J. J., and Breidenthal, R. E., "Ignition and Flame Propagation Process with Rotating Hot Jets in a Simulated Wave Engine Test Cell," AIAA Paper 98-3399, 1998.
25. Akbari, P., Baronia, D., and Nalim, M. R., "Single-Tube Study of a Semi-Intermittent Pressure-Gain Combustor," ASME Paper GT2006-91061, 2006.
26. Lewis, B. R., "Engine Having a Rotor with a Plurality of Circumferentially – Spaced Combustion Chambers," United States Patent No. 2,705,867, 1955.
27. Weber, H. E., *Shock Wave Engine Design*, John Wiley and Sons, New York, 1995.

28. Keller, J. J., "Pressure Wave Machine with Constant-Volume Combustion as a Superstage for Gas Turbines," ABB Technical Report CRBT 90-20, 1990.
29. Walraven, F., "Operational Behavior of a Pressure Wave Machine with Constant Volume Combustion," ABB Technical Report CHCRC 94-10, 1994.
30. Paxson, D. E., "Comparison Between Numerically Modeled and Experimentally Measured Wave-Rotor Loss Mechanism," *Journal of Propulsion and Power*, Vol. 11, No. 5, 1995, pp. 908-914.
31. Akbari, P., Nalim, M. R., and Snyder, P. H., "Numerical Simulation and Design of a Combustion Wave Rotor for Deflagrative and Detonative Propagation," AIAA Paper 2006-5134, 2006.
32. Snyder, P. H., Alparslan, B., and Nalim, M. R., "Wave Rotor Combustor Test Rig Preliminary Design," ASME Paper IMECE2004-61795, 2004.
33. Paxson, D. E., and Perkins, H. D., "Thermal Load Considerations for Detonative Combustion-Based Gas Turbine Engines," AIAA Paper 2004-3396, 2004.
34. Hendricks, R. C., Wilson, J., Wu, T., and Flower, R., "Two-Way Brush Seals Catch a Wave," *Mechanical Engineering*, Vol. 120, No. 11, 1998, pp. 78-80.
35. Akbari, P., Nalim, M. R., Donovan, E. C., and Snyder, P. H., "Leakage Assessment of Pressure-Exchange Wave Rotors," AIAA Paper 2006-4449, 2006.
36. Pekkan, K., and Nalim, M. R., "Two-Dimensional Flow and $NO_X$ Emissions in Deflagrative Internal Combustion Wave Rotor Configurations," *Journal of Engineering for Gas Turbines and Power*, Vol. 125, No. 3, 2003, pp. 720-733.
37. Kilchyk, V., Merkle, C., and Nalim, M. R., "Effect of Channel Rotation on the Turbulent Combustion Rates Attained in the Wave Rotor Combustor," AIAA Paper 2007-5053, 2007.
38. Dwyer, H., and Hasegawa, T., "Some Flows Associated with Premixed Laminar Flame Propagation in a Rotating Tube Flow", *Proceedings of the Combustion Institute*, Vol. 29, No. 2, 2002, pp. 1471–1477.
39. Markstein, G. H., *Nonsteady Flame Propagation*, MacMillan, NY, Chapter D, 1964.
40. Khokhlov, A, Oran, E. S., Chtchelkanova, A. Y., and Wheeler, J. C., "Interaction of a Shock with a Sinusoidally Perturbed Flame", *Combustion and Flame*, Vol. 117, No. 1, 1999 , pp. 99-116.
41. Samtaney, R., Ray, J., and Zabusky, N., "Baroclinic Circulation Generation on Shock Accelerated Slow/Fast Gas Interfaces", *Physics of Fluids*, Vol. 10, No. 5, 1998, pp. 1217-1230.

# Modeling and Dynamic Characterization of Thermal Pulse Combustors

**Achintya Mukhopadhyay**
Mechanical Engineering Department, Jadavpur University Kolkata - 700032

Pulse combustors are a class of air-breathing engines that undergo self-sustained or forced oscillations, which lead to low emission and higher heat transfer rates than steady combustors. The essential parts of a pulse combustor consist of the combustion chamber and the tailpipe. A nonlinear dynamic model of a thermal pulse combustor consisting of four coupled ordinary differential equations has been used to investigate the effect of sinusoidal perturbations in mass flow rate. The system of equations was solved using ODE45 function of MATLAB. The dynamics of the system is characterized using time series plots, phase plots, time delay phase plots, Poincare maps and bifurcation diagrams. The results suggest that in absence of external forcing, the system undergoes transition from periodic state to chaotic state through a series of period-doubling bifurcations. The dynamic data is utilized for early prediction of system extinction. Different strategies are developed and explored for this purpose. All the techniques are found to be successful in early prediction of system extinction. Results are also presented for the response of the system to sinusoidal perturbations in inlet mass flow rate. The response is sensitive to both frequency and amplitude of the perturbation. The extinction is accelerated in presence of a sinusoidal perturbation in inlet mass flowrate.

# Controlled Atomization: Needs and Challenges

**Abhijit Kushari**
Department of Aerospace Engineering, Indian Institute of Technology, Kanpur- 208016, India

## 1. INTRODUCTION

The transformation or break up of bulk liquid fuel into a spray is of great importance in liquid fuel combustion that is used in various propulsion systems and industrial processes. The spray, comprising of a multitude of droplets, provides a much larger surface area compared to the bulk liquid, thus greatly enhancing the liquid evaporation rate. Numerous spraying devices, which operate on different principles and are broadly designated as atomizers or injectors, have been developed over the years.

The combustion of liquid fuels involves atomization of a bulk liquid into a multitude of droplets, thus increasing the liquid surface area and enhancing the rate of evaporation of liquid fuel. The formed fuel vapor then mixes with the air to form a combustible mixer. Therefore, the evaporation time, along with the mixing time and chemical reaction time, play a very important role in the performance of liquid-fueled combustion devices, which, in turn, depends on the sizes of the generated droplets.

The quality of combustion processes, e.g., ignition, flame stability, efficiency, combustion stability and emission levels, in liquid fueled engines strongly depend on the quality of the fuel spray [1-5]. There is evidence [1-2] that the performance of the combustion process under a given operating condition may be optimized by generating a spray with a specific mean droplet size, which may not be the minimum size that could be generated by the atomizer. Fine atomization causes substantial reduction in the production of unburned hydrocarbons because it reduces the evaporation time of the droplets and, thus, improves mixing and increases the time available to attain complete combustion [2]. Complete combustion caused by finer atomization leads to a decrease in soot formation, which is problematic in engines employing conventional pressure-swirl atomizers at off-design operating condition [1]. Finer atomization also improves the combustion efficiency by reducing the evaporation time [1]. On the other hand, the reduction in mean droplet size increases the emissions of $NO_x$. This can be attributed to the reduction in evaporation time and consequent increase in the time spent by the combustion products in the hot combustion zone, thus increasing the time available for NO production reactions to proceed towards equilibrium [2].

Furthermore, finer atomization reduces the lean blow out limit of the engine because of the decrease in evaporation time [1]. Therefore, it could be possible to optimize the combustion processes and reduce emissions by using atomizers that could be controlled to produce sprays with optimal properties, e.g., droplet sizes and velocities, for required fuel flow rates.

The new challenges to be met by aircraft engine companies in the years to come are high thrust, low weight engines for fighter aircraft and low $NO_x$, high thrust, low weight engines for the advanced commercial aircraft [4]. Engines for future fighter aircraft are required to be half the size and generate twice the thrust compared to the current engines [4]. On the other hand, future commercial aircraft engines must provide for doubling the size of the aircraft and doubling the distance traveled (without landing), compared to the current aircraft [4]. In addition, the emission of $NO_x$ must be one-tenth of that produced by current commercial aircraft in order to minimize chemical reaction of $NO_x$ with ozone in the stratosphere [4]. There is a growing recognition that effective design of atomizers and control of their sprays can contribute towards meeting the performance and emission requirements for the future aircraft engines, as discussed in the previous paragraph.

There are many other issues in spray combustion that are directly related to the characteristics of the spray. For example, non-symmetrical spray flames and hot-streaks can cause serious damage to combustor liners and have serious impact on combustor exit temperature distributions (i.e., pattern factor) [4]. Therefore, flames must be kept well away from liner surfaces. Since the flame shape, flame location and pattern factor strongly depend on the characteristics of the fuel spray, controlling the spray properties has become an issue of major concern because of potential damages it can cause to the engine.

In conclusion, significant improvements in the performance of liquid fueled combustors can be achieved by having the ability to control the spray characteristics. Controlled atomization can lead to increase in combustion efficiency and reduction in emissions. However, till date, there does not exist an atomizer that can provide controlled sprays for the entire range of operating conditions of a gas turbine engine. With the advent of faster computers, which has made possible the modeling of the atomization process and the development of fast response actuators, it now seems possible to develop a controllable atomizer.

## 2. BRIEF DISCUSSION OF ATOMIZERS

Atomization is often accomplished by discharging the liquid at a high velocity into a relatively slow-moving stream of air or gas or by destabilizing the liquid column using mechanical or electrostatic devices. In this section, the general features of various types of atomizers are described along with a discussion on their ability to provide controlled atomization.

### 2.1 Pressure Atomizers

As their name suggests, pressure atomizers rely on the conversion of pressure into kinetic energy to achieve a high relative velocity between the liquid and the surrounding gas. Most of the atomizers in general use are of this type. They include plain-orifice and simplex nozzles, as well as variable-

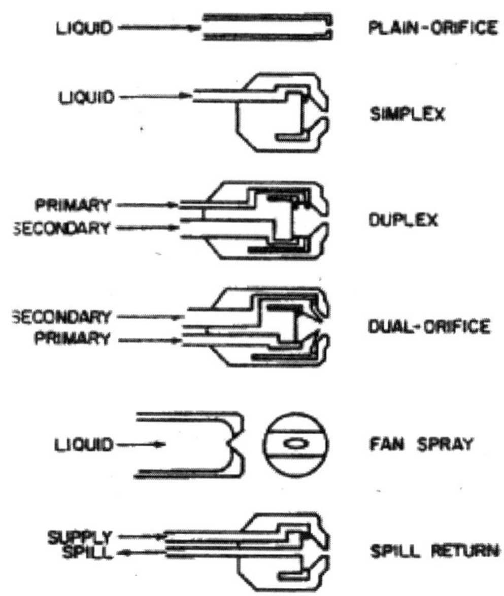


**Figure 1** Schematics of pressure atomizers

geometry, duplex and dual-orifice injectors [6-9]. Schematics of various types of pressure atomizers are presented in Fig. 1. Disintegration of liquid jet, emerging from pressure atomizers, is promoted by an increase in flow velocity, which enhances both the level of turbulence in the liquid jet and the aerodynamic drag forces exerted by the surrounding medium [6-9]. In simplex or pressure-swirl atomizers, a swirling motion is imparted to the liquid so that, under the action of centrifugal force, it spreads out in the form of a conical sheet as soon as it leaves the orifice. This sheet then breaks up into liquid droplets under the influence of the external and internal forces. Since the flow rates and droplet sizes obtained from pressure atomizers depend on the pressure drop across the injector, the quality of atomization is very poor at low flow rates, causing very unsatisfactory combustion. The lower limit of this flow rate depends on the atomizer design [6-9]. At high flow rates the operating pressures are very high, e.g., in the range of 1500 psi. Thus, poor atomization at low flow rates and high operating pressures at high flow rates bound the range of operation of pressure atomizers. Furthermore, since the fuel flow rates and droplet sizes are directly related to each other through the operating pressure, these two parameters cannot be controlled independent of each other.

## 2.2 Rotary Atomizers

In the rotary atomizers [6-9], liquid is supplied through a stationary tube to the inner part of a rotating surface, as shown in the schematic presented in Fig. 2. The friction between the liquid and the rotating surface causes the liquid to rotate with the surface and the centrifugal force causes it to flow towards the periphery. Under the action of this centrifugal force, the liquid leaves the edge of the surface with a high velocity, producing a fine spray. In rotary atomizers, even high flow rates of

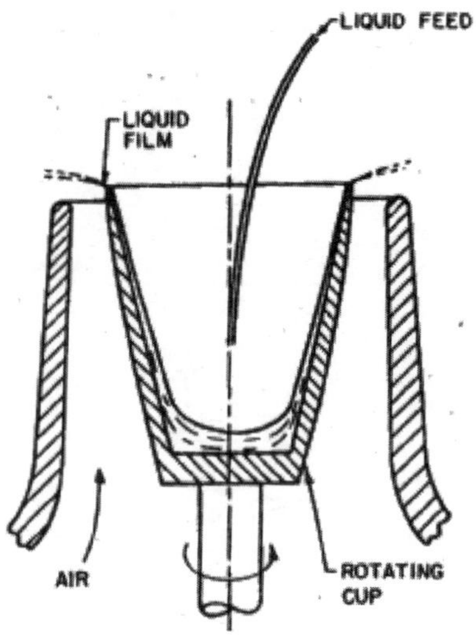


Figure 2 A schematic of a rotary atomizer

liquid can be atomized with relatively low pressure. They can be cleaned readily and display minimal flow blockage. However, the design is complex, including a rotating element with accompanying transmission and lubrication systems. They require high rotational speed of the surface to achieve good quality of atomization. Such high speeds reduce the reliability and require appropriate protections.

## 2.3 Airblast Atomizers

The drive to improve the quality of generated sprays has resulted in an increase in the use of air blast atòmizers [6-10] in modern engines. In such injectors, atomization is induced by blasting of liquid sheets and ligaments by slow moving, high-pressure air from the compressor. However, due to the small pressure drop between the compressor discharge and the combustor, the kinetic energy of the blasting air is limited. Therefore, they require large quantities of air (typically 500 to 600 percent of the liquid mass) to attain satisfactory atomization. But, this additional air is not wasted, because after atomizing the fuel it conveys the drops into the combustion zone and also supplies a fraction of the primary combustion air. A schematic of an air blast atomizer is presented in Fig. 3.

Airblast atomizers have many advantages over pressure atomizers. They require lower fuel pump pressures and produce finer sprays. The atomization process ensures very good mixing of air and fuel, thus, reducing soot formation. The main disadvantage of airblast atomizers is that the large quantities of atomizing air required by these injectors are not available during start up of the engine, idling and high altitude re-light, resulting in poor sprays and combustion processes at those off design operating conditions. Furthermore, the "drowning" and "blasting" of the liquid jet by huge quantities of air leaves very little room for any kind of control of the atomization process [6-9].

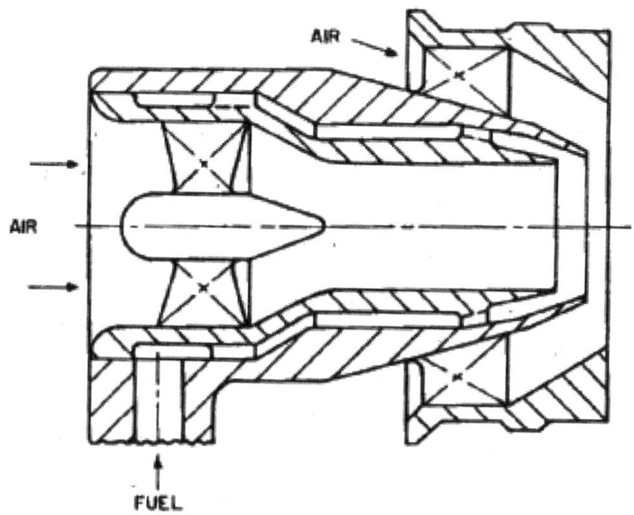


**Figure 3** A schematic of an air-blast atomizer

## 2.4 Effervescent Atomizers

In effervescent atomizers [5, 7, 11-19] a small amount of air or gas is introduced into the bulk liquid in a mixing chamber upstream of the discharge orifice, as shown in Fig. 4. The injected gas forms a bubbly two-phase, gas-liquid flow when it mixes with the liquid in the mixing chamber inside the injector. When the gas bubbles emerge from the injector they "explode". This rapid expansion of gas bubbles shatters the surrounding liquid ligaments, resulting in the formation of very fine droplets [7]. The method of operation of effervescent atomizers is similar to that exhibited by "flashing" atomizers [20], where the atomization is caused by the explosion of cavitation bubbles. Quality of atomization obtained from effervescent atomizers is very good even at very low injection pressures. They have large holes and passages and can thus atomize "dirty" fuels without plugging of the injectors. These injectors are reliable, simple and cheap. However, effervescent atomizers are sensitive to acceleration owing to the sensitivity of air bubbles to acceleration. Consequently, they cannot be safely employed in airborne applications.

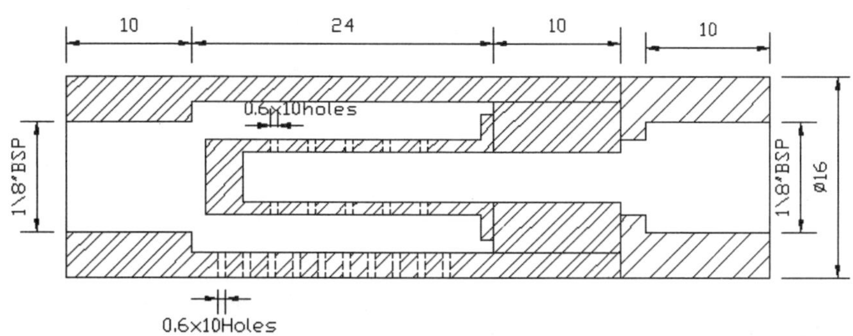


**Figure 4** A schematic of an effervescent atomizer

## 2.5 Advanced Atomizers

In order to rectify the flaws of the conventional atomizers, many advanced concepts of atomization were used in various process industries over the years. These atomizers, e.g., ultrasonic atomizers [21-26], acoustic atomizers [27-28], induced atomizers [29], electrostatic atomizers [30-31], use different physical mechanisms to produce liquid sprays. For example, ultrasonic atomizers use capillary break up of liquid sheets vibrating at ultrasonic frequencies to produce fine liquid spray [21]. In acoustic atomizers, atomization is achieved by de-stabilizing the liquid column in an acoustic field [27-28]. In electrostatic atomizers, the bulk liquid is charged and the droplets are believed to be produced by the instabilities induced by the repulsion of charged molecules [30]. These advanced atomization concepts provide very good control of the atomization process and they produce very fine sprays with mono-sized droplets. However, all of them suffer from the same drawbacks. They need additional supply of electrical energy and additional electrical components to generate the actions responsible for atomization and the flow rate over which they can operate is severely restricted. They cannot handle the high fuel flow rates required in gas turbine applications, which limits their use to special low flow rate applications such as drug delivery [24], ink-jet printers [26] and spray painting.

## 2.6 Internally Mixed Air-Assisted Atomizers

In internally mixed, air-assisted, atomizers the atomizing air interacts with the liquid inside the injector and assists in the atomization process [32-43]. It is believed that two effects induce atomization in an internally mixed injector. First, as both the liquid and the atomizing air share the same flow passage inside the injector, the liquid is restricted to a smaller available flow area. The reduction in flow area accelerates the liquid and, thus, increases its kinetic energy. This increase in the kinetic energy of liquid induces fine atomization. Second, the relative motion between gaseous and liquid phases produces a shear force at their interface. This force strips liquid droplets from the liquid filaments and thus, induces atomization. Since most of the atomizing energy is supplied by the atomizing air, this injector requires a very small pressure drop of liquid across the injector to produce a fine spray, in contrast to the large pressure drop required in pressure atomizers. Furthermore, internally mixed atomizers need a significantly lower flow rate of atomizing air compared to airblast atomizers [9]. An atomizing air flow rate of only 10 – 20 percent of the liquid flow rate (by mass) can produce very fine atomization in internally mixed atomizers. Therefore, the internally mixed injector appears attractive for airborne gas turbine applications, as it requires a small amount of atomizing air to produce fine atomization with only a small liquid pressure drop across the injector. The disadvantage of this injector is that it requires additional compression of the atomizing air.

Conceptually, internally mixed atomizers are similar to effervescent atomizers. However, instead of forming a bubbly air-liquid mixture inside the injector, a low flow rate of pressurized air is impinged upon the liquid, a short distance upstream of the injector's exit in an internally mixed injector. Since the interactions between air and liquid occur over a very short distance, eliminating the possibilities of bubble formation [13], the resulting atomization process is expected to be less sensitive to acceleration.

Another potential advantage of internally mixed injector is choking of the two-phase flow of liquid and air as it passes through the injector, due to the low sonic velocity of two-phase liquid-air mixture [35, 36]. Therefore, the liquid flow rate is relatively insensitive to the variations in the combustion chamber pressure and, thus, the fuel flow rate is not likely to respond to combustor disturbances, which will reduce the likelihood of combustion instabilities by reducing the coupling of combustor pressure and fuel flow oscillations.

## 3. CONTROLLED ATOMIZATION USING INTERNALLY MIXED ATOMIZERS

As has been discussed in the previous section, internally mixed atomizers can be used to control the atomization quality. Some examples of such control are discussed in this section.

### 3.1 A simple Internally Mixed Air-Assisted Atomizer

A cross-sectional view of a simple Internally mixed atomizer [40] is shown in Fig. 5. Liquid enters through the wide opening of a tube that narrows down into a long tube. Air is radially injected into the liquid flow through multiple holes just upstream of the injector's exit. Figures 6 and 7 describe the variations in the spray characteristics when the water supply pressure was kept constant at 20 psi (137.41 kPa) while the amount of atomizing air was varied. It should be noted that although the water supply pressure was kept constant in these experiments, the flow rate of the water was not constant. In fact the water flow rate decreases as the air flow rate increases, which is evident from the nonlinear relationship between the Air Liquid Ratio (ALR) and the air flow rate in Figs. 6 and 7.

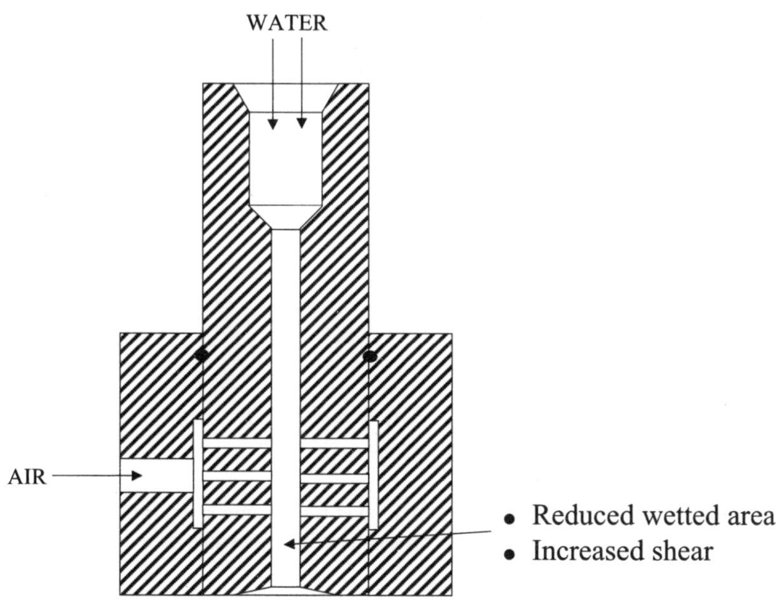


**Figure 5** Schematic of a simple Internally Mixed Air-Assisted Atomizer

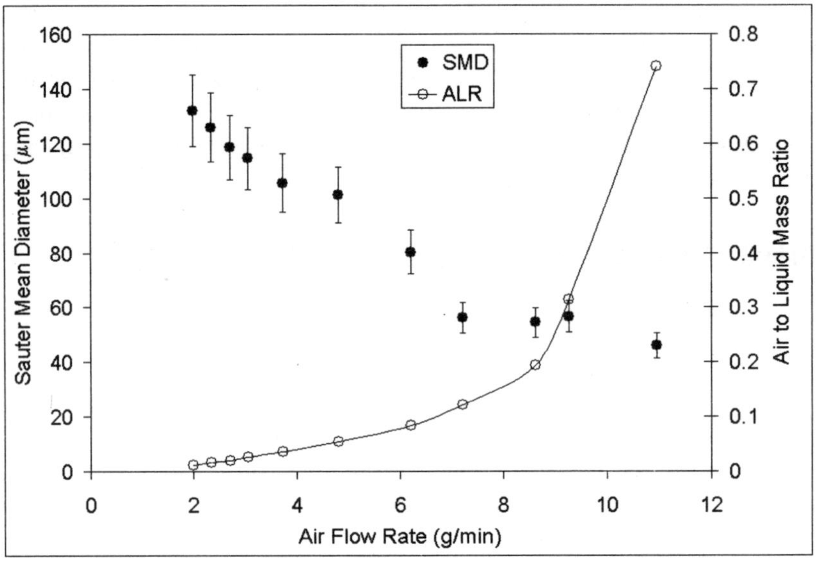


**Figure 6** Dependence of droplet SMD and ALR upon the air flow rate, for a constant water supply pressure of 20 psi (137.41 kPa) and atmospheric back-pressure, at the center of the spray

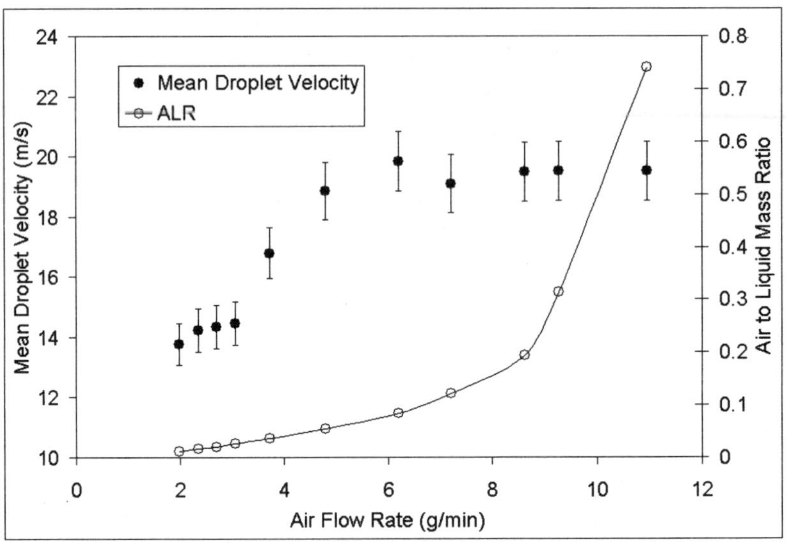


**Figure 7** Dependence of mean droplet velocity and ALR upon the variations in air flow rate, at a constant water supply pressure of 20 psi (137.41 kPa) and atmospheric back pressure, at the center of the spray

This occurs because the pressure inside the atomizer increases as the air flow rate increases, thus reducing the water flow rate through the injector. Consequently, when the flow rate of air was varied from 1.998 to 10.96 g/min, the ALR increased from 0.0123 to 0.741, which corresponds to a decrease from 162.8 g/min to 14.8 g/min in the water flow rate.

Figure 6 shows that the Sauter Mean Diameter (SMD) of the spray droplets decreases from 132 to 45 μm as the air flow rate increases from 1.998 to 10.96 g/min. It is believed that atomization in the investigated injector is induced by two effects. First, the increase in air flow-rate is accompanied by increased restriction on the water passage, thus accelerating the water flow. This, in turn, increases the kinetic energy of the water, resulting in improved atomization. Second, the increase in air flow-rate is accompanied by an increase of the air velocity and, thus, the shear force that it exerts upon the water. Consequently, the number of droplets being stripped from the water filaments increases, resulting in finer atomization. The acceleration of the water due to the introduction of air is evident in Fig. 7. It shows that the mean droplet velocity at the center of the spray, 2.5 inches (6.35 cm) downstream of the injector exit, increased from 13.5 m/sec to 20 m/sec as ALR increases from 1.2 to 10 percent. Correlating the ALR with the SMD and mean droplet velocity in Figs. 7 and 8 reveals that significant variations in the spray's SMD and velocity occur as the ALR increases from 1 to 10 percent. However, further increase in the ALR (up to 74 percent) has almost no effect on the spray's SMD and velocity. Thus, the data in Figs. 6 and 7 suggest that when the investigated injector is operated with an atmospheric back pressure, its ALR need not exceed 10 percent. While the reasons for the limited effect of increased ALR (beyond 10 percent) upon the droplet SMD and average velocity remain unknown, it is speculated that it is related to the different characteristics of the two phase flow within the injector at low and high ALR.

A performance map for the investigated injector was obtained by repeating the above described tests with water supply pressures between 10 psi (68.7 kPa) and 60 psi (412.24 kPa) and the results are presented in Fig. 8. For low water flow rates, the lines in Fig.8 "coalesce" around 40 μm, thus suggesting that this is the smallest obtainable SMD with the investigated injector. It is noteworthy that when hot (i.e., low density) air was supplied to the injector directly from a compressor rather than room temperature air from an accumulator, an SMD of 15 μm was easily obtained.

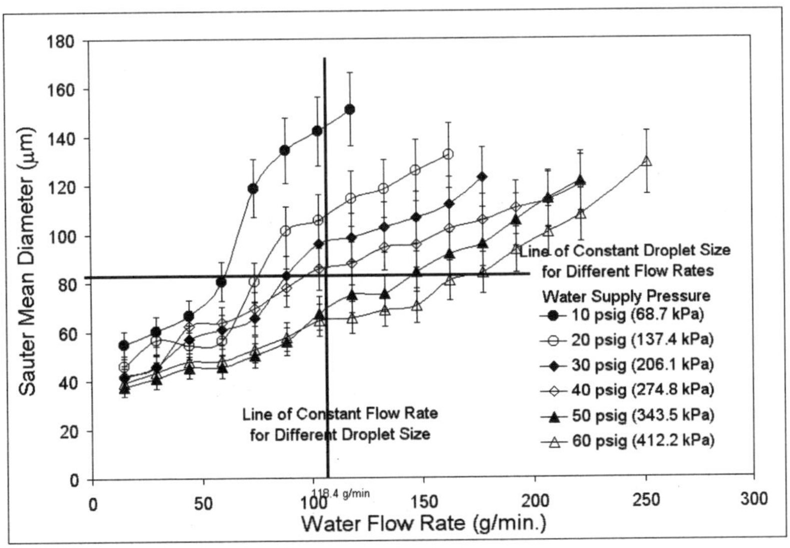


**Figure 8** Performance map of the injector showing the dependence of centerline SMD upon water flow rate at different water supply pressures and for zero back pressure. Water flow rate was varied by changing the amount of atomizing air

The data in Fig. 8 indicate that in contrast to the behavior exhibited by pressure and air blast atomizers, the investigated injector can atomize very small water flow rates into very fine sprays. This observation suggests that the investigated atomizer can be effectively used to obtain good atomization during engine start up and idling.

The performance map in Fig. 8 shows that varying the water supply pressure while keeping the water flow rate fixed (which requires varying the air flow rate as shown in Fig. 3), provides means for controlling the spray's SMD. This can be achieved by moving along a vertical line in Fig. 8. Similarly, changing the water supply pressure along, e.g., the horizontal line in Fig. 8 provides means for changing the water flow rate while keeping the spray SMD constant.

## 3.2 An Internally Mixed Twin-Fluid Swirl Atomizer

Although a simple internally mixed atomizer can provide control over the droplet size and liquid flow rate, the spray produced by such atomizers are always solid cone sprays with no control over the spray angle. It should be pointed out that some applications of liquid atomization require formation of hollow cone sprays due to their large coverage areas. But, some applications require solid cone sprays with allows for local droplet injection. However, most of the commercial atomizers can provide either a hollow cone spray or a solid cone spray but not both. The internally mixed twin-fluid swirl atomizer features the combined phenomena of both conventional swirl atomization as well as an internally mixed air assisted atomization. It uses the tangential momentum provided by the swirl as well as the additional axial momentum provided by the assisting air to produce droplets. Therefore this atomizer is able to provide both a hollow cone spray and a solid cone spray. Due to this feature, this atomizer can find applications in various processes without any design change.

The schematic cross-sectional view of the twin-fluid swirl atomizer is shown in Fig. 9. This internally mixed, air assisted, swirl atomizer, consists of a liquid inlet port, air inlet holes, air settling chamber, a helical passage, a spin chamber and an atomizer orifice. The liquid is supplied to the atomizer from the liquid inlet port. A small amount of air is introduced into the liquid stream through multiple radial holes on the circular wall of the tube in which the liquid flows. Before interacting with the liquid the air is allowed to settle in the settling chamber. This chamber ascertains the uniform distribution of air through the holes in the circular tube. The pressure in this chamber is the total air pressure, which is kept at a slightly higher value compared to static liquid pressure in order to avoid flow of liquid in the air passage. The air coming out of the air inlet holes interacts with the liquid and creates a two-phase air liquid mixture, which flows through the helical passage with a double threaded acme screw element fitting. This passage imparts the tangential component to the flow velocity. The rotating flow then passes through a swirling chamber. This swirling two- phase flow finally comes out of the orifice of the atomizer at high velocity and spray characteristics are obtained depending upon the ALR and the other flow conditions maintained.

Arguably, there are two components governing the spray evolution characteristics at a particular operating condition in the presented atomizer. They are the tangential or the swirling effect and the axial momentum component of the flow. Therefore, the flow of the two-phase air-liquid mixture through the atomizer is a two-dimensional two-phase flow, unlike the one-

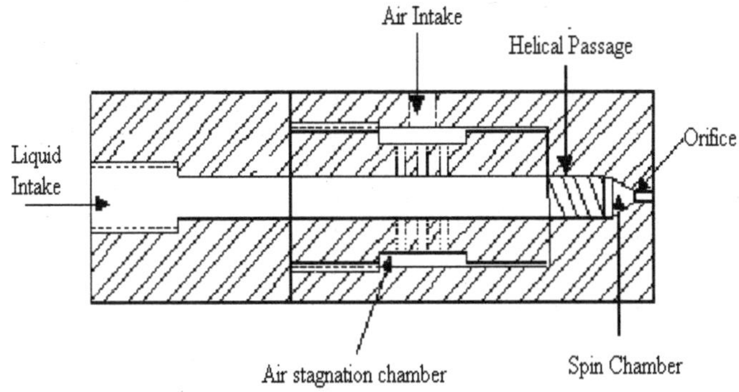


**Figure 9** Schematic cross sectional view of a twin-fluid swirl atomizer

dimensional two-phase flow exhibited by conventional air-assisted atomizers. At low ALR, the tangential component of the liquid flow velocity or the swirling effect dominated the flow, resulting in spreading of the cone as the spray comes out of the atomizer resulting in a hollow cone at low ALR. At high ALR, a solid cone shape is observed.

A performance map for the investigated injector vis-à-vis the spray solidity and liquid flow rate is presented in Fig. 10 for various liquid supply pressures. The data in Fig. 10 indicate that in contrast to the behavior exhibited by pressure atomizers, the investigated injector can atomize very small liquid flow rates into very fine sprays (as seen in the inserted pictures). This observation suggests that the investigated atomizer can be effectively used to obtain good atomization during engine start up and idling.

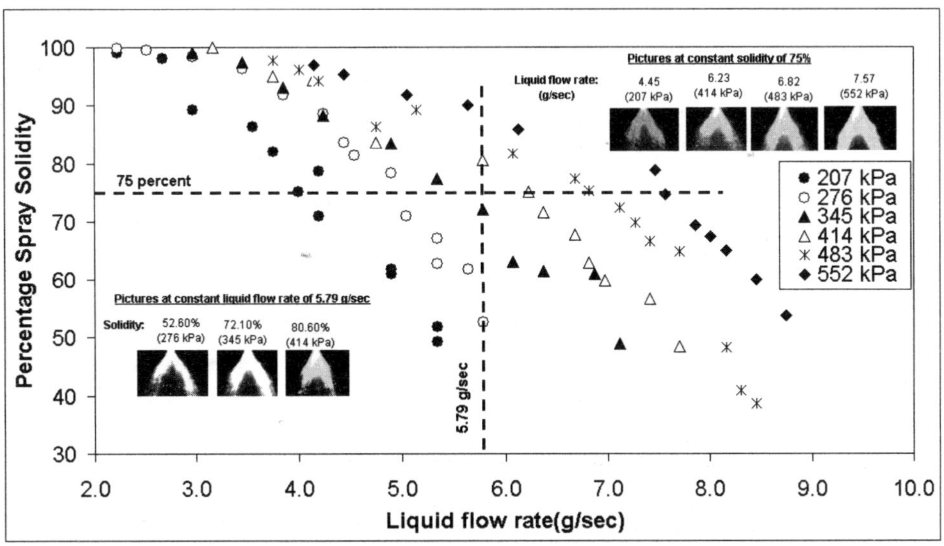


**Figure 10** Solidity versus Liquid flow rate relationship for different liquid supply pressures. The inserted pictures represent various controlled spray conditions

The performance map in Fig. 10 shows that varying the liquid supply pressure while keeping the liquid flow rate constant provides means for controlling the spray's solidity. This can be achieved by moving along a vertical line in Fig. 10. Similarly, changing the liquid supply pressure along the horizontal line in Fig. 10 provides means for changing the liquid flow rate while keeping the spray solidity constant. The inserted pictures in Fig. 10 demonstrate the spray solidity variation at constant liquid flow rate of 5.79 grams per second and the liquid flow rate variation at constant solidity of 75 percent for different liquid supply pressures. The data in Fig. 10 show that in order to increase the liquid flow rate through the presented atomizer, the liquid supply pressure needs to be increased. However, that leads to a decrease in the spray solidity. Therefore, the ALR needs to be increased in order to compensate for this decrease in solidity. However, the required increase in pressure and ALR will lead to an overall increase in the total energy and momentum of the system, causing a spread in the cone angle of the spray.

Another performance map of the investigated atomizer depicting the variation in spray cone angle with the liquid flow rate is presented in Fig. 11. The data in Fig. 11 show that the spray cone angle can be maintained at a constant value and the liquid flow rate can be changed over a range by traversing along a horizontal line in the figure. The pictures inserted in the figure show that indeed the spray cone angle can be kept at 38° angle while the liquid flow rate can be varied from 2.97 g/sec to 4.55 g/sec. The increase in the liquid flow rate for a constant cone angle can be accomplished by simultaneous increase in liquid supply pressure and ALR with a corresponding increase in the spray solidity. Similarly, by traversing along a vertical line, the spray cone angle can be varied while keeping the flow rate constant. The inserted pictures in Fig. 11 show that the spray cone angle can be changed from 53.4° to 70.6° for a constant flow rate of 5.79 g/sec.

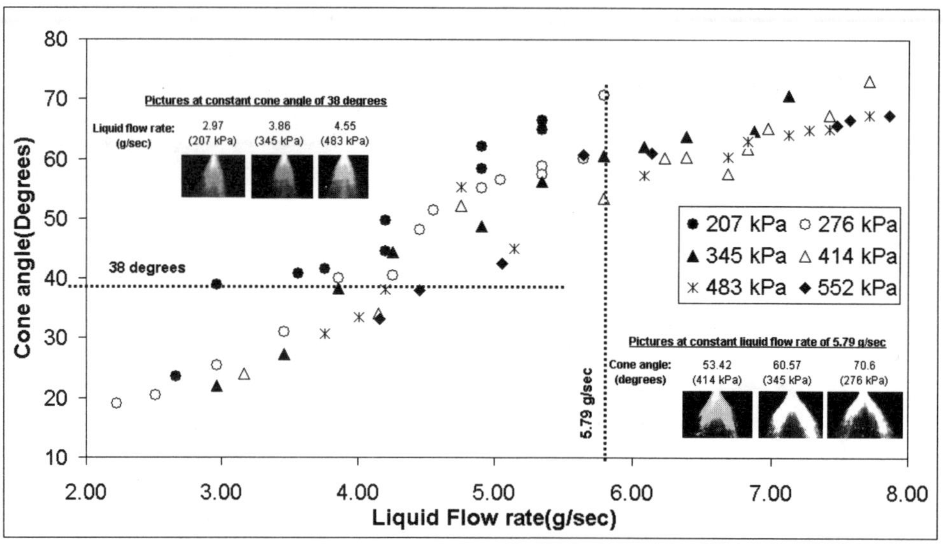


**Figure 11** Spray cone angle versus Liquid flow rate relationship for different liquid supply pressures. The inserted pictures represent various controlled spray conditions

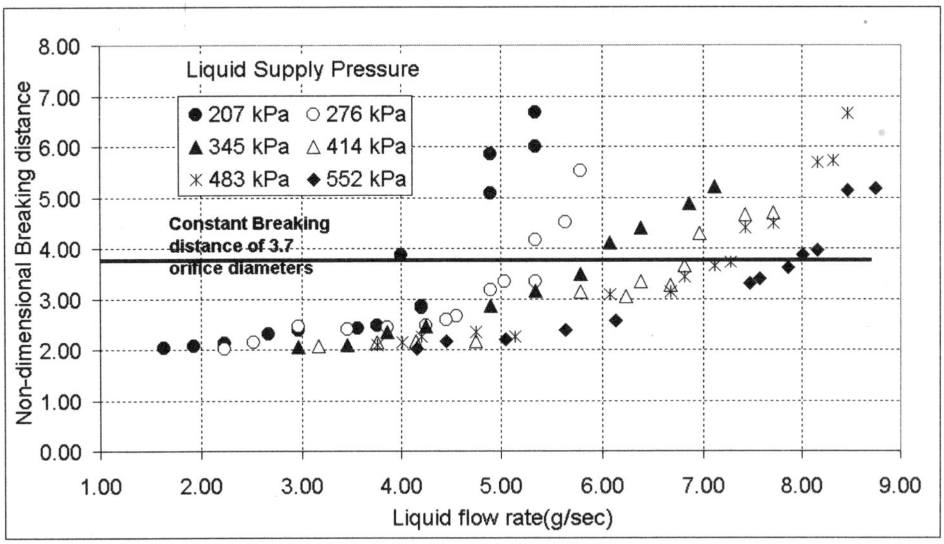


**Figure 12** Breaking distance versus liquid flow rate for different liquid supply pressures

The data in Fig. 12 shows that for a given supply pressure, the breaking distance increases with an increase in the liquid flow rate in a manner similar to that exhibited by conventional swirl atomizers. But that can be kept constant by simultaneously varying the ALR, and thus, a constant breaking distance can be maintained for different liquid flow rates, as exhibited by the horizontal line shown in Fig. 12. This property of the presented atomizer can be used by combustor designers to decide the location of the flame holding device without having to bother about the off-design variation in spray structure. However, it should be pointed out that a constant breaking distance may not guarantee constant flame holding, as the droplet size, spray cone angle and spray solidity all play significant roles in the final evaporation, mixing and combustion.

The variation in SMD with the liquid flow rate at different liquid supply pressures can be seen in Fig. 13. As in the case of other spray properties discussed earlier, the droplet size can also be changed in the present atomizer, independent of the liquid flow rate, by simultaneously varying the liquid supply pressure and ALR. The vertical line in Fig. 13 represents an SMD variation from 110 μm to 66 μm for a constant liquid flow rate of 5 g/sec. The horizontal line in Fig. 13 represents a constant SMD of 70 μm for which the liquid flow rate was varied from 3 g/sec to 8.75 g/sec.

The formation of spray with varied cone angles and solidity ranging from a hollow cone to a perfectly solid cone with an efficient controllability over the breaking distance and droplet size, independent of the liquid flow rate, provides this atomizer discussed with extraordinary flexibility of being efficiently used for different applications. Particularly, the controllability of the spray structure is the feature that adds uniqueness to this novel atomizer design and can make it an effective spray device for the future.

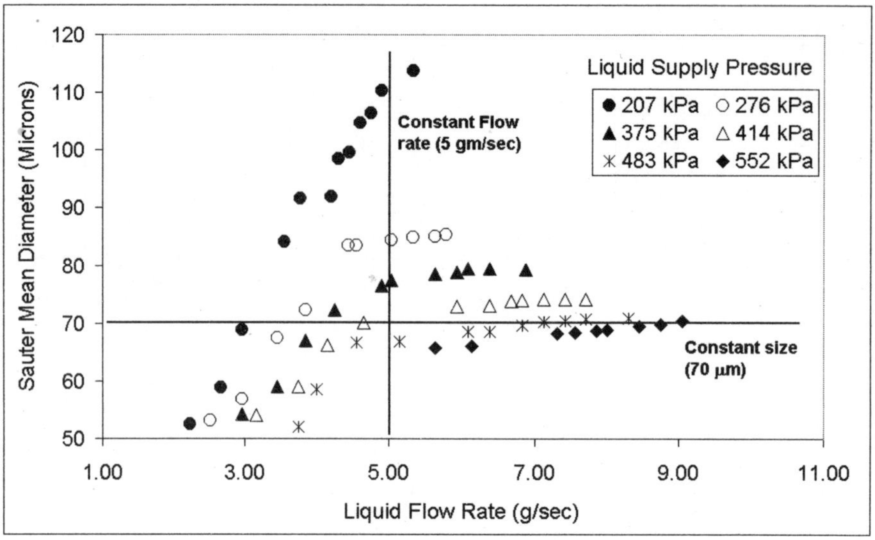


**Figure 13** Variation of spray SMD with the liquid flow rate for different liquid supply pressures

## Conclusions

Atomization plays a significant role in liquid combustion process and the ability to control the atomization process can improve the combustion efficiency and reduce emissions. However, the state of the art atomizers provide very little or no-control over this import process. An internally mixed air-assisted atomizer on the other hand can achieve this control by simultaneously varying the liquid and air supply pressures. However, a simple internally mixed atomizer provides limited control in terms of spray pattern and break distance. A twin-fluid swirl atomizer can control the spray cone angle, solidity, breaking distance as well as the droplet size independent of the liquid flow rate and hence, this type of atomizer can be used for various applications over a wide range of operating conditions with minimal design modifications.

## References

1. A. H. Lefebvre, "Fuel effects on gas turbine combustion - ignition, stability and combustion efficiency", *ASME J. Eng. Gas Turbines and Power*, vol. 107, pp. 24-37, 1985.
2. K. K. Rink and A. H. Lefebvre, "Pollutant formation in heterogeneous mixtures of fuel drops and air", *AIAA J. propulsion,* vol. 3 No. 1, pp. 5-10, Jan-Feb, 1987.
3. C. M. Reeves and A. H. Lefebvre, "Fuel effects on aircraft combustor emissions", *ASME* (paper), 86-GT-212, 1986.
4. N. Chigier, "Spray Science and Technology", *Fluid Mechanics and Heat Transfer in Sprays, ASME Fluid Eng. Div. Publ. FED*, v 178, pp. 1-18, 1993. ASME New York, NY.
5. A. H. Lefebvre, *Gas Turbine Combustion,* McGraw Hill, New York, 1983.
6. E. Giffen and A. Muraszew, *The Atomization of Liquid Fuels,* Chapman & Hall Ltd., London, 1953.
7. A. H. Lefebvre, *Atomization and Spray*, Hemisphere Publishing Corporation, 1989.

8. A. J. Yule and J. J. Dunkley, *Atomization of Melts*, Oxford University Press Inc., New York, 1994.
9. L. Bayvel and Z. Orzechowski, *Liquid Atomization*, Taylor & Francis, 1993.
10. J. S. Chin, N. K. Rizk and M. K. Razdan, "Experimental investigation of hybrid airblast atomizer", *ASME* (paper), 96-GT-464, 1996.
11. T. C. Roesler and A. H. Lefebvre, " Studies on aerated liquid atomization", *Int. J. of Turbo and Jet Engines*, vol. 6, Nos. 3 and 4, pp. 221-230, 1989.
12. J. Li, A. H. Lefebvre and J. R. Rollbuhler, "Effervescent atomizers foe small gas turbines", *ASME* (paper), 94-GT-495, 1994.
13. J. S. Chin, "Effervescent atomization and internal mixing air assisted atomization", *Int. J. of Turbo and Jet Engines*, 12, pp. 119-127, 1995.
14. J. S. Chin and A. H. Lefebvre, "A design procedure for effervescent atomizers", *ASME J. of Engineering for Gas Turbines and Power*, vol. 117, pp. 266-271, April 1995.
15. S. K. Chen and A. H. Lefebvre, "Influence of ambient pressure on effervescent atomization", *AIAA J. of Propulsion and Power*, vol. 9, No. 1, pp. 10-15, Jan.-Feb. 1993.
16. M. T. Lund, C. Q. Jian, P. E. Sojka, J. P. Gore and M. V. Panchagnula, "The influence of atomizing gas molecular weight on low mass flowrate effervescent atomizer performance", FED-Vol. 178/HTD-Vol. 270, *Fluid Mechanics and Heat Transfer in Sprays, ASME*, pp. 123-128, 1993.
17. W. Y. Lee and P. E. Sojka, "The influence of fluid viscoelasticity on low mass flow rate effervescent atomization", FED-Vol. 178/HTD-Vol. 270, *Fluid Mechanics and Heat Transfer in Sprays, ASME*, pp. 129-135, 1993.
18. S. G. Bush and P. E. Sojka, "Entrainment by effervescent sprays at low mass flow rates", FED-Vol. 178/ HTD-Vol. 270, *Fluid Mechanics and Heat Transfer in Sprays, ASME*, pp. 117-121, 1993.
19. S. C. Geckler and P. E. Sojka, "High mass flow rate effervescent atomization of viscoelastic fluids", FED-Vol. 178/HTD-Vol. 270, *Fluid Mechanics and Heat Transfer in Sprays, ASME*, pp. 109-115, 1993.
20. K. A. Nielsen, C. W. Glancy, K. L. Hoy and K. M. Perry, "A new atomization mechanism for airless spraying: the supercritical fluid spray process", *ICLASS 91*, paper 37, pp. 367-374, 1991.
21. V. N. Bindal, S. K. Jain and Y. Kumar, " Performance characteristics of an ultrasonic atomizer", *Indian J. of Technology*, vol. 24, March 1986, pp. 153-156.
22. N. Maehara, S. Nakane, K. Yamamoto and K. Iga, "A pinhole-plate ultrasonic atomizer", *Ultrasonics*, vol. 22, no. 6, pp 259-260, Nov. 1984.
23. Y. Ohyama, S. Ueno and S. Yanagisawa, "Powertrain control system for lean combustion engines", *Hitachi Review*, vol. 35, No. 3, pp. 141-144, 1986.
24. M. N. Topp and P. Eisenklam, " Industrial and medical uses of ultrasonic atomizers", *Ultrasonics*, May 1972, pp. 127-133.
25. H. L. Berger, "Characteristics of a class of widely applicable ultrasonic nozzles", *Proceedings of 3rd International Conference on Liquid Atomization and Spray Systems*, July 1985, pp. 1A/2/1-13.
26. V. I. Bezrukov, "Theoretical and experimental design principles for jet printers with high-pressure drop emission", *Sov. Surf. Eng. Appl. Electrochem.*, no. 5, pp 44-55, 1985.
27. N. R. Ramesh, K. Sridhara and R. Natarajan, "Studies on the atomization and combustion performance of a twin-fluid acoustic atomizer", *Fuel*, vol. 64, pp. 1677-1680, December 1985.
28. N. R. Ramesh and R. Natarajan, "Atomization and combustion performance of a stem-jet type twin-fluid acoustic atomizer", *ICLASS 91*, paper 42, pp. 407- 414, 1991.
29. S. P. Lin and D. R. Woods, "Induced atomization", *ICLASS 91*, paper 39, pp. 383-387, 1991.

30. T. Sakai, M. Sadakata, M. Sato, K. Kimura and M. Saito, "Production of uniformly-sized duel concentric droplets from coaxial smooth jet under applied AC electric field", *ICALSS 91*, paper 40, pp. 389-396, 1991.
31. A. Gomez and K. Tang, "Atomization and dispersion of quasi-monodisperse electrostatic sprays of heptane", *ICALSS 91*, paper 90, pp. 805-812, 1991.
32. P. J. Mullinger and N. A. Chigier, "The design and performance of internal-mixing multijet twin-fluid atomizers, *J. Inst. Fuel*, vol.47, pp. 251-261, 1974.
33. J. S. Chin and L. X. Wang, "Experimental study on internal mixing sonic flow air assisted atomizer for heavy oils", *ASME* (paper), 90-GT-6, 1990.
34. T. Sakai, M. Kito, M. Saito and T. Kanbe, "Characteristics of internal mixing twin-fluid atomizer", *Proceedings of the 1st International Conference on Liquid Atomization and Spray Systems*, Tokyo, August 1978, paper 10-1, pp. 161-168, 1978.
35. M. N. Biswas, "Atomization in two-phase critical flow", *Proceedings of the 2nd International Conference on Liquid Atomization and Spray Systems* (ICLASS-82), pp. 145-151, 1982.
36. J. M. Chawla, "Atomization of liquid employing the low sonic velocity of liquid/gas mixture", *Proceedings of the 3rd International Conference on Liquid Atomization and Spray Systems* (ICLASS-85), pp. LP/1A/5/1-LP/1A/5/7, 1985.
37. J. F. Hurley and B. W. Doyle, "Design of two-phase atomizers for use in combustion furnaces", *Proceedings of the 3rd International Conference on Liquid Atomization and Spray Systems* (ICLASS-85), pp. IA/3/1-IA/3/13, 1985.
38. T. Inamura and N. Nagai, "The relative performance of externally and internally mixed twin-fluid atomizers", *Proceedings of the 3rd International Conference on Liquid Atomization and Spray Systems* (ICLASS-85), pp. IIC/2/1-IIC/2/11, 1985.
39. A. Kushari, Y. Neumeier, O. Israeli, A. Peled and B. T. Zinn, "An internally mixed injector for active control of atomization process in liquid fueled engines", AIAA 99-0329, *37th AIAA Aerospace Sciences Meeting and Exhibit*, Jan. 11-14, 1999, Reno, NV.
40. A. Kushari, Y. Neumeier, O. Israeli, E. Lubarsky and B. T. Zinn, "Internally Mixed Liquid Injector for Active Control of Atomization Process", *Journal of Propulsion and Power*, Vol. 17, No. 4, 2001, pp. 878-882.
41. A. Kushari, "A phenomenological two-phase flow model of atomization in an Internally Mixed Liquid Atomizer", *International Journal of Turbo and Jet Engines*, vol. 24, no. 3, 2007, pp 183-194.
42. J. Karnawat and A. Kushari, "Controlled Atomization using a twin-fluid swirl atomizer" *Experiments in Fluids* (2006) 41: 649-663.
43. J. Karnawat and A. Kushari" Spray Evolution in a Twin-Fluid Swirl Atomizer", *Atomization and Sprays,* vol. 18, no. 5, 2008, pp. 449-470.